Experiments in General Chemistry Featuring MeasureNet

Guided Inquiry, Self-Directed, and Capstone

Second Edition

Bobby Stanton
University of Georgia

Lin Zhu
Indiana University
Purdue University at Indianapolis

Charles H. Atwood
University of Georgia

BROOKS/COLE
CENGAGE Learning™

Australia • Brazil • Japan • Korea • Mexico • Singapore • Spain • United Kingdom • United States

Experiments in General Chemistry Featuring MeasureNet: Guided Inquiry, Self-Directed, and Capstone
Second Edition
Bobby Stanton, Lin Zhu, and Charles H. Atwood

Executive Editor: Mary Finch
Acquisitions Editor: Lisa Lockwood
Development Editor: Stefanie Beeck
Assistant Editor: Stefanie Beeck
Editorial Assistant: Elizabeth Woods
Media Editor: Lisa Weber
Marketing Manager: Nicole Hamm
Marketing Assistant: Kevin Carroll
Marketing Communications Manager: Linda Yip
Content Project Manager: Michelle Cole
Creative Director: Rob Hugel
Art Director: John Walker
Print Buyer: Judy Inouye
Rights Acquisitions Account Manager, Text: Bob Kauser
Rights Acquisitions Account Manager, Image: Mandy Groszko
Cover Designer: Denise Davidson
Cover Image: © Image Source/Corbis
Compositor: Macmillan Publishing Solutions

Library of Congress Control Number: 2009920729

ISBN-13: 978-0-495-56179-8

ISBN-10: 0-495-56179-7

Brooks/Cole
10 Davis Drive
Belmont, CA 94002-3098
USA

Cengage Learning is a leading provider of customized learning solutions with office locations around the globe, including Singapore, the United Kingdom, Australia, Mexico, Brazil, and Japan. Locate your local office at **www.cengage.com/international.**

Cengage Learning products are represented in Canada by Nelson Education, Ltd.

To learn more about Brooks/Cole, visit **www.cengage.com/brookscole**

Purchase any of our products at your local college store or at our preferred online store **www.ichapters.com.**

Printed in the United States of America
1 2 3 4 5 6 7 13 12 11 10 09

Contents

Preface

Laboratory Experiments in General Chemistry Featuring MeasureNet

The second edition of this manual offers several new features: new digital pictures and illustrations, all concept/technique experiments converted to a *guided inquiry format*, the addition of three new *self-directed* experiments, and one new *Capstone* experiment.

Virtually all the artwork and illustrations, in the first edition of the manual, have been replaced with digital pictures and illustrations in the second edition of the manual.

Second, all concept/technique oriented experiments in the first edition of the manual have been converted to a *guided inquiry* format. This includes guided inquiry experimental procedures and laboratory reports that are written in the guided inquiry format. We have inserted numerous *"What is the purpose of,* or *Why were you instructed to . . ."* questions throughout the experimental procedures to encourage students to think more in depth about why they are being ask to add certain chemicals or perform a particular experimental technique. Most of the sample calculations included in the concept/technique experiments in the first edition of the manual have been removed, and in there place are text descriptions describing the approach needed to determine or calculate certain experimental values. New pre- and post-laboratory questions have been written for all the concept/technique experiments.

Third, we have included three new self directed experiments in the second edition of the manual. A *self-directed* experiment is one in which students write their own experiment to solve an assigned problem. Every third or fourth week during the semester, student teams are required to solve a multi-component problem via a *self-directed* experiment. This multi-component problem requires the use of knowledge, skills, and techniques learned in the previous weeks' *concept/technique* experiments. Students submit a *Procedure Proposal* two weeks before the *self-directed* experiment is to be performed. The *ProcedureProposal* is modeled on the work of Thomas Greenbowe at Iowa State University. Dr. Greenbowe's research into the science writing heuristic shows that giving students a set of questions that guides them in the design and implementation of their experiment markedly improves their understanding of the experimental material. Consequently, the *Procedure Proposal* asks the student teams to answer several focusing questions to help solve the multi-component problem. The procedure proposals are graded, corrected, and returned to each team one week before they perform the *self-directed* experiment. Once the *self-directed* experiment is completed, the student teams submit a *Formal Lab Report* detailing the experimental techniques used, all data and observations collected, and a thorough discussion of the results and conclusions drawn from the experiment. We have also relied on Dr. Greenbowe's science writing heuristic to provide focusing questions and guiding principles for the *Formal Lab Report*.

Finally, we have inserted one new *Capstone* experiment, Experiment 32. This experiment encompasses all techniques, skills, concepts, and principles learned in a two semester general chemistry laboratory course.

Acknowledgments

A project of this magnitude requires the assistance of numerous people. We would like to express our deep appreciation and gratitude to the people that have encouraged and assisted us with the production of this lab manual.

First, we would like to thank the excellent staff at Brooks/Cole. We wish to thank our editor Lisa Lockwood. During the development of the second edition, we were assisted by an outstanding editorial assistant Stefanie Beeck. We greatly appreciate her assistance. The authors owe a great deal of gratitude to Bill Cornett, our Brooks/Cole Representative. Bill was instrumental in making Brooks/Cole aware of our interest in initiating this project. He realized that this project could have national appeal and he convinced Brooks/Cole that our ideas were worthwhile. We are indebted to his insight and persistence.

We would like to express our deep appreciation to each of the reviewers of the first edition of this manual for their comments and suggestions. We were fortunate to have a conscientious, diligent, and hard working group of reviewers. The names and affiliations of the *reviewers* of the *first edition* are listed below.

Patricia Amateis - *Virginia Tech*
Charles Baldwin - *Union University*
Michael Denniston - *Georgia Perimeter College*
Michelle Douskey - *University of California-Berkely*
Roy Garvey - *North Dakota State University*
Michael Nichols - *John Carroll University*
Richard Petersen - *The University of Memphis*
Ruth Ritter - *Agnes Scott College*
Phillip Squattrito - *Central Michigan University*

We would like to express our deep appreciation to John Moody for *reviewing* the *second edition* of the manual. His comments and suggestions were extremely useful and beneficial.

John Moody - *The University of Georgia*

Our lab manual features the MeasureNet system that was developed at the University of Cincinnati by Estel Sprague and Robert Voorhees. Subsequently, they have started their own company, MeasureNet Technologies Incorporated. They have been instrumental in helping us prepare this manual: developing new devices and probes, giving us ideas for experiments, allowing us to take part in MeasureNet workshops, financial support for the development of new experiments, and actively working with Brooks/Cole to develop the manual. We would also like to thank Michael

Kurutz, the Marketing Director for MeasureNet Technologies, Ltd. We look forward to collaborating with MeasureNet for many years to come.

At the University of Georgia, we would like to thank our undergraduate and graduate students in the General Chemistry program. They have made numerous suggestions and comments that the authors believe will improve the learning experience for the thousands of students who will use this lab manual in the future.

Common Laboratory Glassware and Equipment

electronic balance

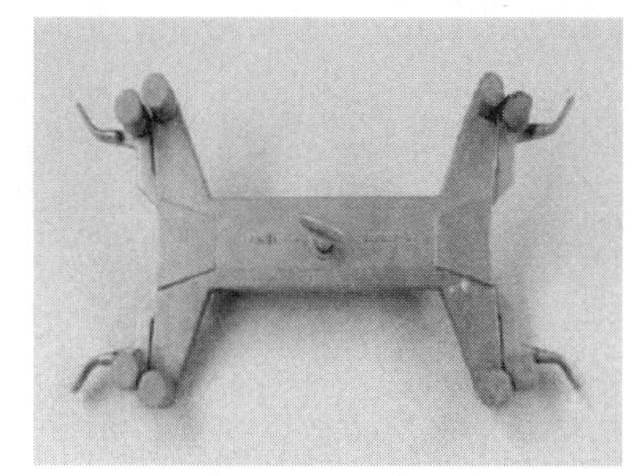

buret clamp

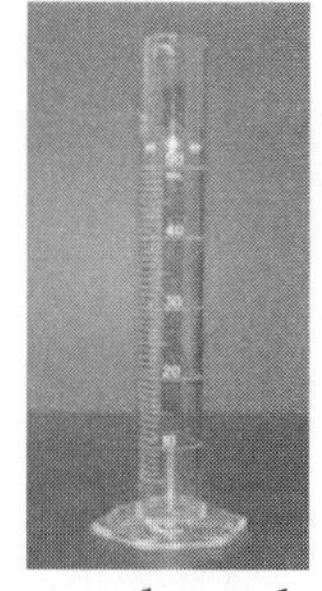

graduated cylinder

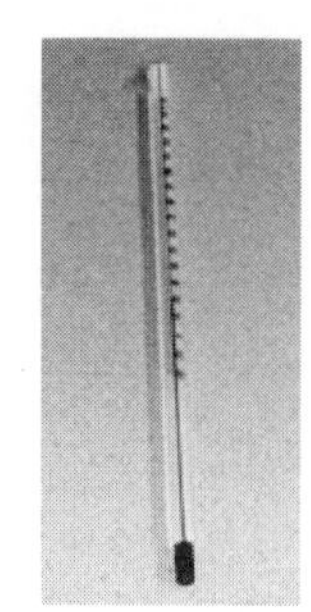

thermometer

beaker

Bunsen burner

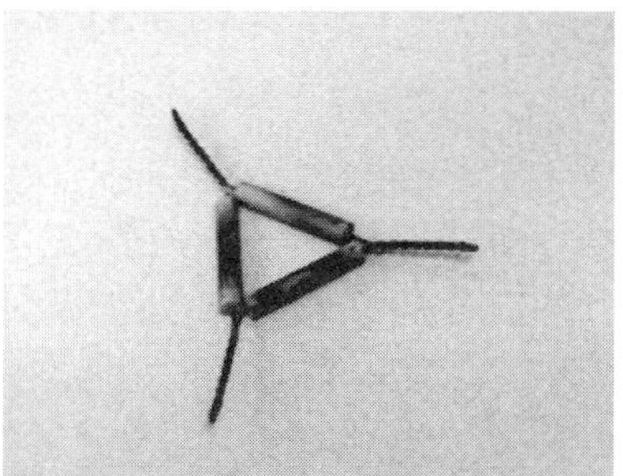

clay triangle

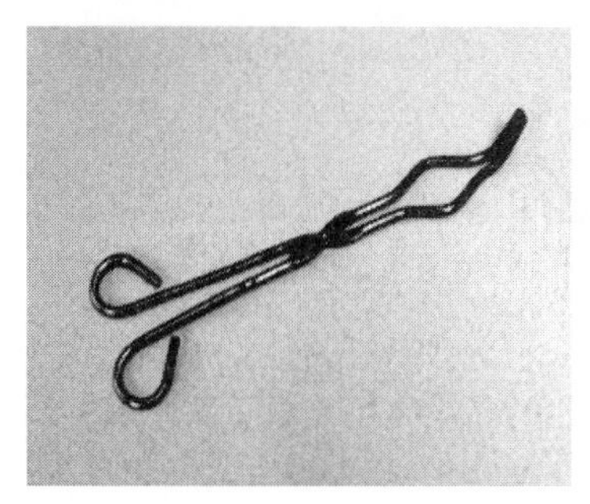

crucible tongs

crucible and lid

Erlenmeyer flask

evaporating dish

funnel

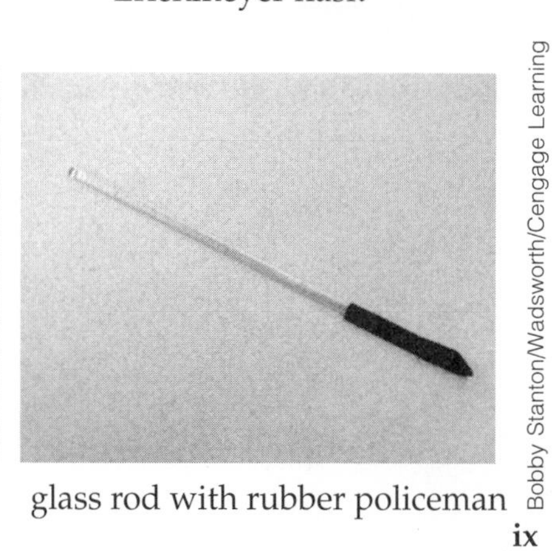

glass rod with rubber policeman

ring stand and iron ring

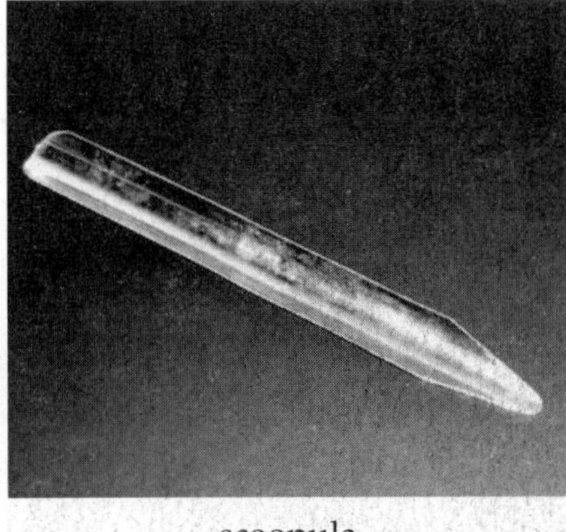

scoopula

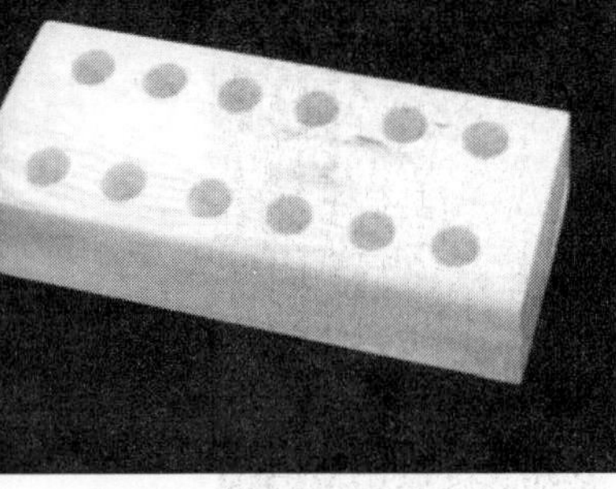

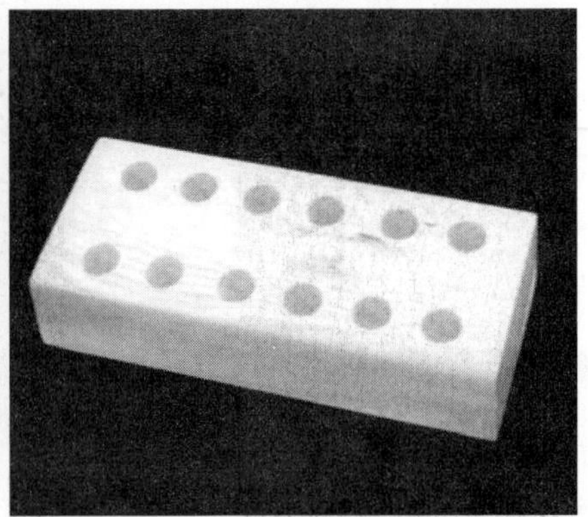

test tube rack

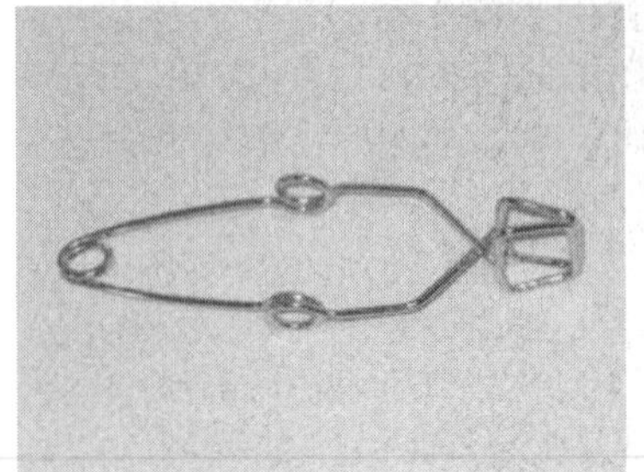

test tube clamp

test tube

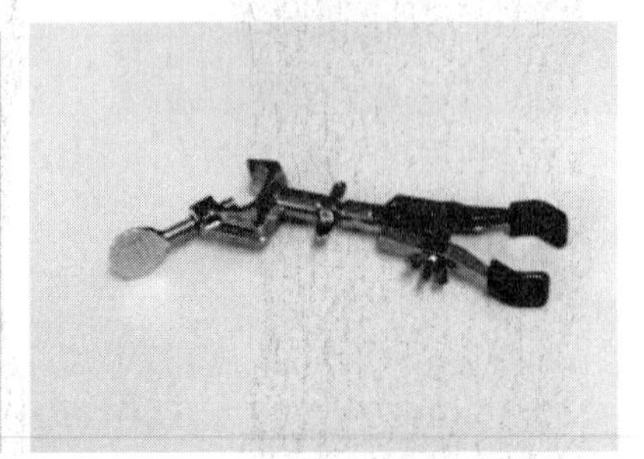

utility clamp

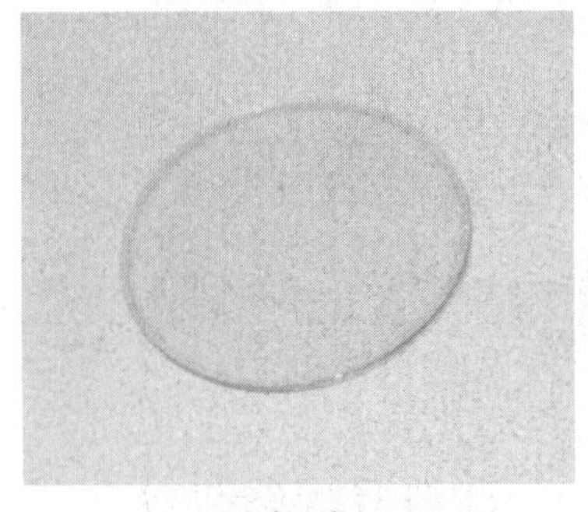

watch glass

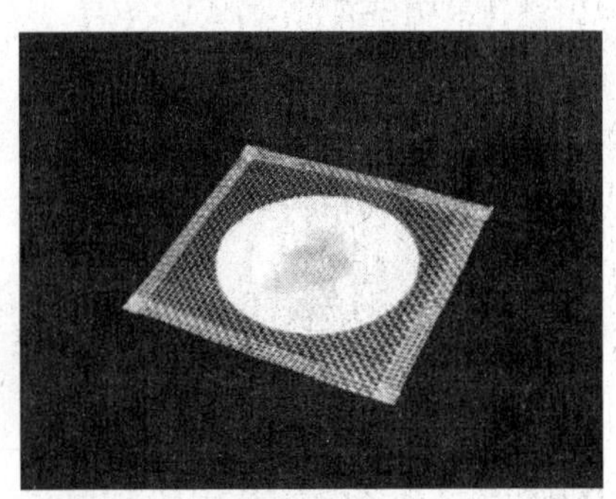

wire gauze

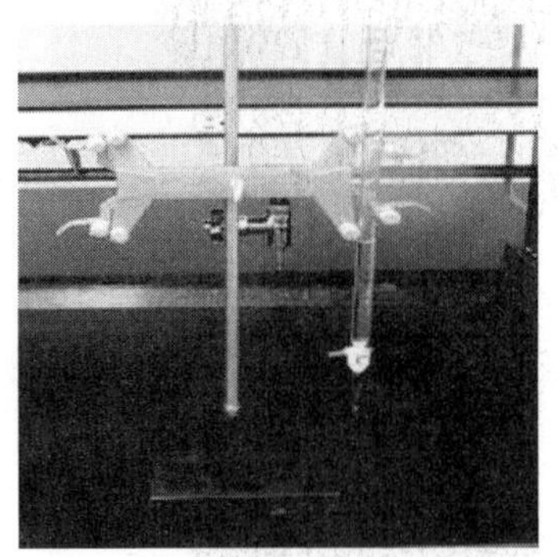

buret

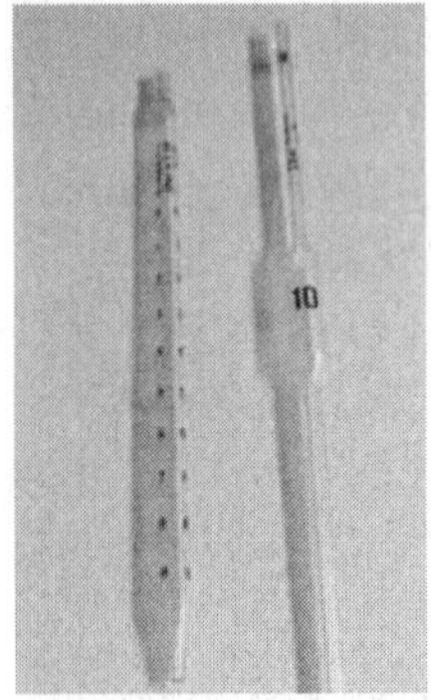

pipets

volumetric flask

Safety Rules

Each student is required to read and fully comprehend the following safety rules that govern laboratory behavior. Each student must complete a laboratory "safety quiz" before entering the laboratory.

1. Safety goggles must be worn in the laboratory *at all times*. Safety glasses are to be worn over eyeglasses. Contact lenses are not permitted in the laboratory. Certain chemical vapors may cause eye irritations when contact lenses are worn.
2. Never eat, drink, chew gum, or smoke in the laboratory. These foods can absorb toxic chemicals.
3. Never inhale vapors or fumes produced in chemical reactions or from bottled chemicals in the laboratory. Use fume hoods when conducting chemical reactions that produce toxic vapors or fumes. Waft non-toxic or non-irritating fumes toward your nose with your hand when instructed to do so.
4. Never taste chemicals used in the laboratory.
5. Never perform an unauthorized experiment (i.e., arbitrarily mix chemicals).
6. Never work in the laboratory unless your instructor is present.
7. Never remove equipment, glassware, reagents or other items from the laboratory.
8. Never return unused chemicals to a reagent bottle. Doing so may contaminate the reagent bottle.
9. Never pour water into concentrated acids. To dilute a concentrated acid, always pour acid slowly into water, while stirring the solution.
10. Never heat a liquid in a test tube with the test tube's open end pointed toward another person.
11. When inserting glass tubing, a glass funnel, a glass rod, or a thermometer into a stopper, always lubricate the stopper and the glass item with glycerin. Before attempting this procedure, your instructor will demonstrate the proper technique.
12. Never leave lighted Bunsen burners unattended. They are a fire hazard.
13. Never dispose of solid materials or filter paper in the sink or troughs. Instead, dispose of them in the trash can or specially provided solid waste containers.
14. Report mercury spills from broken thermometers to your instructor immediately. Because mercury is toxic, it must be disposed of carefully.
15. Never add boiling chips to a hot solution. Adding boiling chips to a hot solution may cause it to boil over rapidly.

16. Short pants, mid-riffs, tank tops, skirts, and open-toed shoes cannot be worn in the laboratory. Free-flowing long hair should be tied in a pony tail to prevent it from accidentally contacting the Bunsen burner flames.
17. Do not store hazardous or flammable chemicals in your locker.
18. You must learn to locate and operate all safety devices provided in the laboratory (i.e., fire extinguishers, fire blankets, eyewash stations, and safety showers).
19. At the end of your laboratory session, return all reagents and equipments to their proper places and clean your work area.
20. Report any laboratory accident, no matter how minor, to your instructor immediately.
21. Obey all fire drill instructions given to you by your instructor.
22. Always use a pipet pump or bulb to fill a pipet. Never pipet by mouth. Do not blow out the last bit of liquid in the pipet.
23. Should chemicals contact your skin, wash the affected area with copious quantities of water and report the incident to your laboratory instructor.
24. After removing chemicals from a reagent bottle, replace the cap on the reagent bottle. After pouring chemical waste into the Waste Container, replace the cap on the container. Not replacing caps on reagent bottles and waste containers is a violation of Federal Law.
25. When heating liquids in a test tube, hold the test tube with a test tube clamp at a 45 degree angle over the flame, and heat the side of the test tube. Do not directly heat the bottom of the test tube.

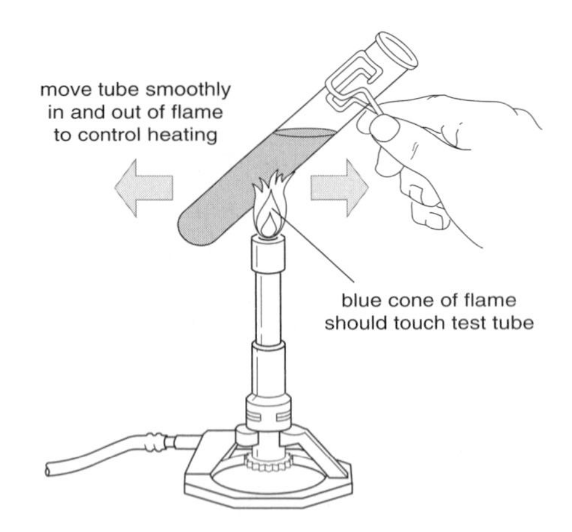

Name ______________________ Section __________ Date __________

Instructor ______________________

Safety Quiz

Indicate whether each of the following statements is true or false by writing the word **TRUE** or **FALSE** in the space provided.

_____ **1.** If chemicals come into contact with your skin, immediately wash the affected area with copious quantities of water.

_____ **2.** Fume hoods are used in the chemical laboratory when using volatile or poisonous chemicals.

_____ **3.** It is permitted to leave a lit Bunsen burner unattended.

_____ **4.** Always return unused chemicals to a reagent bottle to avoid wasting chemicals, you will not contaminate the entire reagent bottle.

_____ **5.** When heating a liquid in a test tube, always point the test tube in a direction away from any other person in the laboratory.

_____ **6.** Always add boiling chips to a hot solution.

_____ **7.** The wearing of shorts, tank tops, mid-riffs and sandals is permitted in the laboratory.

_____ **8.** Drinking soda in the lab is permitted as long as the soda can is at least 10 feet away from all chemicals.

_____ **9.** I am not required to wear safety goggles while in the laboratory unless I am actually performing an experiment.

_____ **10.** It is a violation of Federal Law to leave a Waste Container uncapped.

Densities of Some Liquids and Solids

OBJECTIVES

Familiarize students with some of the common laboratory equipment and determine the density of several liquids and solids.

INTRODUCTION

Chemistry is the study of matter and the changes matter undergoes. Chemists observe matter by determining, measuring, and monitoring physical and chemical properties of matter. A **property** is any characteristic that can be used to describe matter (e.g., size, color, mass, density, solubility, etc.). In this experiment, we will determine the density of liquids and solids. The density of a substance can be used to identify a liquid or solid because density is an intensive property. **Intensive properties** are properties that do not depend on the quantity of the substance. For example, gold, which is relatively dense, can be separated from sand, silt, and rock by panning for it in a stream because of its greater density.

Density is the ratio of the mass of a substance to its volume.

$$\text{Density} = \frac{\text{Mass}}{\text{Volume}} \qquad \text{(Eq. 1)}$$

The units of density are normally expressed as g/mL or g/cm^3. (A mL and a cm^3 are different expressions of the same unit, 1 mL = 1 cm^3).

The density of a liquid can be determined by weighing a known volume of the liquid. For example, to determine the density of a sodium chloride solution, first, weigh a clean, dry 10-mL graduated cylinder (Figure 1). (The lab instructor will demonstrate how to weigh an object using a digital balance by the following methods: a) differences in mass and b) taring the balance.) Next, sodium chloride solution is added to the graduated cylinder and it is reweighed (Figure 2).

The volume of sodium chloride solution in the graduated cylinder is recorded by reading the level of liquid in the cylinder at the bottom of the meniscus.

Figure 1
Weighing an empty 10.0 mL graduated cylinder

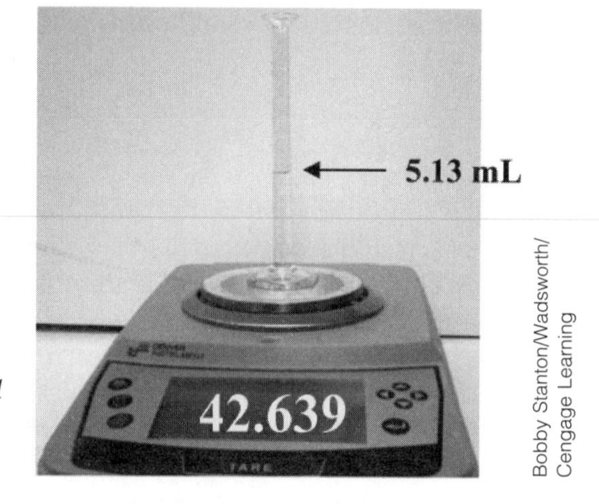

Figure 2
Weighing a 10.0 mL graduated cylinder containing sodium chloride solution.

Knowing the mass (42.639 g − 37.198 g = 5.441 g) and volume of the sodium chloride solution (5.13 mL) in the graduated cylinder, the density of the solution is calculated to be 1.06 g/mL (5.441 g/5.13 mL = 1.06 g/mL).

Solids can be regularly shaped (cylindrical, cubical, spherical, etc.) or irregularly shaped (metal turnings or shot). An object's shape determines which of the following two methods is used to determine its volume.

NOTE: The concave or convex surface of a liquid is referred to as the ***meniscus***. For consistency, a concave meniscus is read at the bottom of the curvature at eye level. A convex meniscus, which liquid mercury displays, is read at the top of the curvature.

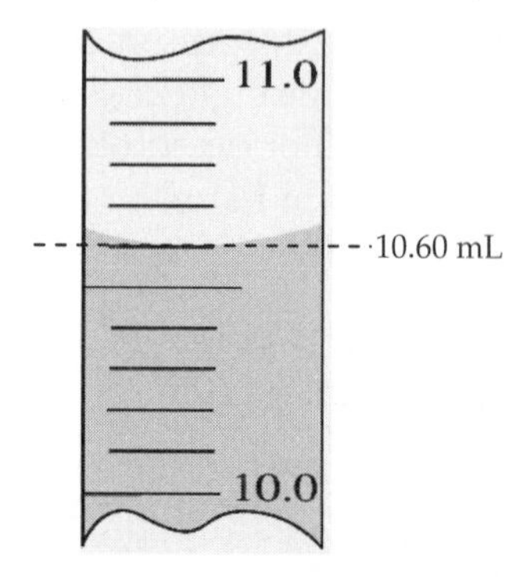

The correct number of significant figures for volumetric glassware is determined from the smallest calibration unit of the graduated container plus one additional, estimated number. The pictured graduated cylinder is calibrated to 0.1 mL. (Each line on the cylinder is 0.1 mL more than the previous line). Thus, the liquid's correct volume is 10.60 mL.

The dimensions of a regularly shaped object can be measured with a ruler and the volume calculated using the appropriate geometric equation. A cylindrical rod of aluminum metal with its length and radius is depicted in Figure 3.

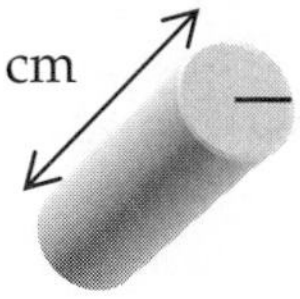

L = 4.67 cm

r = 0.84 cm

Figure 3
Cylindrical aluminum rod

The metal's volume can be calculated by measuring its length (L) and its radius (r) using the geometric relationship for the volume of a cylinder, $V = \pi r^2 L$, where V is volume and π is 3.1416. If the aluminum rod has a mass of 29.7 g, the density of aluminum can be calculated to be 2.7 g/cm^3.

$$V = \pi r^2 L = (3.1416)(0.84\ \text{cm})^2(4.76\ \text{cm}) = 11\ \text{cm}^3$$

$$\text{Density} = \frac{\text{Mass}}{\text{Volume}} = \frac{29.7\ \text{g}}{11\ \text{cm}^3} = 2.7\ \text{g/cm}^3$$

Table 1 *Volume equations for regularly shaped objects*

cube	L^3
rectangular solid	$L \times W \times H$
sphere	$4/3\pi r^3$
cylinder	$\pi r^2 L$

It can be difficult to measure the radius of a sphere. The radius of a sphere is related to its circumference by the following formula. Knowing the circumference of the sphere, its radius can be calculated.

$$\text{circumference} = 2\pi r$$

Irregularly shaped solids cannot easily be measured with a ruler. Instead, their volume is most easily determined by the displacement of water or some other liquid. The solid (which must not react with nor dissolve in the liquid and have a density greater than the liquid) is placed in a calibrated container (usually a graduated cylinder) containing a previously measured volume of the liquid. The solid will displace an amount of liquid equal to its volume. The difference in the liquid's volume in the cylinder before and after the object is added is equal to the volume of the irregularly shaped object (Figure 4).

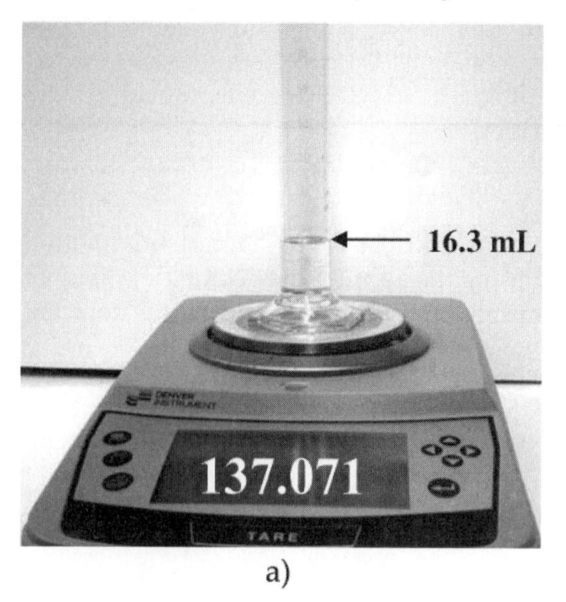

a)

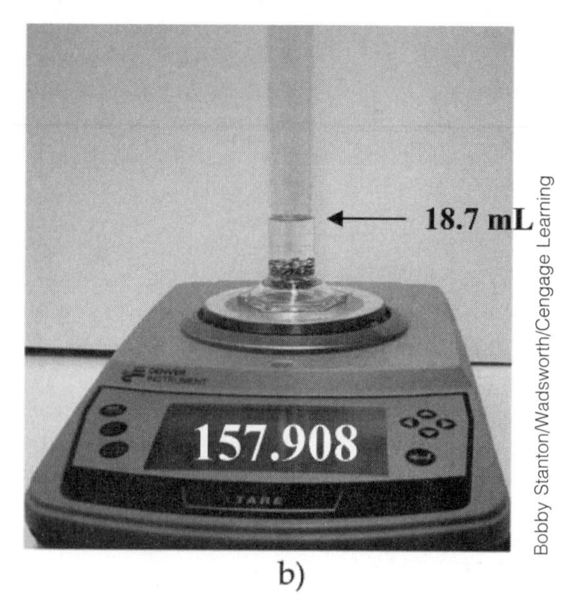

b)

Figure 4
100 mL graduated cylinder with the smallest calibration unit of 1 mL: a) volume of water is 16.3 mL; b) meniscus indicates 18.7 mL after the addition of 20.837 grams of copper shot. Therefore, the volume of the copper shot is 2.4 mL. From this data, the density of copper is determined to be 8.7 g/mL.

PROCEDURE

CAUTION

Students must wear departmentally approved eye protection while performing this experiment. Wash your hands before touching your eyes and after completing the experiment.

1. Determine the density of water using a clean, dry 10-mL graduated cylinder (4-6 mL is a sufficient volume). What experimental data should be recorded in the Lab Report? What calculations should be included to determine the density of water?
2. Determine the density of mineral oil using a clean, dry 10-mL graduated cylinder (4-6 mL is a sufficient volume). What experimental data should be recorded in the Lab Report? What calculations should be included to determine the density of mineral oil?
3. Add similar amounts (~3 mL) of both water and mineral oil to a 10-mL graduated cylinder. Sketch the graduated cylinder containing the liquids. If more than one layer appears, identify each layer with a label on your sketch. Decant the water and mineral oil into the sink. Wash the graduated cylinder with soap and water before proceeding to the next step.
4. Obtain 4-8 mL of unknown sodium chloride solution. Determine the density of the salt solution. What experimental data should be recorded in the Lab Report? What calculations should be included to

determine the density of the salt solution? Decant the sodium chloride solution into the sink.

5. Determine the density of each of the following regularly shaped objects: a cube, a rectangle, a sphere and a cylinder. What experimental data should be recorded in the Lab Report? What calculations should be included to determine the densities of the regularly shaped objects? *Return the regularly shaped objects to your instructor when finished.*
6. Obtain 13-15 grams of an unknown, irregularly shaped metal and record the unknown number on the Lab Report. Obtain a 100-mL graduated cylinder. Determine the density of the metal. What experimental data should be recorded in the Lab Report? What calculations should be included to determine the density of the irregularly shaped metal? Use the information provided in Table 2 to identify your unknown metal.

Table 2 *Densities of some metals*

aluminum	2.70 g/mL
copper	8.92 g/mL
iron	7.86 g/mL
lead	11.3 g/mL
magnesium	1.74 g/mL
zinc	7.14 g/mL

7. Decant the water from the graduated cylinder containing the unknown metal into a beaker. Remove any pieces of metal from the decanted water and combine them with the rest of the wet metal. Pour the decanted water into the sink. *Thoroughly* dry the unknown metal with a towel. *Return the unknown metal to the "Used Metal Container" bearing the same number as your unknown number.*

Name

Section

Date

Instructor

1

EXPERIMENT 1

Lab Report

SHOW ALL WORK TO RECEIVE FULL CREDIT.

1. Determine the density of water.

2. Determine the density of mineral oil.

3. Sketch the graduated cylinder containing water and mineral oil.

4. Determine the density of the sodium chloride solution.

5. Determination of the densities of the regularly shaped objects.

cube

rectangular solid

sphere

cylinder

6. Determine the density of the unknown, irregularly shaped metal.

Unknown number __________ Identity of metal _______________

Name Section Date

Instructor

1 EXPERIMENT 1

Pre-Laboratory Questions

1. Record the volume of water in the graduated cylinder depicted below to the correct number of significant figures.

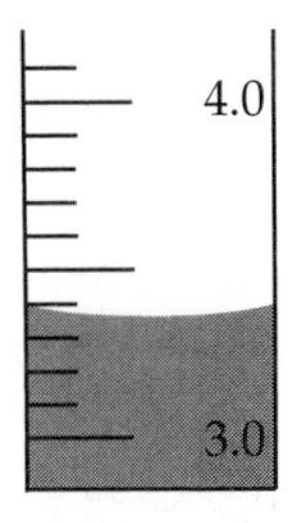

____________ mL

2. Calculate the volumes of the cylinder and the sphere depicted below in cubic centimeters.

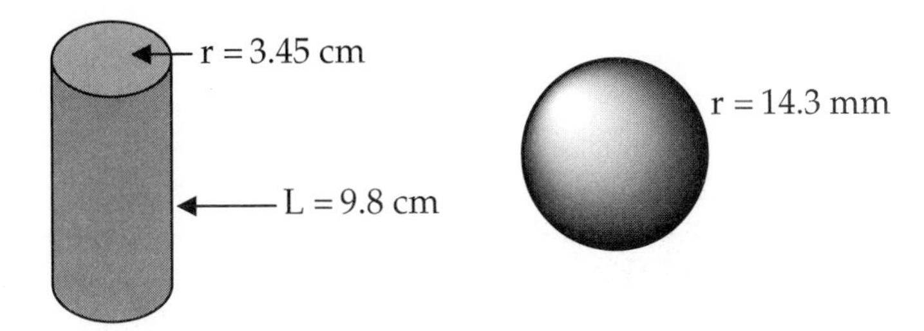

3. The density of copper is 8.92 g/cm^3. A rectangular block of copper is depicted below. Calculate the mass of the copper block in kilograms.

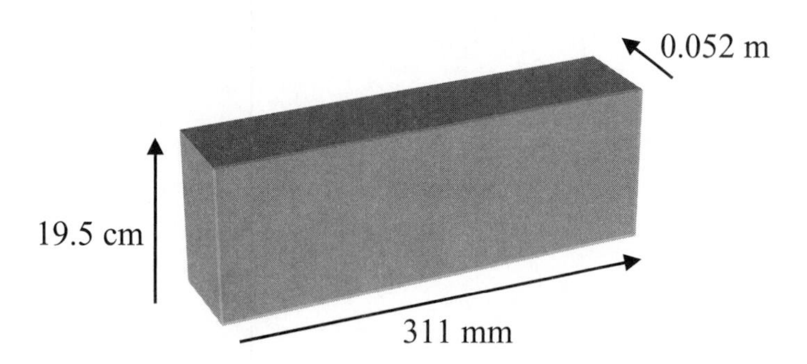

4. A student added some water to a graduated cylinder (Figure a). The student then added zinc shot to the same graduated cylinder to determine the volume of water displaced by the metal (Figure b). Using the density of zinc from Table 2, calculate the mass of iron shot added to the graduated cylinder.

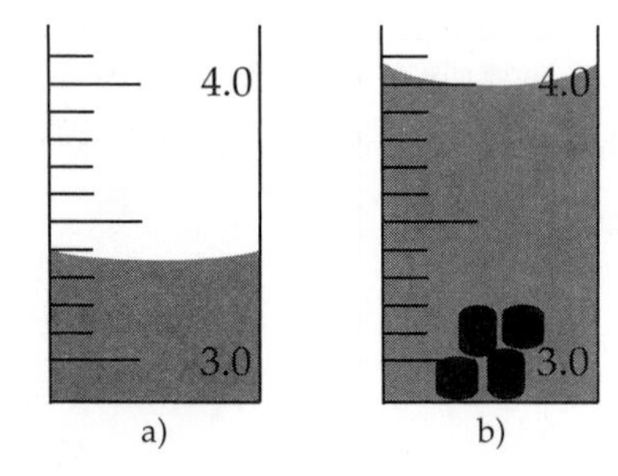

Name ______________________ Section ________ Date ________

Instructor ______________________

1 **EXPERIMENT 1**

Post-Laboratory Questions

1. Three milliliters each of liquids A, B, and C are slowly added to a 10-mL graduated cylinder. Each liquid is immiscible in the other two liquids. Liquid A has a density of 0.69 g/mL. Liquid B has a density of 2.3 g/mL. Liquid C has a density of 1.1 g/mL. Sketch the graduated cylinder containing the mixture. Label each liquid in the cylinder as A, B, or C. Justify your labeling with an explanation.

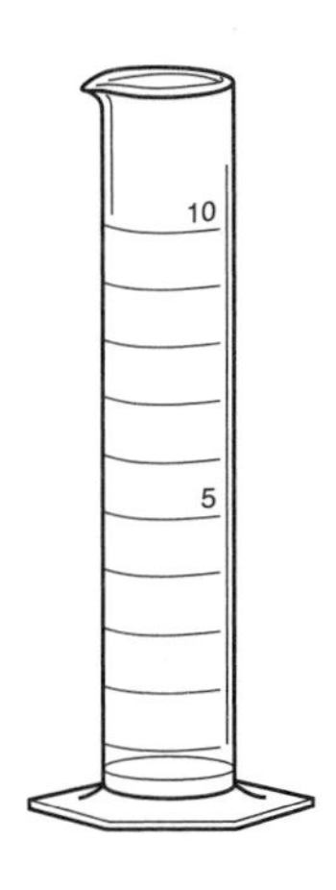

2. A student weighed a wet 10-mL graduated cylinder. The cylinder contained at least 0.5 mL of water when weighed. Next, the student added sodium chloride solution to the cylinder until the bottom of the meniscus read 5.0 mL. How will the 0.5 mL of water affect the calculated density of the sodium chloride solution? Justify your answer with an explanation.

3. In Step 6 of the Procedure, a student measures the volume and mass of an irregular solid. After the irregular solid is added to the graduated cylinder, a few drops of water were accidentally splashed into the graduated cylinder. How will this affect the calculated density of the irregular solid? Justify your answer with an explanation.

Specific Heat of Substances

OBJECTIVES

Determine the specific heat of a solid and understand the relative heat properties of substances having different specific heats.

INTRODUCTION

Substances respond differently upon being heated or cooled. For example, one calorie (4.18 J) of heat raises the temperature of 1.00 gram of water by 1.00 °C. The same amount of heat raises the temperature of 1.00 gram of lead by 32.4 °C. The physical quantity which relates the temperature change and mass of a substance to the amount of heat absorbed is called the **specific heat**. Specific heat is a physical property that is unique to a given substance that can be measured and used to identify an unknown material. In this experiment, the specific heat of an unknown solid will be determined and used to identify the solid.

The specific heat (C) of a substance is the quantity of energy (J or cal) required to change the temperature of 1.00 gram of that substance by 1.00 °C. (A related quantity is the molar heat capacity of a substance which is the quantity of energy required to change the temperature of 1.00 mole of a substance by 1.00 °C.)

$$C = \frac{\text{heat (J or cal)}}{(\text{mass})(\text{change in temperature})} \qquad \text{(Eq. 1)}$$

Specific heat typically has units of cal/g°C or J/g°C. The molar heat capacity has units of cal/mol°C or J/mol°C. The specific heat of an unknown substance can be determined with a device called a **calorimeter**. Calorimeters are used to measure the heat lost or absorbed by a substance in physical or chemical changes. Calorimeters are insulated containers that prevent heat exchange between the process occurring inside the calorimeter and the outside surroundings. This experiment will employ two styrofoam cups, one nested inside the other, to serve as an inexpensive but effective calorimeter.

The First Law of Thermodynamics, or Law of Conservation of Energy, states that energy can neither be created nor destroyed during ordinary physical or chemical changes, but it can be transferred from one system to another. The Second Law of Thermodynamics implies that energy normally flows from substances at higher temperatures to substances at lower temperatures. Both laws will be used in this experiment to determine the specific heat of a substance.

The heat (*energy*) lost or gained by a substance as its temperature changes can be calculated using the equation

$$q = m \times C \times \Delta T \qquad \text{(Eq. 2)}$$

where q is the heat change (units of J or cal) of the substance, m is the mass of the substance in grams, and ΔT is the change in temperature (in degrees °C) of the substance ($\Delta T = T_{final} - T_{initial}$).

One way to determine the specific heat of a substance is to measure the heat lost by a *given mass of a heated substance* after it has been transferred into a *known mass of water* that is initially at room temperature. The aforementioned laws of thermodynamics indicate what will happen as the water and substance contact each other. The total amount of heat energy will be conserved and the cooler substance (water) will experience a temperature rise as the warmer substance cools. After a period of heat exchange, the equilibrium temperature, *or final temperature*, of the system will be attained. Given the specific heat of water (4.18 J/g°C), the mass of the water, and the changes in the temperature of the water and the substance; the heat absorbed by the water can be calculated. From the First Law of Thermodynamics, it is known that the *heat absorbed* by the water must be *equal* in magnitude to the *heat lost* by the substance. Mathematically this statement can be written as

$$-q_{\text{lost by substance}} = q_{\text{gained by water}}$$

(The minus sign, preceding $-$ q above, indicates that heat flows from the substance to the water). From Eq. 2, this can be rewritten as

$$-(m_{substance} \times C_{substance} \times \Delta T_{substance}) = m_{water} \times C_{water} \times \Delta T_{water} \qquad \text{(Eq. 3)}$$

Equation 3 can be rearranged to calculate the specific heat of the substance.

$$C_{substance} = \frac{m_{water} \times C_{water} \times \Delta T_{water}}{-(m_{substance} \times \Delta T_{substance})} \qquad \text{(Eq. 4)}$$

If the heat exchanged in this experiment were solely limited to the metal and the water, Eq. 4 would be sufficient. However, heat will be lost to the surroundings (air, lab bench, etc.) near the mixture. Heat loss to the surroundings can be minimized by performing the experiment in a calorimeter that inhibits heat exchange with the surroundings. However, the calorimeter will also absorb some of the heat from the substance. The heat lost to the calorimeter must also be accounted for in this experiment. Therefore, it will be necessary to calibrate the calorimeter. The **calorimeter constant** (C_{cal}) is defined as the heat energy required to raise the temperature of the calorimeter by 1.00 °C. The heat absorbed by the calorimeter, q_{cal}, can be determined according to the equation

$$q_{calorimeter} = C_{calorimeter} \times \Delta T_{water}$$
or
$$C_{calorimeter} = \frac{q_{calorimeter}}{\Delta T_{water}} \qquad \text{(Eq. 5)}$$

where $C_{calorimeter}$ is the calorimeter constant in J/°C, and ΔT_{water} is the change in the temperature of the water.

If the heat absorbed by the calorimeter is included in Eq. 4, we can derive an accurate expression for the specific heat of a substance.

heat lost by substance = heat absorbed by water + calorimeter

$$-q_{substance} = q_{water} + q_{calorimeter}$$

$$-(m_{substance} \times C_{substance} \times \Delta T_{substance}) = (m_{water} \times C_{water} \times \Delta T_{water}) + (C_{calorimeter} \times \Delta T_{water}) \qquad \text{(Eq. 6)}$$

Equation 6 can be rearranged to yield the specific heat of the substance.

$$C_{substance} = \frac{-[(m_{water} \times C_{water} \times \Delta T_{water}) + (C_{calorimeter} \times \Delta T_{water})]}{m_{substance} \times \Delta T_{substance}} \qquad \text{(Eq. 7)}$$

The mass of the water and the substance are determined by difference (weigh an empty container; add water or substance to the container and reweigh, the difference in the two weights is the weight of water or substance). The specific heat of water is a well-established constant, 4.18 J/g°C. The challenging aspect of this experiment is to accurately determine temperature changes for the water, ΔT_{water}, and the substance, $\Delta T_{substance}$. Temperature changes will be monitored by recording temperature versus time scans (thermograms) using the MeasureNet temperature probe. Figure 1 depicts a thermogram used to determine the calorimeter constant of a styrofoam cup calorimeter.

NOTE: When determining ΔT_{water} or $\Delta T_{substance}$, subtract the initial temperature of the water or substance, T_i, from the final temperature of the solution, T_f.

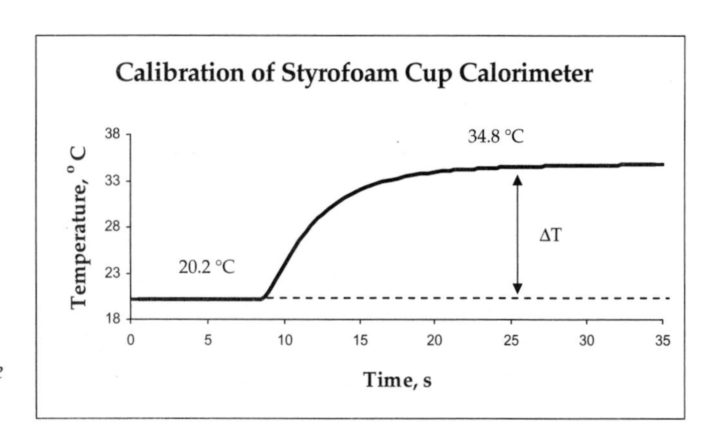

Figure 1
Thermogram for the addition of 31.152 grams of water at 51.0 °C to 31.284 grams of water at 20.2 °C. The equilibrium temperature of the system is 34.8 °C.

SAMPLE CALCULATIONS

Determination of the Calorimeter Constant

In this example, a known amount of warm water is mixed with a known amount of cool water. The hot water loses heat to the cool water until an intermediate, equilibrium temperature is attained. The heat absorbed by the calorimeter is the difference in the heat lost by the warm water and the heat absorbed by the cool water. If 31.152 mL of water (31.152 g assuming the density of water is 1.00 g/mL at ambient temperatures) at 51.0 °C are mixed with 31.284 mL (31.284 g) of water in a calorimeter at 20.2 °C, the equilibrium temperature attained by the system is 34.8 °C (Figure 1). The specific heat of water is 4.18 J/g°C. Calculate the calorimeter constant for this system in J/°C.

$$-q_{\text{warm water}} = q_{\text{cool water}} + q_{\text{calorimeter}}$$

$$q_{\text{calorimeter}} = -q_{\text{warm water}} - q_{\text{cool water}}$$

$$q_{\text{calorimeter}} = -(m_{\text{warm water}} \times C_{\text{warm water}} \times \Delta T_{\text{warm water}}) - (m_{\text{cool water}} \times C_{\text{cool water}} \times \Delta T_{\text{cool water}})$$

$$q_{\text{calorimeter}} = -\left[31.152\ \text{g} \times \frac{4.18\ \text{J}}{\text{g}^\circ\text{C}} \times (34.8 - 51.0\ ^\circ\text{C})\right] - \left[31.284\ \text{g} \times \frac{4.18\ \text{J}}{\text{g}^\circ\text{C}} \times (34.8 - 20.2\ ^\circ\text{C})\right]$$

$$q_{\text{calorimeter}} = 2110\ \text{J} - 1910\ \text{J}$$

$$q_{\text{calorimeter}} = 200.\ \text{J}$$

$$C_{\text{calorimeter}} = \frac{q_{\text{calorimeter}}}{\Delta T_{\text{cool water}}}$$

$$C_{\text{calorimeter}} = \frac{200.\ \text{J}}{(34.8\ ^\circ\text{C} - 20.2\ ^\circ\text{C})} = 13.7\ \text{J}/^\circ\text{C}$$

The calorimeter constant for the styrofoam calorimeter indicates that the calorimeter absorbs 13.7 J of heat for every 1.00 °C change in the temperature of the system. The value of the calorimeter constant will be used in the following sample calculation to determine the specific heat of an unknown substance.

When calculating $C_{\text{calorimeter}}$, always use $\Delta T_{\text{cool water}}$ because the calorimeter is initially at the same temperature as the cool water and both absorb energy from the warm substance.

Determination of the Specific Heat of a Solid

7.050 grams of an unknown solid is heated to 100.0 °C. The warm solid is added to a calorimeter containing 50.00 grams of water at 24.00 °C. The system attains an equilibrium temperature of 24.90 °C. Use this information and Eq. 7 to determine the specific heat of the solid.

$$C_{\text{solid}} = \frac{-[(m_{\text{water}} \times C_{\text{water}} \times \Delta T_{\text{water}}) + (C_{\text{calorimeter}} \times \Delta T_{\text{water}})]}{m_{\text{solid}} \times \Delta T_{\text{solid}}}$$

$$C_{\text{solid}} = \frac{-[(50.00\ \text{g})\ (4.18\ \text{J/g}^\circ\text{C})\ (24.90 - 24.00\ ^\circ\text{C}) + (13.7\ \text{J}/^\circ\text{C}) \times (24.90 - 24.00\ ^\circ\text{C})]}{7.050\ \text{g} \times (24.90 - 100.0^\circ\text{C})}$$

$$C_{\text{solid}} = \frac{-[190\ \text{J} + 12\ \text{J}]}{-529\ \text{g}^\circ\text{C}} = 0.38\ \text{J/g}^\circ\text{C}$$

Table 1 *Specific heat values for selected substances*

Solids	
Aluminum	0.897 J/g°C
Calcium Carbonate	0.837 J/g°C
Copper	0.385 J/g°C
Copper(II) Oxide	0.533 J/g°C
Graphite, $C_{(s)}$	0.709 J/g°C
Iron	0.450 J/g°C
Lead	0.129 J/g°C
Zinc	0.389 J/g°C
Liquids	
1,5-Pentanediol	3.08 J/g°C
1-Hexene	2.18 J/g°C
1-Propanol	2.40 J/g°C
Acetone	2.17 J/g°C
Ethanol	2.45 J/g°C
Glycerol	2.38 J/g°C
Water	4.18 J/g°C

Table 1 lists the specific heat values of several solids and liquids. From the data presented in Table 1, the solid could be either copper or zinc metal.

From Table 1, it is evident that solids generally have lower specific heat values than liquids. In particular, water has a very high specific heat even compared to the other liquids in the table. That is one reason why water is commonly chosen as a coolant for automobile engines.

PROCEDURE

CAUTION

Students must wear departmentally approved eye protection while performing this experiment. Wash your hands before touching your eyes and after completing the experiment.

Part A - Determination of the Calorimeter Constant

1. See Appendix A-1 – **Instructions for Initializing the MeasureNet Workstation to Record a Temperature versus Time Scan.** Complete all steps in Appendix A-1 before proceeding to Step 2 below.
2. See Appendix A-2 – **Instructions for Recording a Temperature versus Time Thermogram to Determine the Styrofoam Cup Calorimeter Constant.** Complete all steps in Appendix A-2 before proceeding to Step 3.

3. *Steps 4–6 are to be completed after the laboratory period is concluded (outside of lab).* Proceed to Step 8, *Determining the Specific Heat of a Solid*
4. From the tab delimited files you saved, prepare plots of the temperature versus time data using Excel (or a comparable spreadsheet program). Instructions for plotting temperature versus time thermograms using Excel are provided in Appendix B-1.
5. How do you determine the equilibrium temperature of the hot-cold water mixture in the calorimeter? How do you determine the temperature changes for the warm water and the cold water? Should these temperatures be recorded in the Lab Report and to how many significant figures?
6. What calculations are required to determine the calorimeter constant in J/°C? From the results of Trials 1 and 2, determine the average calorimeter constant.
7. *Use the same two styrofoam cups in Part B of the experiment that were used in Part A. The calorimeter constant determined in Part A will be used in Part B of this experiment to determine the specific heat of the unknown solid.*

Part B - Determination of the Specific Heat of a Solid

8. 45–50 grams of tap water should be added to the calorimeter to serve as the cool water. Should the exact mass of the cool water be recorded in the Lab Report? If so, to what number of significant figures should the mass be recorded?
9. Place a stir bar in the bottom of the calorimeter and turn the power on to a low to medium setting. Be sure the stir bar does not contact the temperature probe or the walls of the container when spinning.
10. Obtain an unknown solid and record the *Unknown Number* in the Lab Report.
11. 15 to 20 grams of the unknown solid must be added to a 25 × 200 mm test tube (a large test tube), and nested inside of a 250 mL beaker. Must the exact mass of the unknown solid be determined? Should the mass be recorded in the Lab Report, and to how many significant figures?
12. Place the beaker containing the test tube with solid on a hot plate, and add ~ 125 mL of water to the beaker (Figure 2). Bring the water to a *gentle* boil. Do not allow any water to splash into the test tube containing the solid. Secure the test tube to a ring stand with a utility clamp.
13. Insert a thermometer into the solid (inside the test tube, see Figure 2). **The thermometer must not touch the bottom or the walls of the test tube** (tip of thermometer should be at least 1 cm from the bottom of the test tube). The test tube containing the unknown metal should remain in *boiling* water for 10 minutes.
14. Note the temperature on the MeasureNet workstation display. How should the temperature of the cool water be determined? Should this temperature be recorded in the Lab Report and to what number of significant figures?
15. Leave the test tube containing unknown solid in the boiling water for at least 10 minutes. How should the temperature of the solid be determined? Should this temperature be recorded in the Lab Report, and to what number of significant figures?

Figure 2
Heating solid inside a test tube immersed in boiling water

16. Press **Start** on the MeasureNet workstation to start the scan. After 5–10 seconds have elapsed, raise the calorimeter lid. Remove the thermometer from the test tube containing the hot solid. Using the utility clamp to hold the hot test tube, *quickly, but carefully*, transfer all of the hot solid from the test tube to the calorimeter. Do not splash water from the calorimeter as the solid sample is added. Immediately replace the lid on the calorimeter.
17. Be sure the stir bar is still turning. If a stir bar is not available, stir the contents of the calorimeter with the wire stirrer. Be sure the probe stays submerged in the water at all times. Do not slosh any water out of the calorimeter.
18. Once the water temperature has risen and stabilized at a constant temperature (reaching equilibrium), press **Stop** to end the scan.
19. Press **File Options**, then press **F3** to save the scan. Enter a 3 digit code when prompted, then press **Enter**. Save the file to a disk, flash drive, or email the files to yourself via the internet.
20. Record the file name in your Lab Report. Note what type of information is contained in the file in the Lab Report.
21. Press **Display** to clear the previous scan. You are now ready to press **Start** to record a subsequent scan for a new experimental trial.
22. Remove the stir bar from the calorimeter with a magnetic rod. Decant the water in the calorimeter into a beaker. Remove any pieces of metal from the beaker, and pour the water into the sink.
23. Thoroughly dry the unknown solid with a towel. Pour the solid into the "Used Unknown Solid" container *bearing the same number as your unknown*. Thoroughly dry the calorimeter.
24. Perform a second trial to determine the specific heat of the unknown solid by repeating Steps 8–23.
25. When you are finished with the experiment, transfer the files to a flash drive, or email the files to yourself via the internet.
26. Return the dry styrofoam cups and lid to your instructor.
27. From the tab delimited files you saved, prepare plots of the temperature versus time data using Excel (or a comparable spreadsheet program). Instructions for plotting temperature versus time thermograms using Excel are provided in Appendix B-1.
28. How do you determine the equilibrium temperature of the hot solid-cold water mixture in the calorimeter? How do you determine the temperature changes for the hot solid and the cold water? Should these temperatures be recorded in the Lab Report and to how many significant figures?
29. What calculations are required to determine the specific heat of the unknown solid? From the results of Trials 1 and 2, determine the average specific heat of the unknown solid.
30. Using the information provided in Table 1, identify the unknown solid.

Name

Section

Date

Instructor

2 EXPERIMENT 2

Lab Report

Submit all thermograms along with your lab report.

Part A – Determination of the Calorimeter Constant

Experimental data and calculations – Trial 1

Experimental data and calculations – Trial 2

Average calorimeter constant _________ J/°C

Part B – Determination of the Specific Heat of a Solid

Unknown number _________

Experimental data and calculations – Trial 1

Experimental data and calculations – Trial 2

Average specific heat of the solid ________ J/g°C

Using the specific heat values given in Table 1, identify the unknown solid.

____

Name

Section

Date

Instructor

2

EXPERIMENT 2

Pre-Laboratory Questions

1. 30.109 grams of cool water are placed in a styrofoam cup calorimeter. 30.789 grams of water are heated in a beaker to 53.11 °C. The warm water is added to the cool water in the calorimeter, and the temperature change of the system is monitored (see plot below). Calculate the calorimeter constant. *This calorimeter constant will be used in Question 2.*

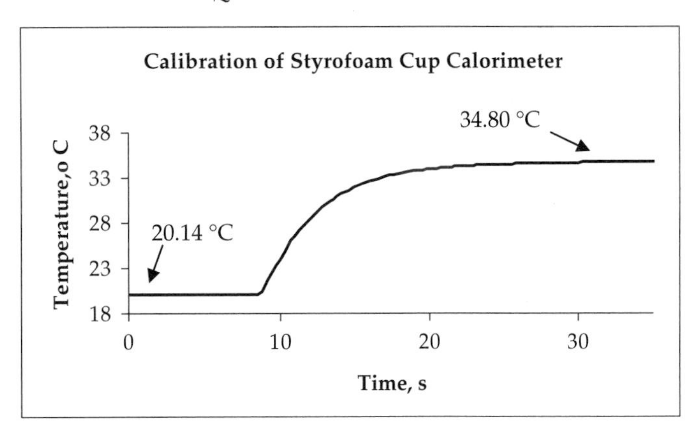

2. 25.000 grams of water are added to a styrofoam cup calorimeter. 41.397 grams of an unknown solid are heated in a large test tube to 99.60 °C. The hot solid is added to the water in the styrofoam cup calorimeter, and the temperature change of the system is monitored. Calculate the specific heat of the unknown solid. Using the information from Table 1, identify the unknown solid.

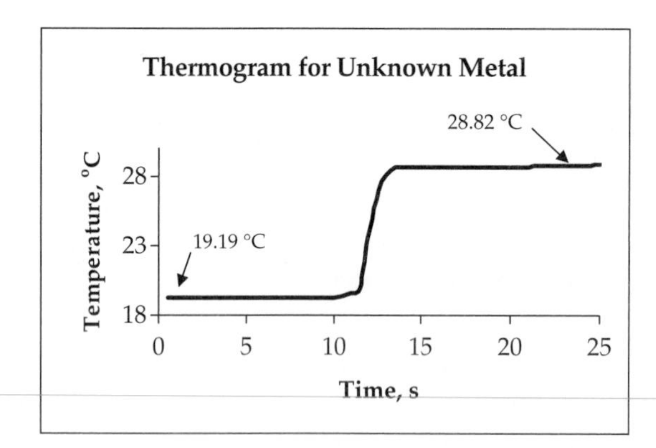

Name

Instructor

Section

Date

2 EXPERIMENT 2

Post-Laboratory Questions

1. In Step 13 of the procedure, you were instructed not to let the thermometer touch the bottom or the walls of the test tube when heating the unknown solid. Why was this instruction given?

2. In an experiment, 10.0 grams of solid A and 10.0 grams of solid B are each heated to 200 °C. Each solid is added to different beakers containing 25 grams of water at 25 °C. The equilibrium temperature of solid A is significantly higher than that of solid B. Which solid has the higher specific heat? Justify your answer with an explanation.

3. If a student incorrectly calibrated the temperature probe so that all subsequent temperature readings are 2.0 °C to high, will it affect the student's specific heat value determined in this experiment? Justify your answer with an explanation.

4. The percent error in this experiment is calculated by taking the absolute value of the difference between your experimental specific heat value and the actual value of the specific heat (from Table 1) for the solid you identified as your unknown, dividing that number by the specific heat value of the unknown solid from Table 1, then multiplying that number by 100%. Calculate the % error in the experimental specific heat value for your unknown solid.

Chromatography

OBJECTIVES

To use paper chromatography to separate the dyes in colored ink markers and identify the transition metal ions present in an unknown solution of transition metal ions by paper chromatography.

INTRODUCTION

Many substances found in nature or synthesized in the laboratory are mixtures. Often chemists must separate the individual components of a mixture to identify and characterize each component. **Chromatography** is a technique in which a **mobile phase** (a gas or a liquid) carries the components of a mixture across a **stationary phase** (a solid or a liquid). Separation of the mixture is based upon differences in migration rates between the mixture's components due to their different absorptive tendencies for the mobile or the stationary phases.

Paper chromatography, first developed in 1943 for the analysis of amino acids, is a useful technique to separate very small quantities of samples. In paper chromatography, a small sample of a mixture is spotted on a piece of chromatography paper. Chromatography paper usually consists of high quality filter paper. Filter paper is composed of cellulose, a polymeric chain of repeating glucose molecules (Figure 1).

Cellulose contains many hydroxyl groups, an -OH group attached to a C atom. The electron charge density on the oxygen atoms in hydroxyl groups is relatively large (oxygen atoms bear a partial negative charge). In contrast, the electron charge density on the hydrogen atoms in the hydroxyl groups is relatively low (hydrogen atoms bear a partial positive charge). Molecules containing large differences in electron density within a given region of a molecule are said to be **polar molecules**. Because of the high and low regions of electron density, polar molecules are attracted to other polar molecules (i.e., partially positive charged hydrogen atoms are attracted to partially negative charged oxygen atoms in adjacent molecules). In this experiment, the polar water molecules in the solvent are attracted to the cellulose. The stationary phase in paper chromatography consists of water molecules adsorbed (bonded to) onto the cellulose fibers of the paper.

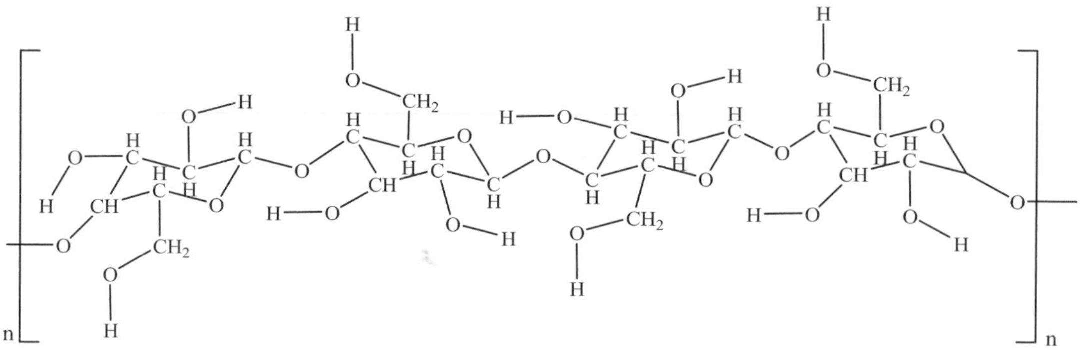

Figure 1
Polymeric structure of cellulose consisting of repeating glucose molecules. The "n" at the end of either side of the chain indicates that the chain continues in either direction. Cellulose is composed of thousands of glucose molecules.

The edge of the paper is placed in the mobile phase, a solvent which is frequently a mixture of water and one or more organic compounds. As the solvent moves over the spotted sample, the mixture's components move up the paper along with the solvent. Chromatography takes advantage of the fact that the mixture's components interact differently with the stationary and mobile phases. Some components are more strongly attracted to (adsorbed onto) the stationary phase, while other components are more attracted to (soluble in) the mobile phase. As the mobile phase moves up the paper, the components that are more soluble in the mobile phase move greater distances. Meanwhile, those components more attracted to the stationary phase move shorter distances. The components in the mixture separate based upon their attraction to the stationary and mobile phases. By adjusting the amounts of organic solvents dissolved in water, it is possible to adjust the mobile phase's polarity, which can improve the chromatographic separation.

The pattern of separated components produced on the chromatography paper forms a **chromatogram**. The point on the chromatogram where the sample is "*spotted*" is the **origin** of a chromatogram. The leading edge of the mobile phase is referred to as the **solvent front**.

The relative strength of a component's interactions with the mobile and stationary phases can be quantitatively measured using a **retention factor, R_f**, defined by Equation 1.

$$R_f = \frac{\text{distance the component travels from the origin}}{\text{distance from the origin to the solvent front}} \qquad \text{(Eq. 1)}$$

R_f values are functions of temperature and the chemical nature of the components in the mixture and the solvent. Under a given set of specific experimental conditions, each component should have a unique R_f value. Thus, the components in an unknown mixture can be identified by comparing their experimental R_f values to known R_f values of pure samples of the components under the same experimental conditions.

The R_f values reflect the solubility of the substance in the mobile phase versus the stationary phase. The charge density of a transition metal ion is dependent upon its charge and size (i.e., ionic radius). Small, highly charged transition metal ions have a large charge density. Large ions with

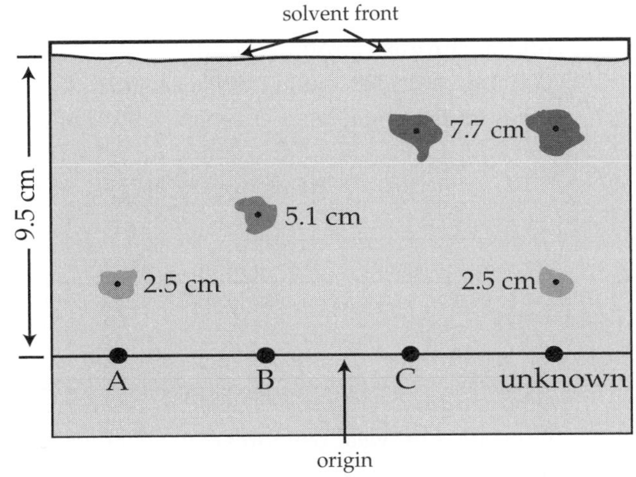

Figure 2
A sample chromatogram

low charges have small charge densities. If the mobile phase is more polar than cellulose (the stationary phase), ions with high charge densities are more soluble in the mobile phase and move greater distances from the origin. Ions with small charge densities are less soluble in the mobile phase (they adhere more strongly to cellulose, the stationary phase) and move shorter distances from the origin. Thus, the separation of transition metal ions in a chromatographic process is highly dependent upon the charge densities of these ions.

Figure 2 depicts a typical chromatogram. Spots of pure substances A, B, and C, and an unknown mixture (containing some of the substances A, B, and C) are placed on the origin of the chromatography paper. The bottom edge of the paper is placed in the solvent. As the solvent moves up the paper, the substances separate. The distance each substance travels is determined by measuring the distance from the origin to the center of the component spot. The R_f values of substances A, B, and C, (designated as $R_{f,A}$, $R_{f,B}$, *and* $R_{f,C}$, respectively) and the components in the unknown mixture ($R_{f,\text{component 1}}$, and $R_{f,\text{component 2}}$) are calculated as follows.

$$R_{f,A} = \frac{2.5\text{ cm}}{9.5\text{ cm}} = 0.26$$

$$R_{f,B} = \frac{5.1\text{ cm}}{9.5\text{ cm}} = 0.54$$

$$R_{f,C} = \frac{7.7\text{ cm}}{9.5\text{ cm}} = 0.81$$

$$R_{f,component\ 1} = \frac{2.5\text{ cm}}{9.5\text{ cm}} = 0.26$$

$$R_{f,component\ 2} = \frac{7.7\text{ cm}}{9.5\text{ cm}} = 0.81$$

Based on the calculated R_f values, the unknown sample is composed of substances A and C.

In Part A of this experiment, we will separate and determine the components in inks from red, blue, green, and black nonpermanent markers using paper chromatography. In Part B of the experiment, we will identify which of four transition metal ions (Ag^{+}, Co^{2+}, Cu^{2+}, or Fe^{3+}) are present in an unknown mixture. To enable us to see transition metal ions on the chromatogram, we will subject them to ammonia vapor to produce colored, complex ions. For example, Cu^{2+} ions reacts with ammonia, forming the royal blue colored $[Cu(NH_3)_4]^{2+}$ ion.

$$Cu^{2+} + 4\ NH_{3(g)} \rightarrow [Cu(NH_3)_4]^{2+} \qquad \text{(Eq. 2)}$$

PROCEDURE

CAUTION

Students must wear departmentally approved eye protection while performing this experiment. Wash your hands before touching your eyes and after completing the experiment.

The solvents used in this experiment are corrosive and flammable. If they contact your skin, wash the affected area with copious quantities of water and inform your lab instructor. There must be no open flames in the lab during this experiment.

Part A. Dye Separation of Nonpermanent Marker Inks

1. Obtain 15 mL of a 1:1:1 (*by volume*) water, glacial acetic acid, and 1-propanol solution. While holding a stirring rod in the center of the 600-mL beaker, *slowly* pour the solvent solution down the stirring rod. *Be careful not to wet the walls of the beaker with the solvent.* Cover the beaker with plastic wrap or aluminum foil.
2. Obtain a 10 × 15 cm piece of chromatography paper (Whatman® 1). *Place it on a clean sheet of plain (notebook) paper, do not place the chromatography paper directly on the lab bench.*
3. Using a pencil (***do not use ink***), draw a line, 1.5 cm from the bottom edge, across the length of the paper as shown in Figure 3.
4. Beginning 3.0 cm from the left edge of the paper, draw four 1 mm ovals, 3 cm apart, on the line (see Figure 3). Label (*with a pencil*) one oval red, one green, one blue, and one black.
5. Obtain one red, one blue, one green, and one black non-permanent fine point markers.
6. Lightly touch the tip of each marker to the corresponding sample spot on the chromatography paper. The spot size should be less than 1mm

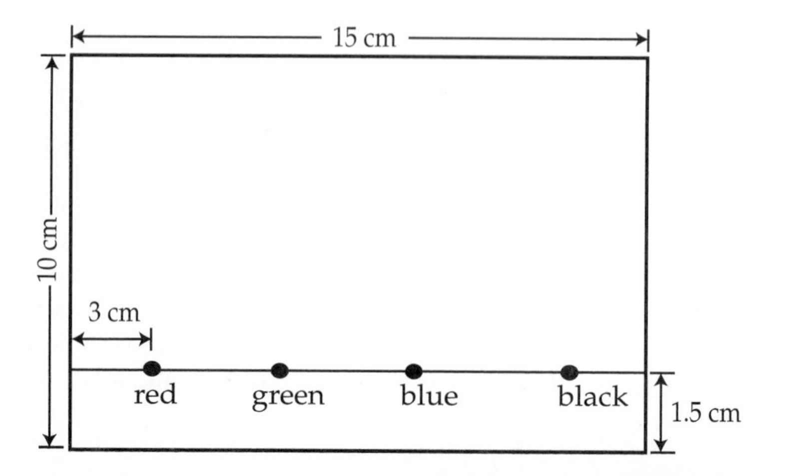

Figure 3
Chromatography Paper Preparation

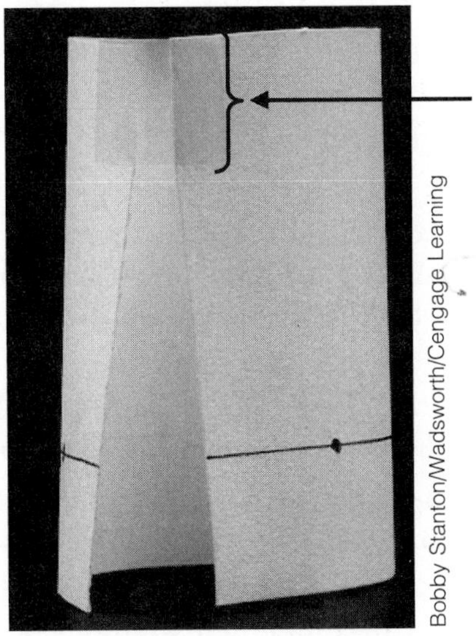

adhesive tape

Bobby Stanton/Wadsworth/Cengage Learning

Figure 4
Chromatography paper folded into a cylinder, note the placement of adhesive tape

in diameter. You may find it necessary to practice making spots on a piece of scratch paper until your spots are reproducible.

7. Allow the spots to dry.
8. Carefully roll the chromatography paper into a cylinder, with the ink spots facing out (visible to you) and the origin line at the bottom of the cylinder (see Figure 4). Tape the *top* corners of the paper cylinder together with a 2 cm strip of adhesive tape *leaving a 1–1.5 cm gap between thetwo edges of the paper,* see Figure 4. Alternatively, you can staple the top and bottom corners of the paper to form a similar cylinder.
9. Remove the cover from the 600-mL beaker containing the solvent solution, and carefully place the paper cylinder inside of the 600-mL beaker. The ink spots should be at the bottom of the paper, just above the surface of the solvent. Be certain that the paper does not touch the walls of the beaker. Cover the beaker with plastic wrap or aluminum foil, being careful not to splash any solvent on the paper.
10. Leave the beaker undisturbed until the solvent front rises to within 1.0–1.5 cm from the top of the paper.
11. At that point remove the paper cylinder from the beaker. Carefully remove the adhesive tape. Using a pencil, trace a line along the solvent front. Allow the paper to dry.
12. After the paper has dried, circle the sample spots with a pencil, and place a dot at the center of each spot. Identify the colors of the dyes (components) that constitute each type of ink in the lab report. Which inks are composed of only one dye? Which inks are composed of multiple dyes?
13. Staple your chromatogram to your lab report. Discard the solvent into the designated Waste Container.

Part B. Identification of Transition Metal Ions Present in an Unknown Solution

14. Obtain 15 mL of a 1:1:1 (*by volume*) 1-butanol, ethanol, and 6 *M* hydrochloric acid solution. While holding a stirring rod in the center of the 600-mL beaker, *slowly* pour the solvent solution down the stirring rod. *Be careful not to wet the walls of the beaker with the solvent.* Cover the beaker with plastic wrap or aluminum foil.

15. Obtain a 10 × 15 cm piece of chromatography paper (Whatman® 1). *Place it on a clean sheet of plain (notebook) paper, do not place the chromatography paper directly on the lab bench.*
16. Using a pencil (*do not use ink*), draw a line, 1.5 cm from the bottom edge, across the length of the paper as shown in Figure 3.
17. Beginning 3.0 cm from the left edge of the paper, draw five 1 mm ovals, 3 cm apart on the line. Use a pencil to label the marks in the following sequence (left to right across the paper): Ag^+, Co^{2+}, Cu^{2+}, Fe^{3+}, and unknown.
18. Obtain an unknown transition metal solution from your instructor. Record the Unknown Number in the Lab Report.
19. Using a plastic Beral pipet, transfer five drops of the Ag^+ known solution to one well of a clean 24 well-plate.
20. Repeat Step 19 for each of the three remaining known transition metal solutions and the unknown solution. Be certain to use a clean Beral pipet for each solution to avoid contamination of the solutions.
21. Practice the "spotting" procedure using water and a new piece of filter paper (not the one prepared in Steps 15–17). Dip the end of a paper clip or a toothpick into distilled water and lightly touch the filter paper. The spot should not exceed 4 mm in diameter. Practice spotting the paper until you can reproduce a 4 mm spot size.
22. Dip the end of a paper clip or toothpick into one of the transition metal solutions and lightly touch the tip of the paper clip or the toothpick to the corresponding spot on the chromatography paper.
23. Using different paper clips or toothpicks, repeat Step 22 for each of the remaining transition metal solutions and the unknown metal solution.
24. Let the paper air dry (place it under a heat lamp if one is available).
25. Repeat steps 22–24 several times in succession to produce a spot size of 3–4 mm for each metal ion on the paper.
26. Carefully roll the chromatography paper into a cylinder, with the metal spots facing out (visible to you) and the origin line at the bottom of the cylinder (see Figure 4). Tape the *top* corners of the paper cylinder together with a 2 cm strip of adhesive tape (*do not allow the two edges of the paper to touch,* see Figure 4).
27. Remove the cover from the 600-mL beaker containing the solvent solution, and carefully place the paper cylinder inside the 600-mL beaker. The metal ion spots should be at the bottom of the paper, just above the surface of the solvent. Be certain that the paper does not touch the walls of the beaker. Cover the beaker with plastic wrap or aluminum foil, being careful not to splash any solvent on the paper.
28. Leave the beaker undisturbed until the solvent front rises to within 1.0–1.5 cm from the top of the paper.
29. Remove the paper cylinder from the beaker. Carefully remove the adhesive tape. Using a pencil, trace a line along the solvent front.
30. Allow the paper to dry. Notice any metal ion spots that are visible on the paper. *Carefully* circle the visible spots with a *pencil*.

a)

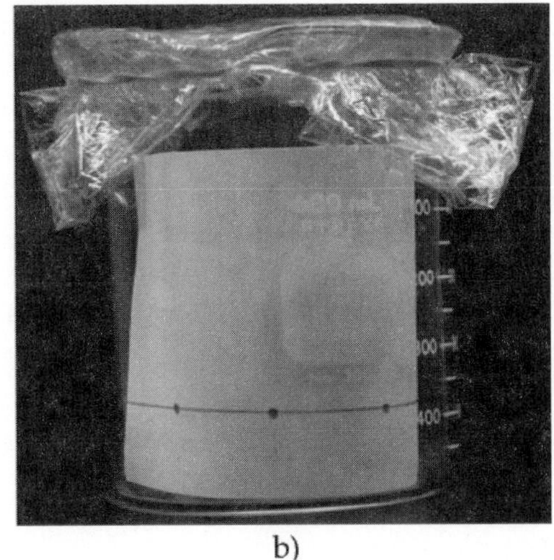

b)

Figure 5

Ammonia Vapor Generator: a) a 50-mL beaker containing 10–15 mL of 3 M aqueous ammonia nested inside a 600-mL beaker; b) the chromatogram surrounding the 50-mL beaker inside the ammonia generator.

31. Under a fume hood, place a 50-mL beaker inside of a 600-mL beaker (see Figure 5a). Add 10–15 mL of 3 M aqueous ammonia to the 50-mL beaker. Be careful not to splatter any of the ammonia out of the 50-mL beaker, the 600-mL beaker must remain dry.

CAUTION

Ammonia is an irritant to the skin and the respiratory system. Do not inhale ammonia vapor. The ammonia chamber must be placed in the fume hood and covered.

32. Place the paper cylinder inside the ammonia chamber (spots facing out), surrounding the 50-mL beaker as depicted in Figure 5b. Cover the chamber with plastic wrap or parafilm.
33. After the chromatogram has been inside the ammonia chamber for approximately 10 minutes, remove the paper. Notice any new spots that are visible. Circle the spots with a *pencil* and place a dot at the center of each spot.
34. Determine the R_f values for each metal ion and for each component in the unknown solution.
35. Identify the metal ions present in the unknown solution.
36. Staple your chromatogram to your Lab Report. Discard the solvent into the designated waste container.

Name

Instructor

Section

Date

3 EXPERIMENT 3

Lab Report

SHOW ALL WORK TO RECEIVE FULL CREDIT.

Part A. Dye Separation of Nonpermanent Marker Inks

Indicate the dyes present in each ink: red, green, blue, and black.

Which inks are composed of only one dye? Which inks are composed of multiple dyes?

Staple your chromatogram here.

Part B. Identification of an Unknown Metal Nitrate Solution

Unknown number ____________________

Determine R_f values for each metal ion and each component of the unknown solution.

Identify the metal ions present in the unknown solution.

____

Staple your chromatogram here.

Name

Section

Date

Instructor

3 EXPERIMENT 3

Pre-Laboratory Questions

SHOW ALL WORK TO RECEIVE CREDIT.

1. What is the purpose of the mobile phase in chromatography?

2. What is the R_f value for a component in a sample?

3. A student developed chromatograms for pure substances A, B, and C, and an unknown mixture (containing some of the substances A, B, and C). Calculate the R_f values of A, B, and C, and each component of the unknown mixture. Identify the substances present in the unknown mixture.

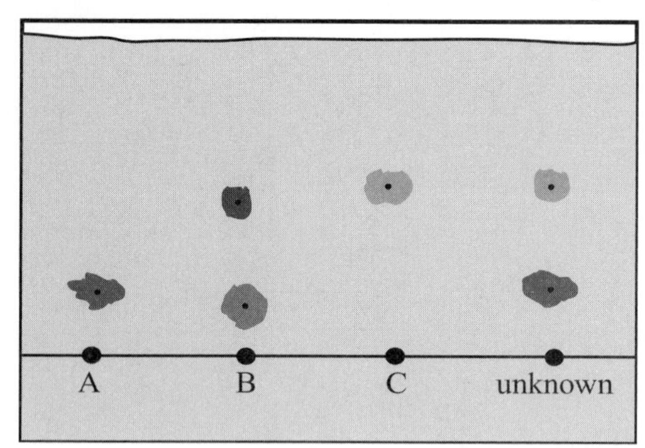

Name

Section

Date

Instructor

3 EXPERIMENT 3

Post-Laboratory Questions

1. What compound servers as the stationary phase in chromatography paper. Why does this compound adhere well to cellulose?

2. Why were you instructed to place the chromatography paper on a clean sheet of paper (instead of placing it directly on the lab bench) when spotting the samples?

3. Addition of which of these compounds to water would produce the *more* polar solution? Justify your answer.

H_3C–O–H

methanol

H_3C–CH_2–CH_2–CH_3

butane

4. A spot from a green, nonpermanent ink marker and a spot of green food dye are placed on a piece of chromatography paper. The chromatogram is placed in a mobile phase consisting of 90% water 10% propanol for 15 minutes. After 15 minutes, the green marker and green food dyes each separated into two components, a yellow component and a blue component. The R_f values for each of the yellow and blue components were identical.

   A. Do these data conclusively prove that the ink marker and the food dye were both made from the same green dye? Justify your answer.

   B. What modifications to the experimental procedure would you suggest to assist you in answering question 4. A) above.

Determination of the Percent by Mass of the Components in a Mixture by Thermal Gravimetric Analysis

OBJECTIVES

Determine the percent by mass of sodium hydrogen carbonate and potassium chloride in a mixture using thermal gravimetric analysis.

INTRODUCTION

Thermal gravimetric analysis is used to determine the percent by mass of a component in a mixture. When a mixture is heated to a sufficiently high temperature, one component in the mixture decomposes to form a gaseous compound(s). The other component(s) in the mixture will not decompose upon heating. The mass of the component that decomposes is stoichiometrically related to the mass of the gaseous compound liberated. For example, $BaCl_2{\cdot}2H_2O$ (barium chloride dihydrate) loses water and forms anhydrous $BaCl_2$ when heated above 100 °C.

$$BaCl_2{\cdot}\ 2H_2O_{(s)} \xrightarrow{\Delta} BaCl_{2(s)} + 2\ H_2O_{(g)} \qquad \text{(Eq. 1)}$$

The balanced equation indicates that 2 moles of H_2O vapor are liberated per mole of $BaCl_2{\cdot}2H_2O$ decomposed. If 0.326 grams of H_2O is liberated when a sample of $BaCl_2{\cdot}2H_2O$ is heated, the original mass of $BaCl_2{\cdot}2H_2O$ can be calculated as follows.

$$0.326\ \text{g}\ H_2O \times \frac{1\ \text{mole}\ H_2O}{18.0\ \text{g}\ H_2O} \times \frac{1\ \text{mole}\ BaCl_2{\cdot}2H_2O}{2\ \text{mole}\ H_2O} \times \frac{244.3\ \text{g}\ BaCl_2{\cdot}2H_2O}{1\ \text{mole}\ BaCl_2{\cdot}2H_2O} = 2.21\ \text{g}\ BaCl_2{\cdot}2H_2O$$

When a mixture containing potassium hydrogen carbonate ($KHCO_3$) and sodium chloride (NaCl) is heated to 300 °C in a porcelain crucible,

$KHCO_3$ decomposes to form solid potassium carbonate (K_2CO_3), carbon dioxide gas (CO_2), and water vapor. Sodium chloride does not decompose at this temperature.

$$2\ KHCO_{3(s)} \xrightarrow{\Delta} K_2CO_{3(s)} + CO_{2(g)} + H_2O_{(g)} \qquad \text{(Eq. 2)}$$

Any mass lost by the mixture is attributed to the decomposition of $KHCO_3$. Equation 2 indicates that 1 mole of CO_2 gas and 1 mole of H_2O vapor are liberated when 2 moles of $KHCO_3$ completely decomposes. The mass of CO_2 and H_2O lost per gram of $KHCO_3$ is calculated in the following manner.

Mass of CO_2 lost per gram of $KHCO_3$.

$$1.000\ g\ KHCO_3 \times \frac{1\ mole\ KHCO_3}{100.1\ g\ KHCO_3} \times \frac{1\ mole\ CO_2}{2\ mole\ KHCO_3} \times \frac{44.0\ g\ CO_2}{1\ mole\ CO_2} = 0.220\ g\ CO_2$$

Mass of H_2O lost per gram of $KHCO_3$.

$$1.000\ g\ KHCO_3 \times \frac{1\ mole\ KHCO_3}{100.1\ g\ KHCO_3} \times \frac{1\ mole\ H_2O}{2\ mole\ KHCO_3} \times \frac{18.0\ g\ H_2O}{1\ mole\ H_2O} = 0.0899\ g\ H_2O$$

The total mass of CO_2 and H_2O lost is 0.310 grams per gram of $KHCO_3$ decomposed. This relationship can be expressed as a combined stoichiometric ratio that includes both the CO_2 and H_2O. The ratio tells us that if we heat 1.000 gram of pure $KHCO_3$, 0.310 grams of CO_2 and H_2O will be liberated.

$$\frac{1.000\ g\ KHCO_3}{0.310\ g\ CO_2\ \text{and}\ H_2O}$$

The product of the stoichiometric ratio and the mass of CO_2 and H_2O liberated in the reaction yields the mass of $KHCO_3$ in a mixture.

$$\text{mass of}\ CO_2\ \text{and}\ H_2O\ \text{liberated} \times \frac{1.000\ g\ KHCO_3}{0.310\ g\ CO_2\ \text{and}\ H_2O} = \text{mass}\ KHCO_3\ \text{in mixture}$$

In this experiment, the percent by mass of sodium hydrogen carbonate ($NaHCO_3$) and potassium chloride (KCl) in a mixture will be determined. Potassium chloride does not decompose upon heating. The thermal decomposition of $NaHCO_3$ yields sodium carbonate (Na_2CO_3), carbon dioxide gas, and water vapor.

$$2\ NaHCO_{3(s)} \xrightarrow{\Delta} Na_2CO_{3(s)} + CO_{2(g)} + H_2O_{(g)} \qquad \text{(Eq. 3)}$$

Heating to Constant Mass

The progress of a thermal gravimetric reaction is monitored by loss of mass from the sample mixture. When there is no longer any detectable loss of mass from the sample after a heating cycle, the reaction has gone to completion and is said to be **heated to constant mass.**

The sample is generally heated for 15–20 minutes for the first heating-cooling cycle. The sample is cooled to room temperature. The sample and its container are weighed. The sample is heated a second time for approximately 5 minutes, and cooled to room temperature. The sample and its container are reweighed. If the mass difference between the first and second heating-cooling cycles is no greater than ± 0.005 g or ± 0.01 g (depending on whether the balance measures to the nearest 0.001 g or 0.01 g), the sample has reached constant mass. If the mass difference is greater than ± 0.005 g or ± 0.01 g, a third 5 minute heating-cooling cycle is required. Heating-cooling cycles are continued until the mass difference between two consecutive cycles differ by no more than ± 0.005 g or ± 0.01 g.

PROCEDURE

CAUTION

Students must wear departmentally approved eye protection while performing this experiment. Wash your hands before touching your eyes and after completing the experiment.

Do not touch a hot crucible, clay triangle, or iron ring.

Do not place a hot crucible on the bench top, lab manual, cloth or paper towels to cool. Place it on wire gauze to cool.

1. Obtain a crucible and lid. Clean the crucible by adding 2–3 milliliters of 6 *M* hydrochloric acid (HCl). Leave the HCl in the crucible for 2–3 minutes. Use a spatula to remove any residual material in the crucible. Pour the acid and residual into the "Waste Container."

Chemical Alert

Avoid splattering 6 M hydrochloric acid from the crucible. Hydrochloric acid is very corrosive. If HCl contacts your skin, immediately wash the affected area with copious quantities of water and inform your lab instructor.

2. Rinse the crucible with copious quantities of water. Decant the water into a laboratory sink and thoroughly dry the crucible.
3. Place a clay triangle on an iron ring supported by a ring stand (Figure 1). Suspend the crucible and lid inside the clay triangle. Place the lid on the crucible *slightly ajar*, do not completely cover the crucible.

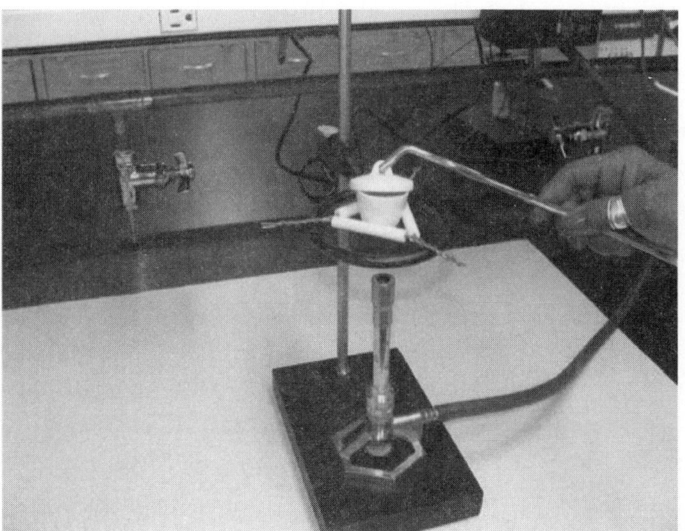

Bobby Stanton/Wadsworth/Cengage Learning

Figure 1
A crucible supported by a clay triangle and iron ring with the placement of the crucible lid slightly ajar

4. Heat the crucible with a Bunsen burner for 5 minutes; the bottom of the crucible should glow red.
5. Using crucible tongs completely cover the crucible with the lid, and *carefullyplace the hot crucible and lid on wire gauze*. Let the crucible cool to room temperature (see Note).

NOTE: Always weigh an object when it is at room temperature. Warm objects cause convection currents around the balance that may cause erroneous readings. In addition, warm objects are more susceptible to absorbing moisture from the atmosphere, which can increase the mass of the object.

---

6. Should you determine the mass of the cool crucible and lid? If so, to what number of significant figures should the mass be recorded in the Lab Report?
7. Add approximately 1 gram (it does not have to be exactly 1 g) of an unknown mixture to the crucible. Record the Unknown Number in the Lab Report.
8. Should the exact mass of the unknown mixture, crucible and lid be determined? If so, to what number of significant figures should the mass be recorded in the Lab Report?
9. Place the crucible containing the mixture in the clay triangle as shown in Figure 1.
10. Heat the mixture *gently* for five 5 minutes. Adjust the air in-take of the burner to remove the inner blue cone of the flame to reduce the flame temperature. Periodically, use crucible tongs to raise the lid and view the mixture. If the sample begins to melt, reduce the flame temperature (see Note).

NOTE: Do not melt the sample. Molten Na_2CO_3 can react with silica in the crucible, releasing additional CO_2, which will lead to an erroneous percent by mass of $NaHCO_3$ in the mixture.

---

11. After 5 minutes of gentle heating, adjust the burner air in-take to produce a distinct inner blue cone of flame and *vigorously* heat for 15 minutes.
12. **Heat the mixture to constant mass**. Should the exact mass of the unknown mixture, crucible, and lid be determined after every heating-cooling cycle? If so, to what number of significant figures should each mass be recorded in the Lab Report.
13. Repeat Steps 1–12 with a second 1 gram sample of the mixture to perform a second trial to determine the percent by mass of $NaHCO_3$ in the mixture.
14. Clean the crucible with 6 *M* HCl (see Step 1). Pour the acid and residual material into the "Waste Container." Rinse the crucible with water, decant the water into a sink, and thoroughly dry the crucible and lid.
15. Determine the stoichiometric ratio for the mass loss of CO_2 and H_2O per gram of $NaHCO_3$ decomposed.
16. Determine the mass of CO_2 and H_2O liberated in the experiment.
17. Determine the percent by mass of $NaHCO_3$ and the percent by mass of KCl in the mixture for each trial.
18. Determine the average percent by masses of $NaHCO_3$ and KCl in the mixture from trials 1 and 2.

Name Section Date

Instructor

4 EXPERIMENT 4

Lab Report

SHOW ALL WORK TO RECEIVE FULL CREDIT.

Experimental data –Trial 1

Experimental data –Trial 2

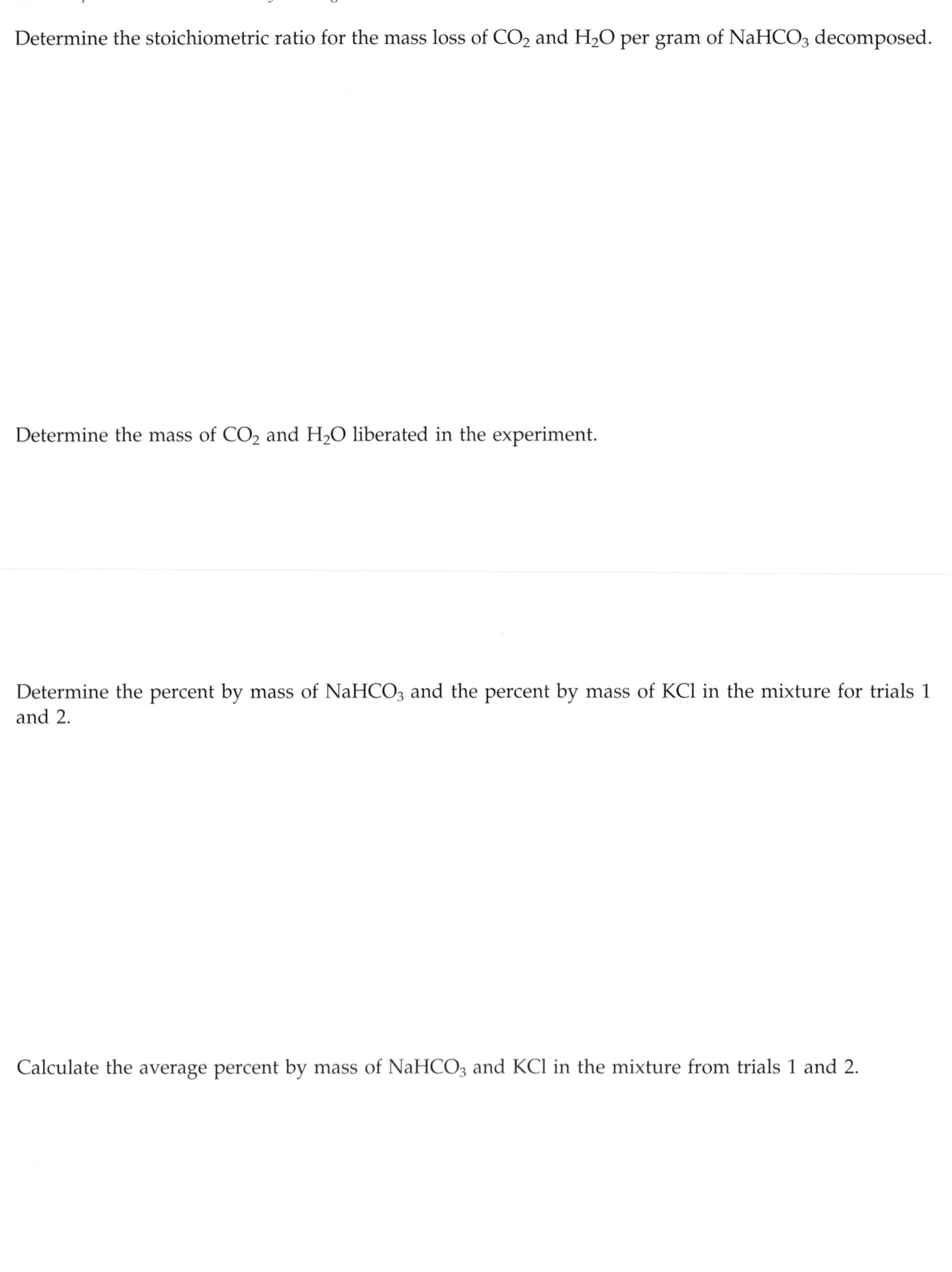

Determine the stoichiometric ratio for the mass loss of CO_2 and H_2O per gram of $NaHCO_3$ decomposed.

Determine the mass of CO_2 and H_2O liberated in the experiment.

Determine the percent by mass of $NaHCO_3$ and the percent by mass of KCl in the mixture for trials 1 and 2.

Calculate the average percent by mass of $NaHCO_3$ and KCl in the mixture from trials 1 and 2.

Name | Section | Date

Instructor

4 EXPERIMENT 4

Pre-Laboratory Questions

SHOW ALL WORK TO RECEIVE CREDIT.

A clean, dry porcelain crucible is weighed and the mass is recorded. A mixture containing $NaHCO_3$ and KCl is added to the crucible and the crucible is reweighed.

$$2\ NaHCO_{3(s)} \rightarrow Na_2CO_{3(s)} + CO_{2(g)} + H_2O_{(g)}$$

Mass of crucible and lid	18.498 g
Mass of crucible, lid, and mixture	19.734 g

The crucible and mixture are heated for 15 minutes, then allowed to cool to room temperature. The crucible and residue are weighed and the mass is recorded.

Mass of crucible, lid, and residue after 1st heating	19.476 g

The crucible and residue are reheated for 5 minutes, then allowed to cool to room temperature. The crucible and residue are weighed and the mass is recorded.

Mass of crucible, lid, and residue after 2nd reheating	19.368 g

The crucible and residue are reheated for 5 minutes, then allowed to cool to room temperature. The crucible and residue are weighed and the mass is recorded.

Mass of crucible, lid, and residue after 3rd reheating	19.369 g

1. Determine the mass of the mixture used in the experiment.

2. Calculate the stoichiometric ratio for the mass loss of CO_2 and H_2O per gram of $NaHCO_3$.

3. Determine the total mass loss of CO_2 and H_2O after three heating-cooling cycles.

4. Determine the mass of $NaHCO_3$ in the mixture.

5. Determine the percent by mass of $NaHCO_3$ and KCl in the mixture.

Name | Section | Date

Instructor

4 EXPERIMENT 4

Post-Laboratory Questions

SHOW ALL WORK TO RECEIVE CREDIT.

1. A student cleaned a crucible with HCl. After pouring the HCl into the waste container, the crucible was neither rinsed with water nor dried before the student added the $NaHCO_3$-KCl mixture. The crucible contained some residual HCl. How would this experimental error affect the calculated percentage of $NaHCO_3$ in the mixture? Justify your answer with an explanation. (*HINT*: $NaHCO_3$ is a base).

2. A student dried a wet crucible with a Kimwipe™. A piece of Kimwipe™ was left clinging to the inside wall of the crucible. The student did not see the piece of Kimwipe™ when adding the $NaHCO_3$-KCl mixture to the crucible. The crucible and its contents were weighed and subjected to three heating-cooling cycles (each time the temperature reached 600 °C). Could this experimental error significantly affect the calculated percentage of $NaHCO_3$ in the mixture? Justify your answer with an explanation.

3. The laboratory technician accidentally mixed $NaHCO_3$ with $BaCl_2{\cdot}2H_2O$ instead of KCl when preparing unknown mixtures for this experiment.

A. How would this mistake affect the calculated percentage of $NaHCO_3$ in the mixture? Justify your answer with an explanation.

B. Is it still possible to determine the percentage of $NaHCO_3$ in the mixture given the error made by the lab technician? Justify your answer with an explanation.

Thermal Insulating Materials: A Self-Directed Experiment

OBJECTIVES

Determine which of several common substances are the best thermal insulating materials.

INTRODUCTION

Heat is thermal energy that flows between two samples of matter at different temperatures. Sometimes it is desirable to minimize heat or thermal energy transfer. This is achieved by establishing a thermal barrier between hot and cold objects. For example, in winter we may wish to keep heat from escaping from our bodies to the cold surroundings by wearing a jacket. We may also want to keep hot coffee from becoming cold by putting it in a thermos bottle. Thermal insulating materials are used to minimize heat transfer by reducing conduction (transfer of heat between two objects in direct contact), convection (transfer of heat by gases or liquids in motion), and radiation (transfer of heat by electromagnetic waves).

In Part A of this experiment, student teams will determine the best material to be used as a coffee-cup calorimeter. In Part B of the experiment, student teams will determine which of several common substances is the best thermal insulating material.

Teams of students will rely on chemical knowledge learned and the laboratory skills and techniques acquired so far in this course to design and perform this self-directed experiment. Your lab instructor will specify the number of students per team. Each team will submit a **Procedure Proposal** to their laboratory instructor two weeks before the experiment is performed. The Procedure Proposal must address several important questions:

1. What is the central question to be answered in this experiment?
2. What experimental techniques will be utilized to answer the central question?

3. What are the criteria for identifying the best coffee-cup calorimeter?
4. What are the criteria for determining the best insulating material?
5. What calculations should be provided in the Procedure Proposal?
6. What safety precautions must be addressed in the Procedure Proposal?
7. What additional questions should be addressed by each team?

Each team may use their lab manual, textbook, reference books, and reliable internet resources for assistance in writing the Procedure Proposal. The format for writing the Procedure Proposal is provided in Appendix C-1. The graded and annotated Procedure Proposal will be returned to each team one week before the lab period during which the experiment is to be performed.

When the experiment is completed, each team will submit one **Formal Lab Report** addressing the following questions:

1. What conclusions can be drawn from the experimental data colleted?
2. Does the data answer the central question?
3. Are the team's conclusions in agreement with the information obtained from reference resources, textbook, or the internet?
4. What are some possible sources of error in the experimental data?
5. What modifications could be made to the experimental design and procedures to improve the accuracy of the data?

The format for writing the Formal Lab Report is provided in Appendix C-2. Your lab instructor will inform you of the submission date for the Formal Lab Report.

PROCEDURE

In Part A of this experiment, each team will be given four different types of cups: metal, ceramic, plastic, and styrofoam. Using the heat transfer principles studied in previous experiments, students will determine which of these types of cups makes the best calorimeter.

Part B of this experiment is a collaborative project. Each team will be provided with one of the following insulating materials: styrofoam peanuts (i.e, packing material), paper towels, cloth towels, aluminum foil, or sea sand. Each team will use the assigned material to construct an insulated Dixie® cup in a 600-mL beaker. A cardboard lid will be provided to cover the Dixie® cup.

Teams will share their data for a given type of insulating material with other teams in the lab section by exchanging data files and experimental results. Using the heat transfer principles studied in previous experiments, each team of students will determine which of these materials is the best insulator.

At the end of the experiment, return the Dixie cup, insulating materials (cloth, paper, aluminum foil, styrofoam peanuts, sea sand, if dry), cardboard lid, and the cups to the laboratory instructor.

LIST OF SPECIAL EQUIPMENT

A. MeasureNet workstation and temperature probe
B. Thermometer
C. Balance
D. 600 mL beaker
E. Dixie® cup
F. Cardboard lid (for calorimeters)

Reaction Stoichiometry

OBJECTIVES

Using reaction stoichiometry, determine the molar mass and the identity of an unknown, powdered metal, and to identify an unknown metal nitrate solution.

INTRODUCTION

A balanced chemical equation provides several important pieces of information about a chemical reaction: 1) the substances consumed (reactants) and formed (products) in the reaction; 2) the physical states of the reactants and products; 3) the reaction conditions; and 4) the relative amounts of reactants and products involved in the reaction. **Reaction stoichiometry** is the study of the amounts of materials consumed and produced in chemical reactions.

At the macroscopic level, the coefficients in a balanced chemical equation represent the molar ratios of substances involved in the reaction. These molar ratios can be written as stoichiometric factors. A **stoichiometric factor** is a fraction expressing the molar relationship between two reactants, between a reactant and a product, or between two products.

For example, consider the complete combustion of propane (C_3H_8) in air to produce carbon dioxide and water.

$$C_3H_{8(g)} + 5\ O_{2(g)} \rightarrow 3\ CO_{2(g)} + 4\ H_2O_{(g)} \qquad \text{(Eq. 1)}$$

The balanced chemical equation indicates that 5 moles of O_2 are required for the complete combustion of 1 mole of C_3H_8. Two stoichiometric factors (one the reciprocal of the other) express this relationship.

$$\frac{5 \text{ mol } O_2}{1 \text{ mol } C_3H_8} \text{ or } \frac{1 \text{ mol } C_3H_8}{5 \text{ mol } O_2}$$

Similarly, 3 moles of CO_2 and 4 moles of H_2O can be produced per mole of C_3H_8 reacted. The stoichiometric factors for these relationships are written as follows.

$$\frac{3 \text{ mol } CO_2}{1 \text{ mol } C_3H_8} \text{ or } \frac{1 \text{ mol } C_3H_8}{3 \text{ mol } CO_2} \text{ and } \frac{4 \text{ mol } H_2O}{1 \text{ mol } C_3H_8} \text{ or } \frac{1 \text{ mol } C_3H_8}{4 \text{ mol } H_2O}$$

Frequently, quantities of reactants are not exactly equal to the stoichiometric factors that appear in chemical reactions. For example, the moles of oxygen required for the complete combustion of 2.13 moles of C_3H_8 is calculated in the following manner.

$$2.13 \text{ mol } C_3H_8 \times \frac{5 \text{ mol } O_2}{1 \text{ mol } C_3H_8} = 10.65 \text{ mol } O_2$$

Similarly, the number of moles of CO_2 and H_2O produced by the complete combustion of 2.13 moles of C_3H_8 are calculated in the following manner.

$$2.13 \text{ mol } C_3H_8 \times \frac{3 \text{ mol } CO_2}{1 \text{ mol } C_3H_8} = 6.39 \text{ mol } CO_2$$

$$2.13 \text{ mol } C_3H_8 \times \frac{4 \text{ mol } H_2O}{1 \text{ mol } C_3H_8} = 8.52 \text{ mol } H_2O$$

In this experiment, we will identify an unknown, powdered metal and an unknown, metal nitrate solution using their reaction stoichiometry. In Part A of the experiment, an unknown, powdered sample of a metal, generically designated as M, will react with hydrochloric acid, HCl, producing a salt (MCl_2) and releasing hydrogen gas, H_2, as shown in Equation 2.

$$M_{(s)} + 2\ HCl_{(aq)} \rightarrow MCl_{2(aq)} + H_{2(g)} \qquad \text{(Eq. 2)}$$

Equation 2 indicates that 1 mole of metal is required to completely react with 2 moles of HCl. The solution bubbles vigorously as H_2 gas is liberated. From the moles of metal that react with a given number of moles of acid, and knowing the mass, in grams, of the metal added to the HCl solution, the molar mass of the metal can be calculated and the unknown metal identified.

One definition of an **acid** is a substance that produces hydrogen ions, $H^+_{(aq)}$, in aqueous solution. **Ionization** is a process in which a molecular compound separates into ions in water. Hydrochloric acid is a strong acid. All strong acids ionize essentially 100% in water. For example, HCl gas (a molecular compound containing no ions) *completely* ionizes to form $H^+_{(aq)}$ and $Cl^-_{(aq)}$ ions in aqueous solution as shown in Equation 3.

$$HCl_{(g)} \xrightarrow{H_2O} H^+_{(aq)} + Cl^-_{(aq)} \qquad \text{(Eq. 3)}$$

Free H^+ ions do not exist in aqueous solution. H^+ ions readily associate with water molecules to form hydronium ions, H_3O^+ ions as represented in Equation 4.

$$H^+_{(aq)} + H_2O_{(\ell)} \rightarrow H_3O^+_{(aq)} \qquad \text{(Eq. 4)}$$

Commonly, H_3O^+ and H^+ are used interchangeably when referring to hydrated hydrogen ions in aqueous solution. Equation 5 best represents the complete ionization of HCl in water.

$$HCl_{(g)} + H_2O_{(\ell)} \rightarrow H_3O^+{}_{(aq)} + Cl^-{}_{(aq)} \qquad \text{(Eq. 5)}$$

The pH of an aqueous solution is directly related to its hydrogen ion concentration. The molarity of the H_3O^+ ions in solution is commonly written as $[H_3O^+]$. The **pH** is the negative logarithm of the hydrogen ion concentration in a solution.

$$pH = -\log[H_3O^+] \qquad \text{(Eq. 6)}$$

The common logarithm of a number is the power to which 10 must be raised to equal that number. For example, the logarithm of 1.00×10^{-7} is -7. The pH scale is used as a convenient, short hand method of expressing the acidity or basicity of a solution. The pH of an aqueous solution is a number typically between 0 and 14. A solution with a $pH < 7$ is acidic. A solution with $pH = 7$ is neutral. A solution with a $pH > 7$ is basic. For example, a solution with $[H_3O^+] = 1.00 \times 10^{-2}$ M would have a pH of 2.00 and is acidic. The pH of a solution can be measured with **a pH meter,** which uses an electrode to measure the $[H_3O^+]$ in solution.

In Part A of this experiment, a specified volume of 1.0 *M* HCl and water are added to a beaker. Before the addition of any of the unknown metal, the $[H_3O^+]$ in the HCl solution is high because HCl is a strong acid. This is reflected in the solution's low pH of approximately 1.0 to 1.5. As the metal powder reacts with H^+ ions in the HCl solution, the $[H_3O^+]$ decreases. The pH of the solution, monitored with a pH meter, gradually increases. After the pH of the solution has increased by approximately 2 pH units, 99% of the HCl has been consumed, and the reaction is essentially complete. The moles of acid reacted is calculated by multiplying the volume of HCl used times the molarity of the HCl solution.

$$\text{moles} = (\text{molarity of solution})\ (\text{volume of solution used in L}) \qquad \text{(Eq. 7)}$$

Using the reaction stoichiometry from Eq. 2, the moles of metal reacted can be calculated from the moles of HCl consumed in the reaction. From the moles of metal reacted and the mass of metal added to the HCl solution, the molar mass of the metal can be determined.

$$\text{molar mass of metal} = \frac{\text{mass of metal}}{\text{moles of metal reacted}} \qquad \text{(Eq. 8)}$$

Part B of the experiment involves the reaction of an unknown metal nitrate solution with sodium hydroxide (NaOH) solution. Many transition metal ions form insoluble metal hydroxides upon reaction with NaOH solution. From the stoichiometry of this precipitation reaction, we can identify the transition metal ion present in the unknown solution.

Dissociation is a process in which ionic compounds separate into ions when they dissolve in water. Sodium hydroxide (an ionic compound containing Na^+ and OH^- ions) completely dissociates when dissolved in water. Equation 9 represents the dissociation of NaOH.

$$NaOH_{(s)} \overset{H_2O}{\rightarrow} Na^+_{(aq)} + OH^-_{(aq)} \qquad \text{(Eq. 9)}$$

Transition metal nitrates also completely dissociate in water. For example, nickel(II) nitrate produces Ni^{2+} and NO_3^- ions when dissolved in water. Because of a process called **hydrolysis**, many transition metal ions react with water to produce acidic aqueous solutions having a pH in the 3 to 6 range. These acidic transition metal nitrate solutions, such as $Ni(NO_3)_{2(aq)}$, react with OH^- ions to form insoluble metal hydroxide precipitates as represented in Equation 10.

$$Ni^{2+}_{(aq)} + 2\ OH^-_{(aq)} \rightarrow Ni(OH)_{2(s)} \qquad \text{(Eq. 10)}$$

In Part B of this experiment, a transition metal nitrate solution will be titrated with sodium hydroxide. Generally, a **titration** involves the addition of small increments of a reagent with a *known* concentration to a known volume of a solution of *unknown* concentration until the reaction is complete. In a titration, a **buret** (Figure 1a) is used to dispense small increments of a reagent solution.

Because the smallest calibration unit for a typical buret is 0.1 mL (Figure 1b), the volume dispensed from the buret is estimated to the nearest 0.01 mL. (*It should be noted that the volume of an aqueous solution in a buret is always read at the bottom of the solution's meniscus*).

The **equivalence-point** of a titration occurs when the added quantity of one reactant is the exact amount necessary for stoichiometric reaction with

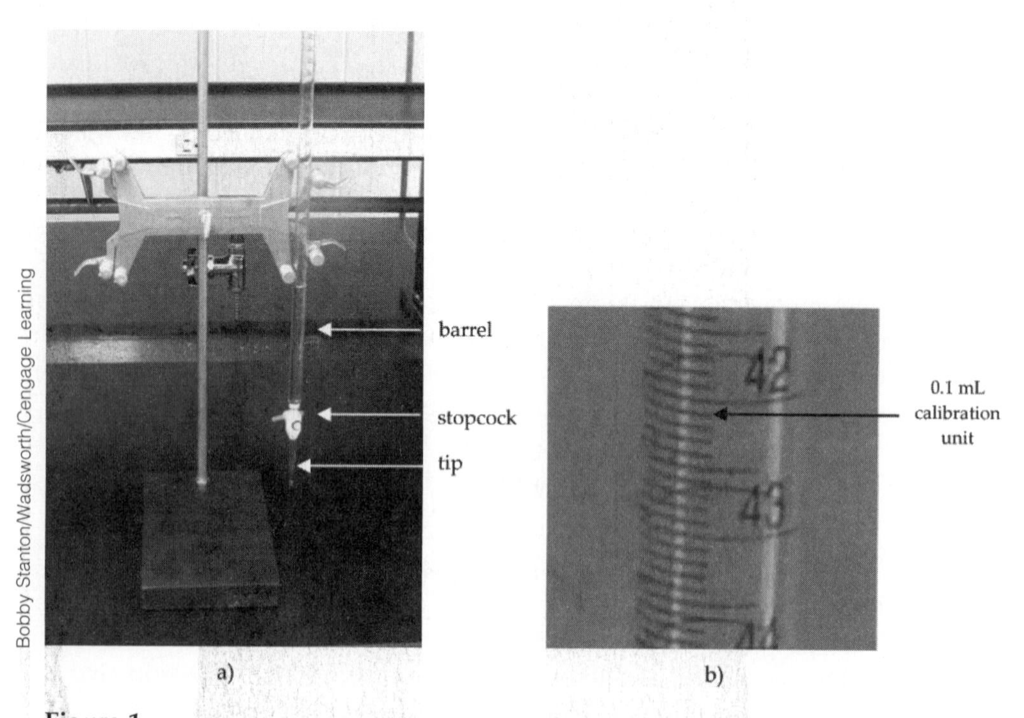

Figure 1
Figure 1a) depicts a typical 50-mL buret. Figure 1b) depicts the smallest calibration for a 50-mL buret, which is 0.1 mL.

the second reactant. Monitoring the equivalence-point for a titration is done with visual indicators, or by measuring a chemical change in the solution (e.g., pH of the solution). In this experiment, we will monitor the pH change as NaOH is added to the solution.

For example, when titrating nickel(II) nitrate solution with 0.10 *M* NaOH, NaOH solution is added drop-wise from a buret to the beaker containing 10.00 mL of 0.050 *M* $Ni(NO_3)_{2(aq)}$. The initial pH of the nickel(II) nitrate solution is in the 2.5 to 5.0 pH range. As NaOH solution is added to the beaker, the solution's pH gradually rises and a precipitate forms. The equivalence-point for the titration occurs when sufficient NaOH has been added to completely precipitate the Ni^{2+} ions in the solution. According to the balanced chemical equation (Eq. 11), 2 moles of NaOH are required to precipitate 1 mole of $Ni(NO_3)_2$.

$$Ni(NO_3)_{2(aq)} + 2\ NaOH_{(aq)} \rightarrow Ni(OH)_{2(s)} + 2\ NaNO_{3(aq)} \quad \text{(Eq. 11)}$$

On a titration curve (Figure 2), a graph of pH versus volume of NaOH added, the equivalence-point occurs at the mid-point of the vertical portion of the curve where the pH rises rapidly. We determine the volume of base required to completely precipitate the Ni^{2+} ions at the equivalence-point. Figure 2 indicates that the equivalence-point occurs when 13.62 mL of NaOH is added, which is the volume of NaOH necessary to precipitate all of the Ni^{2+} ions.

In this experiment the unknown metal nitrate solution will contain either cobalt(II) nitrate or iron(III) nitrate. Both react with NaOH to form insoluble metal hydroxides as shown in Eq. 12 and 13.

$$Co(NO_3)_{2(aq)} + 2\ NaOH_{(aq)} \rightarrow Co(OH)_{2(s)} + 2\ NaNO_{3(aq)} \quad \text{(Eq. 12)}$$

$$Fe(NO_3)_{3(aq)} + 3\ NaOH_{(aq)} \rightarrow Fe(OH)_{3(s)} + 3\ NaNO_{3(aq)} \quad \text{(Eq. 13)}$$

The balanced equations indicate that 2 moles of NaOH are required to completely precipitate 1 mole of $Co(NO_3)_{2(aq)}$ and that 3 moles of NaOH are required to completely precipitate 1 mole of $Fe(NO_3)_{3(aq)}$. The moles of NaOH added in the titration and the moles of metal nitrate in the beaker can be calculated using Eq. 7. The different stoichiometric ratios for the two reactions can be used to identify the unknown metal nitrate solution.

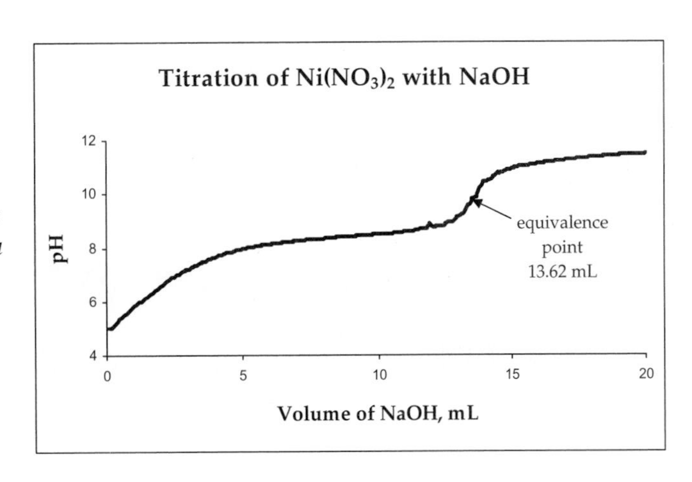

Figure 2
Titration curve for the precipitation of $Ni(NO_3)_{2(aq)}$ with $NaOH_{(aq)}$. The pH of the reaction mixture is plotted as a function of the volume of NaOH added. The graph indicates that the volume of NaOH required to completely precipitate the Ni^{2+} ions is 13.62 mL

PROCEDURE

CAUTION

Students must wear departmentally approved eye protection while performing this experiment. Wash your hands before touching your eyes and after completing the experiment.

Both NaOH and HCl are corrosive. If NaOH or HCl contacts your skin, immediately wash the affected area with copious quantities of water and inform your lab instructor.

Set up the MeasureNet Workstation to Record pH

1. Press the **On/Off** button to turn on the power to the MeasureNet workstation.
2. Press **Main Menu**, then press **F3 pH vs. mVolts**, then press **F1 pH vs. Time**.
3. Press **Calibrate**. The MeasureNet pH probe will be stored in a beaker containing pH 7.00 buffer solution. Using a thermometer, measure the temperature of the pH 7.00 buffer solution and enter it at the workstation. Press **Enter**.
4. Enter 7.00 as the pH of the buffer solution at the workstation, press **Enter**.
5. *Gently* stir the buffer solution with a stirring rod. When the displayed pH value stabilizes, press **Enter**. The pH should be close to 7.00, it does not have to read exactly 7.00.
6. Press **F1** to complete a 1 point calibration of the pH meter using pH 7.00 buffer only. *A 1 point calibration is sufficient for this experiment.*
7. Remove the MeasureNet pH electrode from the pH 7.00 buffer solution, rinse the tip of the probe with distilled water, and dry it with a Kimwipe®.
8. Press **Display** to accept all values.

Part A - Identification of an Unknown Powdered Metal by Reaction with HCl

9. Obtain a beral pipet to prepare a plastic scoopula. Beginning at the center of the tip, use a pair of scissors to cut upward at a 15 degree angle, removing approximately 2–2.5 cm of the top half of the pipet (Figure 3).
10. Obtain an unknown, powdered metal from your lab instructor. Should the Unknown Number be recorded in the Lab Report?
11. Add exactly 10.00 mL of 1.00 *M* HCl solution to a clean 250-mL beaker. Should this volume and concentration be recorded in the Lab Report and to how many significant figures?

Figure 3
The top half of the tip of a beral pipet is removed to prepare a plastic scoopula

Bobby Stanton/Wadsworth/Cengage Learning

12. Add 90–100 mL of distilled water to the acid solution. The $[H_3O^+]$ of distilled water is 0.0000001 *M*. Will this significantly change the moles of H_3O^+ in the 1.00 *M* HCl solution? Should the volume of water be recorded in the Lab Report?

13. Place a stir bar in the beaker, and set the beaker on a magnetic stirrer. (If a magnetic stirrer is not available, use a stirring rod to stir the solution.)

14. Remove the pH electrode from the 7.00 buffer solution; rinse it with distilled water over an empty beaker. *Gently* dry the tip of the electrode with a Kimwipe®. Insert the pH probe into the beaker containing HCl solution. The pH electrode should be approximately 1.5 cm from the bottom of the beaker.

15. If the cut out notch, on the bottom of the probe tip cover, is not completely submerged in the acid solution, add distilled water until it is covered. Should the volume of water be recorded in the Lab Report?

16. Why were you instructed to add distilled water in Step 12, and possibly more distilled water in Step 15, to the acid solution? What purpose does the water serve?

17. Should you record the initial pH of the acid solution in the Lab Report?

18. Turn on the stir bar to a slow to medium speed. The stir bar must not contact the pH electrode when stirring.

19. Place 0.3 grams of unknown metal powder in a weighing boat. Should the exact mass of the unknown metal powder be determined? Should the mass be recorded in the Lab Report, and to how many significant figures?

20. Using the plastic scoopula prepared in Step 9, withdraw approximately 0.02 g of metal powder from the weighing boat (see Note). *Be very careful, you must not spill any of the powder.* Insert the plastic scoopula into the solution, and swirl it to remove the powdered metal. Dry the tip of the scoopula. Repeat Step 20 two more times in succession.

NOTE: It is helpful to use a balance as a visual aid in determining the size of 0.02 g of metal on the tip of the plastic scoopula. Place a weighing boat containing 0.5 g of metal on a balance and repeatedly remove samples of metal until you are proficient at removing 0.02 g of metal.

21. Note the evolution of hydrogen gas (the solution should bubble vigorously), and monitor the pH of the solution. The pH of the acid solution will increase *very slowly* initially, then increase more rapidly). Why does the pH initially increase very slowly?

22. Once the bubbling stops, use the scoopula to add another 0.02 g of metal powder.

23. Repeat Step 22 until the pH of the solution increases by 1.5 pH units higher than the initial pH of the acid solution, or until the addition of metal no longer produces bubbling of hydrogen gas and the pH has ceased to increase (see Note).

NOTE: In theory, the reaction is essentially complete when the pH of the solution increases by 2 pH units (99% complete). In reality, when the pH has risen by 1.2 to 1.5 pH units, bubbling may not be visible. At that point, do not add additional metal to the solution. Wait for the pH of the solution to stop increasing.

Experimental Technique: Determining when to stop adding metal to the acid solution is key to an accurate determination of the metal's molar mass. Knowing when to stop adding metal to the reaction mixture requires *keen observation, good judgment,* and *patience.* Pay very close attention to the reaction mixture and the solution's pH. This can be a difficult decision to make. Completion of this reaction may require 30–45 minutes.

24. When the reaction is complete, should the exact mass of the unknown metal powder remaining in the weighing boat be determined? Should the mass be recorded in the Lab Report, and to how many significant figures?
25. If the pH of the reaction mixture is < 3.0, add 0.1 M NaOH solution, one drop at a time, to the reaction mixture until the pH of the mixture is above 3.0 (but below 12.0). Decant the solution into a laboratory sink.
26. If time permits, perform a second trial by repeating Steps 11–25 to determine the molar mass of the unknown metal.
27. *Complete Steps 28–29 at the end of the laboratory period. Proceed to Part B of the experiment.*
28. How do you determine the molar mass of the metal?
29. Using a periodic table, identify the unknown metal.

Part B - Identification of an Unknown Metal Nitrate Solution via Precipitation with NaOH

30. Add exactly 10.00 mL of 0.050 *M* unknown metal nitrate solution to a clean 250-mL beaker. Should this volume and concentration be recorded in the Lab Report and to how many significant figures?
31. Should the Unknown Number of the metal nitrate solution be recorded in the Lab Report.
32. Add approximately 100 mL of distilled water to the beaker containing the metal nitrate solution. Should the volume of water be recorded in the Lab Report? What purpose does the water serve?
33. See Appendix F – **Instructions for Recording a Titration Curve Using the MeasureNet pH Probe and Drop Counter.** Complete all steps in Appendix F before proceeding to Step 34 below. The acid solution referred to in Appendix F is the unknown metal nitrate solution (*an acidic solution*).
34. Perform a second trial by repeating Steps 30–33 to determine the identity of the metal nitrate solution.
35. From the tab delimited files you saved, prepare titration curves using Excel (or a comparable spreadsheet program). Instructions for preparing titration curves using Excel are provided in Appendix B-4.
36. How do you determine the volume of NaOH required to completely precipitate the metal ions in the unknown solution? Should this volume be recorded in the Lab Report, and to how many significant figures?

37. How do you determine the molar ratio (moles of NaOH per mole of metal ions) for the reaction for each trial.

38. Determine the average molar ratio for the reaction.

39. Is the unknown metal nitrate solution $Co(NO_3)_2$ or $Fe(NO_3)_3$?

Name ______ Section ______ Date ______

Instructor ______

6 EXPERIMENT 6

Lab Report

Part A - Identification of an Unknown Powdered Metal by Reaction with HCl

Unknown number metal powder ________

Experimental data unknown metal powder – Trial 1

Will the 90–100 mL of distilled water added to the acid solution significantly change the moles of H_3O^+ in the 1.00 *M* HCl solution? Why or why not?

Why were you instructed to add distilled water in Step 12, and possibly more distilled water in Step 15? What purpose does the water serve?

In Step 21 why does the pH of the acid solution initially increase very slowly when metal is first added to the acid solution, but gradually increase at a faster rate as the reaction proceeds to completion?

How do you determine the molar mass of the unknown metal?

Using a periodic table, identify the unknown metal. ____________________

Part B - Identification of an Unknown Metal Nitrate Solution via Precipitation with NaOH

Unknown number metal nitrate solution __________

Experimental data and calculations unknown metal nitrate solution – Trial 1

Experimental data and calculations unknown metal nitrate solution – Trial 2

In Step 32 100 mL of distilled water was added to the beaker containing the metal nitrate solution. What purpose does the water serve?

How do you determine the molar ratio (moles of NaOH per mole of metal ions) for the reaction for each trial?

Average molar ratio of NaOH to metal nitrate __________

Identity of the unknown metal nitrate solution __________

Name

Section

Date

Instructor

6

EXPERIMENT 6

Pre-Laboratory Questions

1. If 0.327 g of an unknown metal completely reacts with 10.00 mL of 1.00 *M* HCl according to Eq. 2, calculate the molar mass of the unknown metal. Identify the metal from its molar mass.

   A. Calculate the moles of acid added to the beaker.

   B. Calculate the moles of metal that reacted with the moles of acid in 1A.

   C. Determine the molar mass of the metal.

   D. Identify the unknown metal. ______________________________

2. An unknown 0.0500 *M* metal nitrate solution is either cobalt(II) nitrate or iron(III) nitrate. Both react with NaOH to form insoluble metal hydroxides (see Eqs. 12 and 13). 10.00 mL of 0.0500 *M* an unknown metal nitrate solution is titrated with 0.100 *M* NaOH solution (see titration curve below).

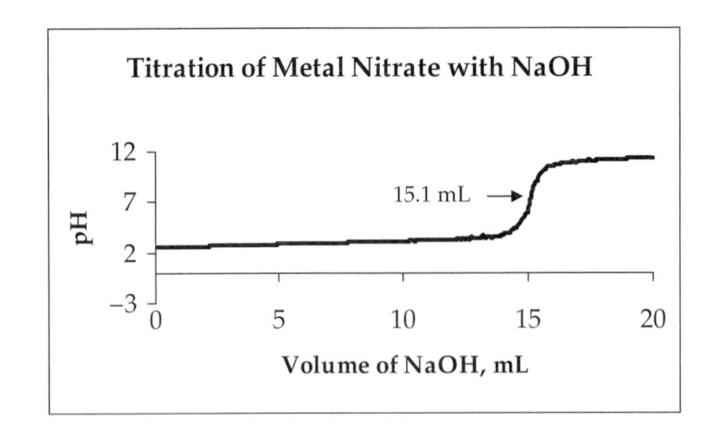

A. Calculate the number of moles of unknown metal ion solution used in the reaction.

B. Calculate the number of moles of NaOH needed to completely precipitate the metal ions from solution.

C. Calculate the ratio of moles NaOH to moles unknown metal nitrate solution.

D. Is the unknown metal nitrate solution $Co(NO_3)_3$ or $Fe(NO_3)_3$? ______________

Name | Section | Date

Instructor

6 EXPERIMENT 6

Post-Laboratory Questions

1. Can calcium hydroxide solution be used to distinguish $Co(NO_3)_2$ and $Fe(NO_3)_3$ solutions using only the titration technique presented in this lab? Justify your answer with an explanation.

2. In Step 32 a student added 150 mL of water to the beaker containing 10 mL of unknown metal nitrate solution instead of 100 mL of water as instructed? Will this experimental error significantly affect the volume of NaOH that must be added in the titration? Justify your answer with an explanation.

3. If a student used an iron scoopula, instead of a plastic one, to add the unknown metal to the HCl solution, will it affect the calculated molar mass of the unknown metal? If so, how? Justify your answer with an explanation.

4. 90.0 mL of distilled water is added to 10.0 mL of 1.00 *M* HCl. Calculate the $[H_3O^+]$ and the pH of the solution before any metal has been added to the solution.

Calculate the $[H_3O^+]$ and the pH of the solution after 99.0% of the H_3O^+ has reacted with and unknown metal.

EXPERIMENT 7

Types of Chemical Reactions

OBJECTIVES

Develop a classification scheme for different types of chemical reactions. Predict products of reactions and write formula unit, total ionic, and net ionic chemical equations.

INTRODUCTION

Chemical reactions can be grouped into categories. The reactions most often encountered in general chemistry experiments fall into two categories: 1) **oxidation-reduction reactions** (redox reactions), and 2) **metathesis** or **double displacement reactions**. Recognizing chemical reaction types helps us predict the reaction products.

I. Redox Reactions

A redox reaction occurs when one or more electrons are transferred between reactant species. These types of reactions are also referred to as oxidation-reduction reactions. In the broadest sense, reactants that lose electrons are oxidized and reactants that gain electrons are reduced. For example, potassium metal reacts with iodine to form potassium iodide.

$$2\ K_{(s)} + I_{2(s)} \rightarrow 2\ KI_{(s)} \qquad \text{(Eq. 1)}$$

Potassium and iodine are neutral species, the reaction product, potassium iodide, contains a potassium ion (K^+) and an iodide ion (I^-). In the course of the reaction, each potassium atom loses one electron and is oxidized ($K \rightarrow K^+ + e^-$), while each iodine atom gains one electron ($I + e^- \rightarrow I^-$) and is reduced. It should be noted that the electrons are transferred from potassium atoms to iodine atoms. Since there are two potassium and two iodine atoms involved in the reaction, the total number of transferred electrons is two.

$$2K \rightarrow 2\ K^+ + 2\ e^- \qquad \text{(Eq. 2)}$$

$$I_2 + 2\ e^- \rightarrow 2\ I^- \qquad \text{(Eq. 3)}$$

Table 1 *Rules for assigning oxidation numbers*

1. The oxidation number of an element in its elemental state is zero. For example, the oxidation number of Na in sodium metal is zero; the oxidation number of F in F_2 is zero.
2. The oxidation number of hydrogen in compounds is generally +1.
3. The oxidation number of oxygen in compounds is generally −2. A notable exception is peroxides containing O_2^{2-} ions, where oxygen has a –1 oxidation number.
4. Monatomic ions have oxidation numbers that are equal to their charge. For example, S^{2-} has an oxidation number of −2.
5. In a molecular compound, the more electronegative element is assigned a negative oxidation number equal to its charge if it were an anion in an ionic compound.
6. The sum of the oxidation numbers of the elements in a compound must equal zero. For example, in Cl_2O, the oxidation number of Cl is +1 and the oxidation number of oxygen is −2.
7. The sum of the oxidation numbers in a complex ion must equal the charge on the ion. For example, the charge on the hydroxide ion is 1−, OH^-. The oxidation number of hydrogen is +1, the oxidation number of oxygen is −2, the sum of the oxidation numbers is $+1 + (-2) = -1$, equal to the charge on the OH^- ion.
8. The oxidation numbers of Group IA and IIA metals (in compounds) are always +1 and +2, respectively.
9. Oxidation numbers are assigned on a per atom basis. In Cl_2O, the sum of the oxidation numbers for Cl is +2 to counterbalance the −2 for O. Since there are two chlorine atoms in the compound, each chlorine has a +1 oxidation number (+2/2 atoms = +1).

By assigning oxidation numbers to each atom involved in a chemical reaction, we can monitor the electron transfer that occurs. An **oxidation number** (or oxidation state) is a hypothetical charge assigned to an atom based on an established set of rules. A redox reaction occurs when one atom's oxidation number increases (oxidation), while simultaneously, a second atom's oxidation number decreases (reduction). Table 1 lists the rules used to assign oxidation numbers.

In this experiment, three types of redox reactions will be examined: 1) combination reactions, 2) decomposition reactions, and 3) single displacement reactions.

In **combination reactions**, two species react together producing a third species as represented by the following general equation.

$$A + B \rightarrow AB$$

The combustion of magnesium metal in the presence of oxygen or burning of iron in the presence of sulfur are examples of combination reactions.

$$2\,Mg_{(s)} + O_{2(g)} \rightarrow 2\,MgO_{(s)} \quad \text{(Eq. 4)}$$

$$8\,Fe_{(s)} + S_{8(s)} \rightarrow 8\,FeS_{(s)} \quad \text{(Eq. 5)}$$

Decomposition reactions occur when one chemical compound breaks apart into simpler compounds and/or elements. The general formula for a decomposition reaction is given below.

$$AB \rightarrow A + B$$

An example of this type of reaction is the decomposition of sodium chlorate into sodium chloride and oxygen gas.

$$2\ NaClO_{3(s)} \xrightarrow{\Delta} 2\ NaCl_{(s)} + 3\ O_{2(g)} \qquad \text{(Eq. 6)}$$

Displacement reactions (sometimes referred to single displacement or single replacement reactions) occur when one element in a compound is replaced by another element. For these reactions to occur, one element must be more chemically reactive than the element it is displacing. The activity series of some elements is given in Table 2, where the most chemically active elements are listed at the top of the table, and the least active elements are listed at the bottom of the table.

The general formula for a displacement reaction is:

$$A + BC \rightarrow AC + B$$

For example, magnesium metal can displace copper from aqueous copper(II) sulfate solution because magnesium is a more active metal than copper. Magnesium metal is dissolved as it displaces Cu^{2+} ions from solution. In the process Mg^{2+} ions and Cu metal (solid) are produced.

$$Mg_{(s)} + CuSO_{4(aq)} \rightarrow MgSO_{4(aq)} + Cu_{(s)} \qquad \text{(Eq. 7)}$$

Table 2 *Chemical activity of some elements*

K	most active
Ca	
Na	
Mg	
Al	
Zn	
Fe	
Co	
Pb	
H	
Cu	
Ag	least active

The progress of this reaction is easily followed via color changes. Magnesium metal is a gray solid. A solution containing Cu^{2+} ions is a blue liquid. The resulting solution is colorless and it contains red copper metal.

Many active metals can displace hydrogen from aqueous acids such as HCl, HBr, or CH_3COOH. For example, magnesium metal displaces hydrogen from aqueous hydrochloric acid. The vigorous bubbling of hydrogen gas is evidence that the reaction is occurring.

$$Mg_{(s)} + HCl_{(aq)} \rightarrow MgCl_{2(aq)} + H_{2(g)} \qquad \text{(Eq. 8)}$$

A less active metal (one lower in Table 2) cannot displace a more active metal from aqueous solution. For example, if copper metal is mixed with aqueous zinc sulfate, no reaction occurs because copper is less active than zinc.

$$Cu_{(s)} + ZnSO_{4(aq)} \rightarrow \text{no reaction} \qquad \text{(Eq. 9)}$$

II. Metathesis Reactions

In some reactions between two compounds dissolved in aqueous solution, the positive and negative ions "change partners" to form new compounds with no change in oxidation number (non-redox reactions). In the hypothetical chemical reaction below, AB exists as A^+ and B^- ions in solution, and CD exists as C^+ and D^- ions in solution.

$$AB + CD \rightarrow AD + CB$$

The cation in AB is attracted to the anion in CD and vice versa. These types of reactions are often referred to as **double displacement** or **metathesis reactions**. The driving force for metathesis reactions is the removal of ions from solution. Ions are removed from solution by forming an insoluble precipitate in precipitation reactions, or by forming predominantly non-ionized (weak or nonelectrolyte) molecules in acid-base reactions.

In a typical **precipitation reaction**, a solid is formed when two aqueous solutions of ionic compounds are mixed. Whether a precipitate will form depends on the solubilities of the products. General rules for the solubilities of ionic compounds in water are given in Table 3.

For example, the reaction of aqueous solutions of calcium chloride and silver nitrate produces silver chloride (a white precipitate) and aqueous calcium nitrate as shown in the equation below.

$$CaCl_{2(aq)} + 2\ AgNO_{3(aq)} \rightarrow 2\ AgCl_{(s)} + Ca(NO_3)_{2(aq)} \qquad \text{(Eq. 10)}$$

It may be more useful to write the equation as a **total ionic equation**, wherein all species are shown as they exist in solution (see Note). The total ionic equation for the reaction of calcium chloride and silver nitrate is:

$$Ca^{2+}_{(aq)} + 2\,Cl^-_{(aq)} + 2\ Ag^+_{(aq)} + 2\ NO_3{}^-_{(aq)} \rightarrow 2\ AgCl_{(s)} + Ca^{2+}_{(aq)} + 2\ NO_3{}^-_{(aq)} \qquad \text{(Eq. 11)}$$

NOTE: When writing total ionic equations, strong acids, strong bases, and water-soluble ionic compounds are written as aqueous ionic species.

Table 3 *Solubilities of some ionic compounds*

Generally Soluble	*Exceptions*
Group IA metals and the ammonium ion (NH_4^+)	None
Nitrates (NO_3^-), Acetates ($C_2H_3O_2^-$), Chlorates (ClO_3^-), and Perchlorates (ClO_4^-)	None
Chlorides (Cl^-), Bromides (Br^-), and Iodides (I^-)	Ag^+, Pb^{2+}, and Hg_2^{2+} halide salts are insoluble
Sulfates (SO_4^{2-})	$CaSO_4$, $SrSO_4$ and Ag_2SO_4 are slightly soluble. $BaSO_4$, $HgSO_4$ and $PbSO_4$ are insoluble
Generally Insoluble	*Exceptions*
Most Hydroxides (OH^-)	Group IA metals and NH_4^+ are soluble. $Ba(OH)_2$ and $Sr(OH)_2$ are moderately soluble. $Ca(OH)_2$ is slightly soluble.
Carbonates (CO_3^{2-}) and Phosphates (PO_4^{3-})	Group IA metals and NH_4^+ are soluble.
Sulfides (S^{2-})	Group IA metals, NH_4^+, MgS and CaS are soluble.

Ions that appear on both sides of the chemical equation in exactly the same form (i.e., no chemical change has occurred) are called **spectator ions**. Spectator ions are removed from the total ionic equation to give the **net ionic equation**. In Eq. 11, the spectator ions are Ca^{2+} and NO_3^-, thus, the net ionic equation is:

$$2\,Cl^-_{(aq)} + 2\,Ag^+_{(aq)} \rightarrow 2\,AgCl_{(s)} \qquad \text{(Eq. 12)}$$

The net ionic equation indicates that silver ions in the presence of chloride ions form the insoluble precipitate silver chloride (see Table 3).

Many acid-base reactions involve transferring a hydrogen ion from an acid to a base. One of the more common acid-base reactions occurs between strong acids (e.g., HCl, H_2SO_4, or HNO_3) and strong bases (Group IA and some Group IIA metal hydroxides). The products of any strong acid-strong base reaction are water and a salt. The H^+ ion from the acid, HA, reacts with the OH^- ion from the base, MOH, to form neutral water. The M^+ ion from the base reacts with the A^- ion from the acid to form a salt. These acid-base reactions are another type of metathesis reaction.

$$HA_{(aq)} + MOH_{(aq)} \rightarrow MA_{(aq)} + H_2O_{(\ell)} \qquad \text{(Eq. 13)}$$

For example, when hydrochloric acid reacts with sodium hydroxide, the products are water and sodium chloride.

$$HCl_{(aq)} + NaOH_{(aq)} \rightarrow NaCl_{(aq)} + H_2O_{(\ell)} \qquad \text{(Eq. 14)}$$

The total and net ionic equations for the reaction of hydrochloric acid and sodium hydroxide are shown below.

$$H^+_{(aq)} + Cl^-_{(aq)} + Na^+_{(aq)} + OH^-_{(aq)} \rightarrow H_2O_{(\ell)} + Na^+_{(aq)} + Cl^-_{(aq)} \quad \text{(Eq. 15)}$$

$$H^+_{(aq)} + OH^-_{(aq)} \rightarrow H_2O_{(\ell)} \quad \text{(Eq. 16)}$$

The net ionic equation for any strong acid-strong base reaction producing a *water soluble salt* is $H^+_{(aq)} + OH^-_{(aq)} \rightarrow H_2O_{(\ell)}$.

A second common type of neutralization reaction is the reaction of an acid (HA) with a metal hydrogen carbonate ($MHCO_3$) or a metal carbonate (M_2CO_3). When an acid reacts with a metal hydrogen carbonate or a metal carbonate, the products of the reaction are a salt, carbon dioxide gas, and water. For example, when nitric acid reacts with potassium hydrogen carbonate, the products of the reaction are water, carbon dioxide gas, and potassium nitrate.

$$HNO_{3(aq)} + KHCO_{3(aq)} \rightarrow KNO_{3(aq)} + H_2O_{(\ell)} + CO_{2(g)} \quad \text{(Eq. 17)}$$

The hydrogen ion (H^+) from nitric acid reacts with the hydrogen carbonate ion (HCO_3^-) from potassium hydrogen carbonate to form an intermediate species, carbonic acid (H_2CO_3), which decomposes into water and carbon dioxide. The net ionic equations for the reactions of an acid with a metal hydrogen carbonate or an acid with a metal carbonate are given in Equations 18 and 19, respectively.

$$H^+_{(aq)} + HCO_3^-{}_{(aq)} \rightarrow [H_2CO_{3(aq)}] \rightarrow H_2O_{(\ell)} + CO_{2(g)} \quad \text{(Eq. 18)}$$

$$2\,H^+_{(aq)} + CO_3^{2-}{}_{(aq)} \rightarrow [H_2CO_{3(aq)}] \rightarrow H_2O_{(\ell)} + CO_{2(g)} \quad \text{(Eq. 19)}$$

Many metals form insoluble hydroxides in the presence of a limited amount of hydroxide ions. For example, when aqueous beryllium nitrate is reacted with a limited amount of potassium hydroxide, insoluble beryllium hydroxide forms according to Equation 20.

$$2\,KOH_{(aq)} + Be(NO_3)_{2(aq)} \rightarrow Be(OH)_{2(s)} + 2\,KNO_{3(aq)} \quad \text{(Eq. 20)}$$

Several insoluble metal hydroxides are amphoteric. **Amphoterism** is the ability of a substance to react with either acids or bases. Beryllium hydroxide is a typical amphoteric metal hydroxide. Beryllium hydroxide behaves as a base when it reacts with acids to form salts and water. In a solution containing excess OH^- ions, $Be(OH)_2$ acts like an acid and dissolves forming the complex ion $[Be(OH)_4]^{2-}$. A list of insoluble metal hydroxides that form soluble complexes in excess OH^- ions is provided in Table 4.

$$2\,KOH_{(aq)} + Be(OH)_{2(aq)} \rightarrow K_2[Be(OH)_4]_{(aq)} \quad \text{(Eq. 21)}$$

$$2\,OH^-_{(aq)} + Be(OH)_{2(aq)} \rightarrow [Be(OH)_4]^{2-}{}_{(aq)} \quad \text{(Eq. 22)}$$

Aqueous ammonia (also known as ammonium hydroxide, NH_4OH) also reacts with some metal ions to form insoluble metal hydroxides (Table 5).

Table 4 *Some amphoteric metal hydroxides*

Metal Ion	*Insoluble in Limited* OH^-	*Soluble Complex Ion in Excess* OH^-
Be^{2+}	$Be(OH)_2$	$[Be(OH)_4]^{2-}$
Al^{3+}	$Al(OH)_3$	$[Al(OH)_4]^-$
Cr^{3+}	$Cr(OH)_3$	$[Cr(OH)_4]^-$
Zn^{2+}	$Zn(OH)_2$	$[Zn(OH)_4]^{2-}$
Pb^{2+}	$Pb(OH)_2$	$[Pb(OH)_4]^{2-}$
Co^{2+}	$Co(OH)_2$	$[Co(OH)_4]^{2-}$
Cu^{2+}	$Cu(OH)_2$	$[Cu(OH)_4]^{2-}$
Sn^{2+}	$Sn(OH)_2$	$[Sn(OH)_4]^{2-}$

Table 5 *Some Metal Hydroxides that Form Ammine Complexes*

Metal Ion	*Insoluble in Limited* NH_3	*Soluble Complex Complex in Excess* NH_3
Ni^{2+}	$Ni(OH)_2$	$[Ni(NH_3)_6]^{2+}$
Ag^+	$AgOH$	$[Ag(NH_3)_2]^+$
Zn^{2+}	$Zn(OH)_2$	$[Zn(NH_3)_4]^{2+}$
Cd^{2+}	$Cd(OH)_2$	$[Cd(NH_3)_4]^{2+}$
Co^{2+}	$Co(OH)_2$	$[Co(NH_3)_6]^{2+}$
Cu^{2+}	$Cu(OH)_2$	$[Cu(NH_3)_4]^{2+}$
Hg^{2+}	$Hg(OH)_2$	$[Hg(NH_3)_4]^{2+}$

For example, if a limited amount of aqueous ammonia is added to an aqueous solution of copper(II) nitrate, copper(II) hydroxide forms.

$$2\,H_2O_{(\ell)} + 2\,NH_{3(g)} + Cu(NO_3)_{2(aq)} \rightarrow Cu(OH)_{2(s)} + 2\,NH_4NO_{3(aq)} \quad \text{(Eq. 23)}$$

If excess aqueous ammonia is added to the solution, $Cu(OH)_2$ dissolves forming the complex ion $[Cu(NH_3)_4]^{2+}$. A list of insoluble metal hydroxides that form soluble ammine complexes in excess aqueous ammonia is provided in Table 5.

$$4\,NH_{3(aq)} + Cu(OH)_{2(s)} \rightarrow [Cu(NH_3)_4]^{2+}{}_{(aq)} + 2\,OH^-{}_{(aq)} \quad \text{(Eq. 24)}$$

In this experiment, you will perform several redox, precipitation, acid-base, and complex-ion formation reactions by mixing various chemicals. For each mixture of chemicals, you must determine whether a chemical reaction occurs. Evidence (indicators) that a chemical reaction occurs include: 1) color changes, 2) precipitate formation, 3) evolution of a gas, or 4) temperature changes. If none of these indicators are observed, you may assume that a chemical reaction did not occur.

PROCEDURE

CAUTION

Students must wear departmentally approved eye protection while performing this experiment. Wash your hands before touching your eyes and after completing the experiment.

Chemical Alert

Avoid splattering acid and bases on your skin. If these chemicals contact your skin, wash the affected area with copious quantities of water and inform your lab instructor.

Redox: Combination Reactions

1. Obtain a small (1 in. × 1 in.) piece of aluminum foil. Hold one end of the aluminum foil over a Bunsen burner flame using crucible tongs. If the foil undergoes a change, record your observations in the Lab Report. If a reaction occurs, write a balanced formula unit equation (also referred to as a molecular equation) for the reaction. If a redox reaction occurs, indicate which element is oxidized and which element is reduced.
2. Let the residue *cool to room temperature* and discard it into the trash can.
3. Obtain a strip of magnesium ribbon that is approximately 1 inch in length. Roll the strip of magnesium into a tight coil and place the coil inside a crucible. Place a clay triangle on an iron ring supported by a ring stand (Figure 1). Suspend the crucible inside the clay triangle.
4. Heat the crucible with a Bunsen burner until the bottom of the crucible glows cherry red for 5 minutes. Does the metal ignite and burn? Observe the residue in the crucible and record your observations in the lab report. If a reaction occurs, write a balanced formula unit equation

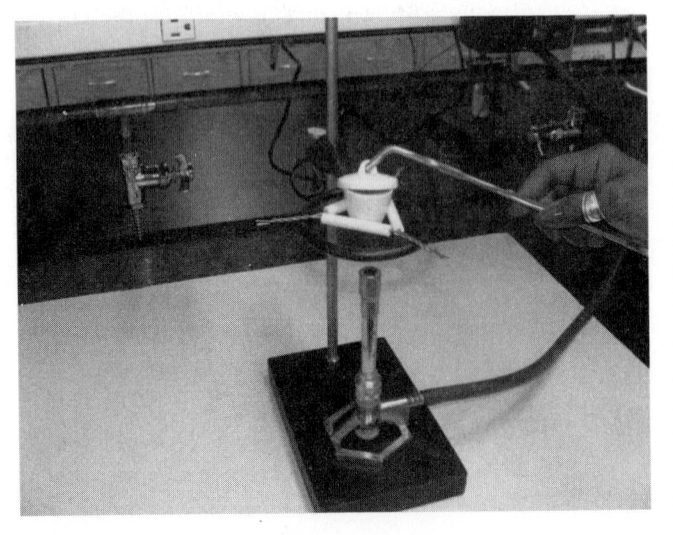

Bobby Stanton/Wadsworth/Cengage Learning

Figure 1
A crucible with lid supported by a clay triangle and iron ring

for the reaction. If a redox reaction occurs, indicate which element is oxidized and which element is reduced.

5. Let the crucible cool and discard any residue into the trash can.

Redox: Decomposition Reactions

6. Place 2 mL of 3% hydrogen peroxide (H_2O_2) in a 13 × 100 mm test tube (4 in test tube). Add 4–5 manganese(IV) oxide (MnO_2) crystals to the test tube. (*MnO_2 is a catalyst. Catalysts are used to increase the rate of a chemical reaction; they are not consumed by the reaction*). Light a wooden splint in a Bunsen burner flame. After the wood ignites, blow out the flame. Insert the glowing splint half way into the test tube. What happens to the glowing splint? Explain your observations. If a reaction occurs, write a balanced formula unit equation for the reaction. If a redox reaction occurs, indicate which element is oxidized and which element is reduced.

7. Discard the residue into the "Waste Container."

Redox: Displacement Reactions

8. Add 2 mL of 0.1 *M* copper(II) nitrate solution and 0.1 g of zinc metal to a 13 × 100 mm test tube. Let the test tube stand for 5 minutes. Does a displacement reaction occur? Record your observations. If a reaction occurs, write a balanced formula unit equation for the reaction. If a redox reaction occurs, indicate which element is oxidized and which element is reduced. If a reaction does not occur, state a reason why the reaction did not occur.

9. Discard the reaction mixture into the "Waste Container."

10. Add 2 mL of 0.1 *M* sodium chloride solution and 0.1 g of copper metal to a 13 × 100 mm test tube. Let the test tube stand for 5 minutes. Does a displacement reaction occur? Record your observations? If a reaction occurs, write a balanced formula unit equation for the reaction. If a redox reaction occurs, indicate which element is oxidized and which element is reduced. If a reaction does not occur, state a reason why the reaction did not occur.

11. Discard the reaction mixture into the "Waste Container."

Unknown – Determine Activity Series for Two Metals

12. Obtain 0.1 g of unknown metal A (solid) and 2 mL of unknown metal A nitrate solution. Obtain 0.1 g of unknown metal B (solid) and 2 mL of unknown metal B nitrate solution. Devise a series of reactions to determine whether unknown A or B is the more active metal. Describe the series of reactions used, your observations, and your reasons for concluding the activity series of the metals in the Lab Report.

Metathesis: Precipitation Reactions (Nonredox)

13. Add 1 mL of 0.1 *M* potassium chloride solution and 1 mL of 0.1 *M* copper(II) nitrate solution to a 13 × 100 mm test tube. Does a displacement reaction occur? Record your observations. If a reaction occurs, write balanced formula unit and net ionic equations for the reaction. If a reaction does not occur, state a reason why the reaction did not occur. Discard the reaction mixture into the "Waste Container."

14. Add 1 mL of 0.1 *M* lead(II) nitrate solution and 1 mL of 0.1 *M* sodium sulfate (Na_2SO_4) solution to a 13 × 100 mm test tube. Does a displacement reaction occur? Record your observations. If a reaction occurs, write balanced formula unit and net ionic equations for the reaction. If a reaction does not occur, state a reason why the reaction

did not occur. Discard the reaction mixture into the "Waste Container."

Metathesis: Acid-Base Reactions (Nonredox)

15. Add 9 mL of 0.1 *M* hydrochloric acid to a 50-mL beaker. Add 2–3 drops of phenolphthalein indicator to the HCl solution. Should the color of the HCl and indicator solution be noted in the Lab Report?

    Chemical Information – A visual indicator is a weak acid used to determine the equivalence point of an acid-base reaction. An indicator has a unique color in acidic solution, but it has a different color in basic solution. When sufficient base has been added to an acid to reach the equivalent point of the reaction, the next drop of base reacts with the indicator and a distinct color change marks the endpoint of the reaction.

16. Add 10 mL of 0.1 *M* sodium hydroxide drop wise to the beaker of HCl and indicator solution. Note your observations in the Lab report. Should the color of the solution be noted in the Lab Report? If a reaction occurs, write complete and balanced formula unit and net ionic equations for the reaction.
17. Discard the residue into the laboratory sink.
18. Add approximately 0.1 gram of solid sodium hydrogen carbonate to a 150-mL beaker. *Slowly and cautiously*, add 1 mL of 1.0 *M* hydrochloric acid drop wise to the beaker. Note your observations in the Lab report. If a reaction occurs, write complete and balanced formula unit and net ionic equations for the reaction.
19. Discard the residue into the laboratory sink.

Metathesis – Amphoteric Hydroxides and Complex Ion Formation

20. Obtain a clean, dry 50-mL beaker. Add 6 drops of 4.0 *M* sodium hydroxide solution and 44 drops of distilled water to the beaker. Stir the solution. Determine the concentration of the new solution (***limited*** OH^- solution).
21. Obtain two clean 20 × 150 mm test tubes. To one of the test tubes, add 2 drops of 0.1 *M* magnesium nitrate solution. Add 2 drops of 0.1 M copper(II) nitrate solution to the second test tube. Add 2 drops of the sodium hydroxide (***limited*** OH^-) solution prepared in Step 20 to each of the test tubes. Watch carefully for any changes in the mixture. Record your observations in the Lab Report. If a precipitate forms, write complete and balanced formula unit and net ionic equations for the reaction. If a reaction does not occur, state a reason why the reaction did not occur.
22. To each test tube in Step 21 in which a precipitate is observed, add 20–25 drops of 4.0 *M* sodium hydroxide solution (***excess*** OH^-). Do any of the precipitates dissolve? Record your observations in the Lab Report. If the precipitate dissolves, write complete and balanced formula unit and net ionic equations for the reaction. If a reaction does not occur, state a reason why the reaction did not occur.
23. Discard the reaction mixtures into the "Waste Container."
24. Obtain a clean, dry 50-mL beaker. Add 6 drops of 4.0 *M* aqueous ammonia solution and 44 drops of distilled water to the beaker. Stir the solution. Determine the concentration of the new solution (***limited*** NH_3 solution).

25. Obtain two clean 13 × 100 mm test tubes. To one of the test tubes, add 2 drops of 0.1 *M* magnesium nitrate solution. Add 2 drops of 0.1 *M* copper(II) nitrate solution to the second test tube. Add 2 drops of the aqueous ammonia (***limited*** NH_3) solution prepared in Step 24 to each of the test tubes. Watch carefully for any changes in the mixture. Record your observations in the Lab Report. If a precipitate forms, write complete and balanced formula unit and net ionic equations for the reaction. If a reaction does not occur, state a reason why the reaction did not occur.
26. To each test tube in Step 25 in which a precipitate is observed, add 3–4 drops of 4.0 *M* aqueous ammonia solution (***excess*** NH_3). Do any of the precipitates dissolve? Record your observations in the Lab Report. If the precipitate dissolves, write complete and balanced formula unit and net ionic equations for the reaction. If a reaction does not occur, state a reason why the reaction did not occur.
27. Discard the reaction mixtures into the "Waste Container."

Unknown – Identifying Ions in Solution

28. Obtain 5 mL of an unknown metal ion solution from your instructor. Should you record the unknown number in the Lab Report? The unknown solution will contain 1 or 2 of the following metal ions: Ag^{+}, Fe^{3+}, or Pb^{2+} ions. Devise a scheme (a series of reactions) that would allow you to selectively precipitate each of these ions to confirm their presence in solution. You may use any of the reagents previously used in this experiment as test solutions. *Use 4–5 drops of unknown solution and 4–5 drops of each test solution per experiment.* Write a balanced net ionic equation for each precipitation reaction. Identify the metal ion(s) present in the unknown solution.

Name | Section | Date

Instructor

EXPERIMENT 7

Lab Report

Redox: Combination Reactions

Observations and reaction for the combustion of aluminum foil.

Observations and reaction for the combustion of magnesium metal.

Redox: Decomposition Reactions

Observations and reaction for the decomposition of H_2O_2. What happens to the glowing splint? Explain.

Redox: Displacement Reactions

Observations and reaction for $Cu(NO_3)_2$ with Zn.

Observations and reaction for NaCl with Cu.

Unknown – Determine Activity Series for Two Metals

Observations, reactions, and reasons for concluding the activity series of unknown metals A and B.

Metathesis: Precipitation Reactions (Nonredox)

Observations and reactions for KCl with $Cu(NO_3)_2$.

Observations and reactions for $Pb(NO_3)_2$ with Na_2SO_4.

Metathesis: Acid-Base Reactions (Nonredox)

Observations and reactions for HCl with NaOH.

Observations and reactions for $NaHCO_3$ with HCl.

Metathesis – Amphoteric Hydroxides and Complex Ion Formation

Determine the concentration of the limited OH^- solution.

Observations and reactions for the addition of limited OH^- ions to $Mg(NO_3)_2$ and $Cu(NO_3)_2$.

Observations and reactions for the addition of excess OH^- ions to $Mg(NO_3)_2$ and $Cu(NO_3)_2$.

Determine the concentration of the limited NH_3 solution.

Observations and reactions for the addition of limited $NH_{3(aq)}$ to $Mg(NO_3)_2$ and $Cu(NO_3)_2$.

Observations and reactions for the addition of excess $NH_{3(aq)}$ to $Mg(NO_3)_2$ and $Cu(NO_3)_2$.

Unknown – Identifying Ions in Solution

Unknown number ______________________________

Observations, reactions, and identity of the ions in the unknown mixture.

Name

Section

Date

Instructor

7 EXPERIMENT 7

Pre-Laboratory Questions

Classify each of the following as a redox, precipitation, acid-base, or complex ion formation reaction. Further indicate whether the reaction is a combination, decomposition, or displacement reaction where applicable.

1. $KHCO_{3(aq)} + HNO_{3(aq)} \rightarrow KNO_{3(aq)} + H_2O_{(\ell)} + CO_{2(g)}$

______________________ ______________________

2. $3\ Ca_{(s)} + 2\ Al^{3+}{}_{(aq)} \rightarrow 3\ Ca^{2+}{}_{(aq)} + 2\ Al_{(s)}$

______________________ ______________________

3. $LiHCO_{3(s)} + \text{heat} \rightarrow Li_2CO_{3(s)} + CO_{2(g)} + H_2O_{(g)}$

______________________ ______________________

4. $Zn(OH)_{2(s)} + 2\ OH^-{}_{(aq)} \rightarrow [Zn(OH)_4]^{2+}{}_{(aq)}$

______________________ ______________________

5. $Ca_{(s)} + H_{2(g)} \rightarrow CaH_{2(s)}$

______________________ ______________________

6. $2\ Cl^-{}_{(aq)} + Pb^{2+}{}_{(aq)} \rightarrow PbCl_{2(s)}$

______________________ ______________________

7. Complete and balance each of the following formula unit or net ionic equations. Write **NR** if no reaction occurs. Include the states (s, ℓ, g, or aq) of all products.

a. $Mg_{(s)} + Al(NO_3)_{3(aq)} \rightarrow$

b. $Ag_{(s)} + FeCl_{3(aq)} \rightarrow$

c. $Ni_{(s)} + 2\ Na^{+}_{(aq)} \rightarrow$

d. $Ba(OH)_{2(aq)} + H_2SO_{4(aq)} \rightarrow$

e. $NH_{3(aq)} + HClO_{4(aq)} \rightarrow$

f. $HCO_3^{-}{}_{(aq)} + OH^{-}{}_{(aq)} \rightarrow$

g. $NH_4Cl_{(aq)} + Li_3PO_{4(aq)} \rightarrow$

h. $Pb(NO_3)_{2(aq)} + MgSO_{4(aq)} \rightarrow$

8. Write a balanced formula unit and a net ionic equation for the following reaction. Include the states (s, ℓ, g, or aq) of all reactants and products.

zinc chloride + lithium hydroxide → zinc hydroxide + lithium chloride

9. When $Co(NO_3)_2$ solution is mixed with limited NH_{3aq}, a precipitate forms. When excess NH_{3aq} is added, the precipitate dissolves. Write balanced formula unit and net ionic equations for these observations. Include the states (s, ℓ, g, or aq) of all reactants and products.

Name

Section

Date

Instructor

7 EXPERIMENT 7

Post-Laboratory Questions

1. Solutions of potassium carbonate, copper(II) chloride, and lead(II) nitrate are each added to a beaker. Will any compound(s) precipitate from solution? Write a balanced **net ionic** equation for any precipitation reaction that occurs.

2. List a series of tests that would determine the order of activity (least active to most active) for the following metals: Zn, Mn, and Ag.

3. The laboratory instructor gives you a test tube containing Ag^{+}, Ba^{2+}, and Cu^{2+} ions. Devise a reaction scheme that would allow you to selectively precipitate each of these ions. Write a balanced **net ionic** equation for each precipitation reaction.

Identification of Metal Ions and Inorganic Compounds by their Chemical Reactions

OBJECTIVES

Use characteristic chemical reactions of metal ions and inorganic compounds to identify an unknown metal ion and an unknown inorganic compound in aqueous solution.

INTRODUCTION

In Experiment 7, a classification scheme was developed for the recognition of different types of chemical reactions. In this experiment, we will use that classification scheme to perform characteristic chemical reactions of some metal ions and inorganic compounds in aqueous solution and develop systems to identify an unknown metal ion and an unknown inorganic compound in aqueous solution.

In this experiment, you must determine whether a chemical reaction occurs for each mixture of chemicals. Evidence (indicators) that a chemical reaction occurs include: 1) color changes, 2) precipitate formation, 3) evolution of a gas, or 4) temperature changes. Many transition metals produce intensely colored aqueous solutions that are characteristic of the metal ion. If none of these indicators are observed, assume that a chemical reaction did not occur (see Note). *You should pay special attention to the color of the original known and unknown solutions.*

In Part A of this experiment, you will establish a system for identifying five common metal ions. You will observe each ion's reactions with three anions and aqueous ammonia. The metal ions and the reagents used in this experiment are listed in Table 1.

NOTE: If two colorless solutions are mixed and a blood-red solution is formed, this indicates that a chemical reaction has occurred. However, if a green solution is mixed with a colorless solution, and the resulting solution is light green, this is not indicative of a chemical reaction (it's an indication of solution dilution).

Table 1 *Metal ions and test reagents*

Metal Ions to Test	*Test Reagents*
Barium ion, Ba^{2+}	Sulfate ion, SO_4^{2-}
Magnesium ion, Mg^{2+}	Hydroxide ion, OH^-
Lead(II) ion, Pb^{2+}	Thiocyanate ion, SCN^-
Cobalt(II) ion, Co^{2+}	Aqueous ammonia, NH_3
Iron(III) ion, Fe^{3+}	

After observing the reactions of the five metals with each of the test reagents, you will identify an unknown metal ion solution using the same test reagents. *The unknown solution will contain one of the five metal cations listed in Table 1.*

From Table 3, Solubilities of Some Ionic Compounds, in Experiment 7, we know that most metal sulfates are soluble in water except $HgSO_4$, $PbSO_4$, and $BaSO_4$. It was also noted that most metal hydroxides are generally insoluble in water. Group IA metal hydroxides are water soluble. Barium hydroxide and strontium hydroxide are moderately soluble in water; calcium hydroxide is sligtly soluble in water.

The amphoteric nature of some insoluble metal hydroxides was also introduced in Experiment 7. Many metal ions, in the presence of limited hydroxide ions (from either a strong base such as sodium hydroxide or a weak base such as aqueous ammonia), form insoluble metal hydroxide precipitates. In excess hydroxide ions, some insoluble metal hydroxides dissolve, forming complex ions. For example, when aqueous zinc nitrate reacts with *limited* potassium hydroxide, insoluble zinc hydroxide forms.

$$2\ KOH_{(aq)} + Zn(NO_3)_{2(aq)} \rightarrow Zn(OH)_{2(s)} + 2\ KNO_{3(aq)} \quad \text{(Eq. 1)}$$

In the presence of *excess* hydroxide ions, $Zn(OH)_2$ dissolves, forming the complex ion $[Zn(OH)_4]^{2-}$.

$$2\ KOH_{(aq)} + Zn(OH)_{2(aq)} \rightarrow K_2[Zn(OH)_4]_{(aq)} \quad \text{(Eq. 2)}$$

When aqueous copper(II) nitrate solution reacts with *limited* aqueous ammonia, copper(II) hydroxide precipitates.

$$2\ H_2O_{(l)} + 2\ NH_{3(g)} + Cu(NO_3)_{2(aq)} \rightarrow Cu(OH)_{2(s)} + 2\ NH_4NO_{3(aq)} \quad \text{(Eq. 3)}$$

In the presence of *excess* aqueous ammonia, $Cu(OH)_2$ dissolves forming the complex ion $[Cu(NH_3)_4]^{2+}$.

$$4\ NH_{3(aq)} + Cu(OH)_{2(aq)} \rightarrow [Cu(NH_3)_4]^{2+}{}_{(aq)} + 2\ OH^-{}_{(aq)} \quad \text{(Eq. 4)}$$

Tables 4 and 5 in Experiment 7 list the amphoteric metal hydroxides in aqueous sodium hydroxide and ammonia solutions.

Metal ions can react with complexing agents to produce more intensely colored solutions. The thiocyanate ion, SCN^-, is a commonly used complexing agent. For example, aqueous copper(II) nitrate solution has a characteristic blue color. However, when $Cu(NO_3)_2$ solution is mixed with SCN^- ions, a green colored complex ion, $[Cu(SCN)]^+$, forms.

$$Cu^{2+}_{(aq)} + SCN^{-}_{(aq)} \rightarrow [Cu(SCN)]^{+}_{(aq)} \qquad \text{(Eq. 5)}$$

In Part B of this experiment, you will determine whether a double displacement reaction occurs when aqueous solutions of the five compounds listed in Table 2 are mixed. You will be assigned an unknown solution containing one of the five compounds listed in Table 2. Based on your observations from Part B, you will identify the unknown inorganic compound.

Table 2 *Test solutions - inorganic compounds*

1. sodium hydroxide, $NaOH$
2. copper(II) sulfate, $CuSO_4$
3. lead(II) nitrate, $Pb(NO_3)_2$
4. zinc nitrate, $Zn(NO_3)_2$
5. barium chloride, $BaCl_2$

PROCEDURE

CAUTION

Students must wear departmentally approved eye protection while performing this experiment. Wash your hands before touching your eyes and after completing the experiment.

Chemical Alert

Avoid splattering 4 *M* aqueous ammonia and 4 *M* sodium hydroxide solutions on your skin. If these chemicals contact your skin, immediately wash the affected area with copious quantities of water and inform your lab instructor. Concentrated sodium hydroxide can cause permanent cornea damage to the eye. *Eye protection must not be removed for any reason during this experiment.*

Part A – Identification of Metal Ions in Aqueous Solution

1. Thoroughly clean and dry five 10 × 75 mm test tubes. Label the test tubes 1, 2, 3, 4 and 5.
2. Add four drops of 0.1 *M* $Ba(NO_3)_2$ solution to test tube 1, four drops of 0.1 *M* $Mg(NO_3)_2$ solution to test tube 2, four drops of 0.1 *M* $Pb(NO_3)_2$ solution to test tube 3, four drops of 0.1 *M* $Co(NO_3)_2$ solution to test tube 4, and four drops of 0.1 *M* $Fe(NO_3)_3$ solution to test tube 5.
3. Should the colors of the metal ion solutions be recorded in the Lab Report?
4. Add four drops of 0.5 *M* $(NH_4)_2SO_4$ to each of the five test tubes in Step 2. Carefully observe the test tubes for any evidence of reaction. Should your observations be recorded in the Lab Report? Should you write the formulas of the substances formed in these reactions in the Lab Report?

5. Empty the contents of the five test tubes into the ''Waste Container.''
6. Repeat Steps 1 and 2 above, adding four drops of each metal nitrate solution to separate clean, dry test tubes.
7. Add four drops of 0.5 *M* NH_4SCN to each of the five test tubes from Step 6. Carefully observe the test tubes for any evidence of reaction. Should your observations be recorded in the Lab Report? Should you write the formulas of the substances formed in these reactions in the Lab Report?
8. Empty the contents of the five test tubes into the ''Waste Container.''
9. Repeat Steps 1 and 2, adding four drops of each metal nitrate solution to separate clean, dry test tubes.
10. Determine the number of drops of distilled water required to dilute ten drops of 4 *M* NaOH to produce a 0.5 *M* NaOH solution. Should your calculations be included in the Lab Report? Prepare the 0.5 *M* NaOH solution in a 50-mL beaker. *Be sure to add the concentrated base to water when mixing.*
11. Add two drops of the freshly prepared 0.5 *M* NaOH solution (*limited* OH^- *ions*) to each of the five test tubes in Step 9. Carefully observe the test tubes for any evidence of reaction. Should your observations be recorded in the table in the Lab Report? Write the formulas of the substances formed in these reactions in the Lab Report.
12. Empty the contents of the five test tubes into the ''Waste Container.''
13. Repeat Steps 1 and 2, adding four drops of each metal nitrate solution to separate clean, dry test tubes.
14. Add 15-20 drops of 4 *M* NaOH solution (*excess* OH^- *ions*) to each of the five test tubes in Step 13. Carefully observe the test tubes for any evidence of reaction. Should your observations be recorded in the Lab Report? Should you write the formulas of the substances formed in these reactions in the Lab Report?
15. Empty the contents of the five test tubes into the ''Waste Container.''
16. Repeat Steps 1 and 2, adding four drops of each metal nitrate solution to separate clean, dry test tubes.
17. Determine the number of drops of distilled water required to dilute ten drops of 4 *M* $NH_{3(aq)}$ to produce a 0.5 *M* $NH_{3(aq)}$ solution. Should your calculations be included in the Lab Report? Prepare the 0.5 *M* $NH_{3(aq)}$ solution in a 50-mL beaker. *Be sure to add the concentrated base to water when mixing.*
18. Add two drops of the freshly prepared 0.5 *M* $NH_{3(aq)}$ solution (*limited* $NH_{3(aq)}$) to each of the five test tubes in Step 16. Carefully observe the test tubes for any evidence of reaction. Should your observations be recorded in the Lab Report? Should you write the formulas of the substances formed in these reactions in the Lab Report?
19. Empty the contents of the five test tubes into the ''Waste Container.''
20. Repeat Steps 1 and 2, adding four drops of each metal nitrate solution to separate clean, dry test tubes.
21. Add 15–20 drops of 4 *M* $NH_{3(aq)}$ solution (*excess* $NH_{3(aq)}$) to each of the five test tubes in Step 20. Carefully observe the test tubes for any evidence of reaction. Should your observations be recorded in the Lab

Report? Should you write the formulas of the substances formed in these reactions in the Lab Report?

22. Empty the contents of the five test tubes into the "Waste Container."

23. Write the *balanced net ionic chemical equations* for all reactions that occurred in Steps 4–21.

24. Obtain 2 mL of an unknown metal ion solution from your laboratory instructor. Should the Unknown Number be recorded in the Lab Report? Should the color of the unknown metal ion solution be recorded in the Lab Report?

25. Label six clean, dry test tubes (10 × 75 mm) as 1, 2, 3, 4, 5, and 6. Add four drops of the unknown metal solution to each of the test tubes.

26. Add four drops of 0.5 *M* $(NH_4)_2SO_4$ solution to test tube 1. Add four drops of 0.5 *M* NH_4SCN solution to test tube 2. Add four drops of 0.5 *M* NaOH solution (prepared in Step 10) to test tube 3. Add four drops of 4 *M* NaOH solution to test tube 4. Add four drops of 0.5 *M* NH_3 solution (prepared in Step 17) to test tube 5. Add four drops of 4 *M* NH_3 solution to test tube 6. Carefully observe the test tubes for any evidence of reaction. Should your observations be recorded in the table in the Lab Report?

27. From your observations of the reactions of the known and unknown metal ions, identify the unknown metal ion. *The unknown metal ion is one of the metal ions listed in Table 1.*

28. Empty the contents of the five test tubes into the "Waste Container." Thoroughly clean and dry the test tubes.

Part B - Identification of Inorganic Compounds in Aqueous Solution

29. Develop and carry out a procedure for determining whether double displacement reactions occur when solutions of the following five compounds are mixed. How many reaction combinations are possible between these 5 solutions? *Use only four drops of each solution for each reaction.* Should the colors of the solutions listed below be recorded in the Lab Report?

1. 0.5 *M* NaOH

2. 0.5 *M* $CuSO_4$

3. 0.5 *M* $Pb(NO_3)_2$

4. 0.5 *M* $Zn(NO_3)_2$

5. 0.5 *M* $BaCl_2$

30. Carefully observe the mixtures in each test tube for any evidence of reaction. Should your observations be recorded in the Lab Report? Should you write the formulas of the substances formed in these reactions in the Lab Report?

31. Empty the contents of the test tubes into the "Waste Container."

32. Thoroughly clean and dry all test tubes.

33. Write *balanced net ionic chemical equations* for all reactions that occur in Steps 29–30.

34. Obtain an unknown inorganic compound solution from your instructor. Should you record the unknown number and color of the solution in the Lab Report?

35. Develop and carry out a procedure for identifying the unknown inorganic compound with each of the test solutions from Step 29. Use four drops of each solution for each reaction.
36. Carefully observe the mixtures in each test tube for any evidence of reaction. Should you record your observations in the Lab Report?
37. Pour the contents of the test tubes into the "Waste Container."
38. Thoroughly clean and dry all test tubes.
39. From your observations of the reactions of the known and the unknown inorganic compounds, identify the unknown inorganic compound. *The unknown is one of the compounds listed in Step 29.*

Name Section Date

Instructor

8 EXPERIMENT 8

Lab Report

Part A – Identification of Metal Ions in Aqueous Solution

Observations of the physical appearance of the metal ion solutions.

Observations and formulas of compounds formed in the reaction of $(NH_4)_2SO_4$ with 0.1 *M* solutions of $Ba(NO_3)_2$, $Mg(NO_3)_2$, $Pb(NO_3)_2$, $Co(NO_3)_2$, and $Fe(NO_3)_3$.

Observations and formulas of compounds formed in the reaction of NH_4SCN with 0.1 *M* solutions of $Ba(NO_3)_2$, $Mg(NO_3)_2$, $Pb(NO_3)_2$, $Co(NO_3)_2$, and $Fe(NO_3)_3$.

Determine the number of drops of distilled water required to dilute ten drops of 4 *M* $NaOH_{(aq)}$ to produce a 0.5 *M* $NaOH_{(aq)}$ solution.

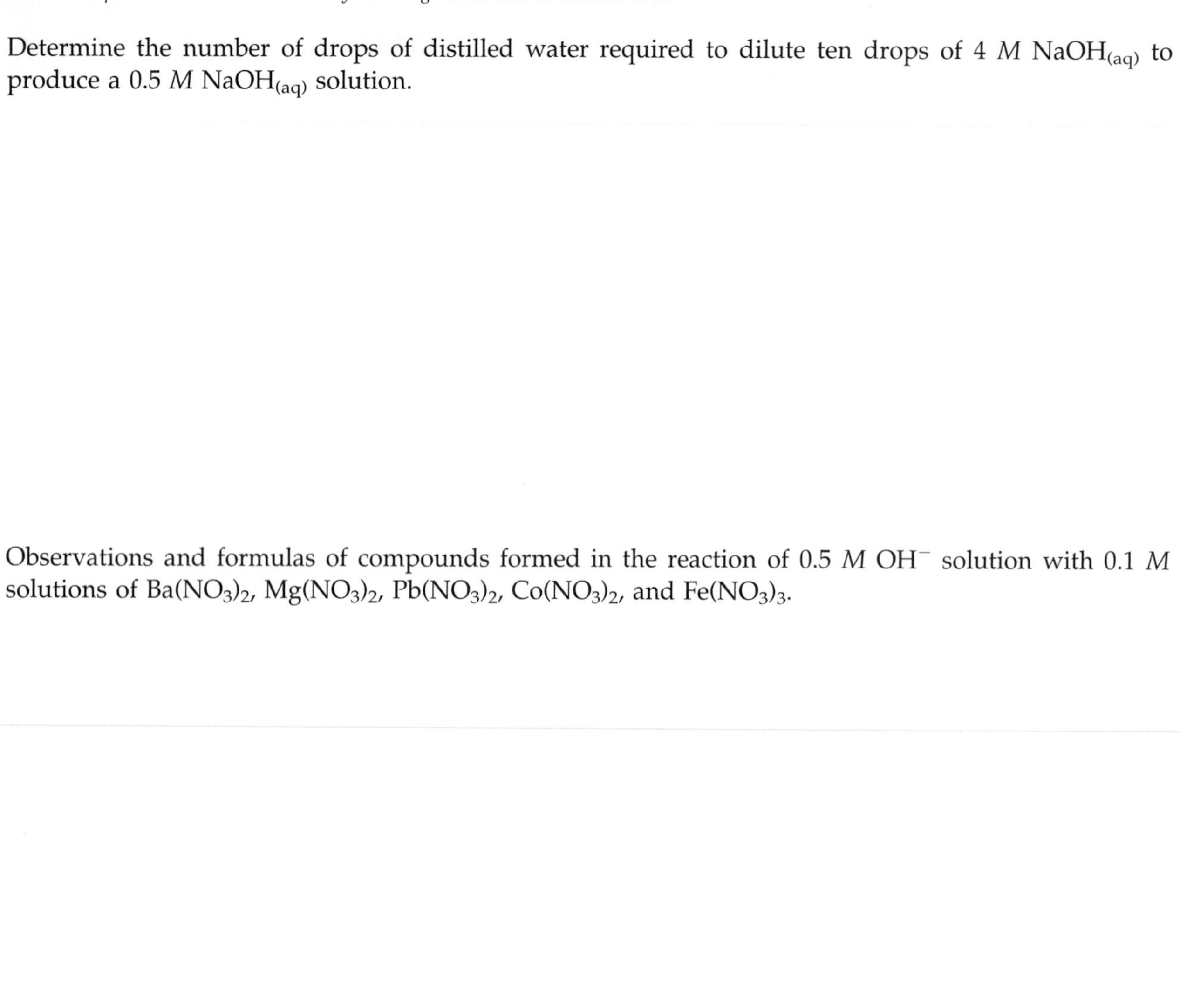

Observations and formulas of compounds formed in the reaction of 0.5 *M* OH^- solution with 0.1 *M* solutions of $Ba(NO_3)_2$, $Mg(NO_3)_2$, $Pb(NO_3)_2$, $Co(NO_3)_2$, and $Fe(NO_3)_3$.

Observations and formulas of compounds formed in the reaction of 4.0 *M* OH^- solution with 0.1 *M* solutions of $Ba(NO_3)_2$, $Mg(NO_3)_2$, $Pb(NO_3)_2$, $Co(NO_3)_2$, and $Fe(NO_3)_3$.

Determine the number of drops of distilled water required to dilute ten drops of 4 *M* $NH_{3(aq)}$ to produce a 0.5 *M* $NH_{3(aq)}$.

Observations and formulas of compounds formed in the reaction of 0.5 *M* $NH_{3(aq)}$ solution with 0.1 *M* solutions of $Ba(NO_3)_2$, $Mg(NO_3)_2$, $Pb(NO_3)_2$, $Co(NO_3)_2$, and $Fe(NO_3)_3$.

Observations and formulas of compounds formed in the reaction of 4.0 *M* $NH_{3(aq)}$ solution with 0.1 *M* solutions of $Ba(NO_3)_2$, $Mg(NO_3)_2$, $Pb(NO_3)_2$, $Co(NO_3)_2$, and $Fe(NO_3)_3$.

Write *balanced net ionic chemical equations* for all reactions that occur in Steps 4–21.

Unknown number of metal ion ______________________________

Observations for the reactions of the unknown metal ion.

Identity of the unknown metal ion. ____________________

Part B – Identification of Inorganic Compounds in Aqueous Solution

Procedure, observations, and formulas of substances formed when displacement reactions are carried out using different combinations of 0.5 *M* solutions of NaOH, $CuSO_4$, $Pb(NO_3)_2$, $Zn(NO_3)_2$, and $BaCl_2$. How many combinations are possible between these 5 solutions?

Write *balanced net ionic chemical equations* for all reactions that occurred in Steps 29–30.

Unknown number of unknown inorganic compound __________

Observations for the reactions of the unknown inorganic compound.

Identity of the unknown inorganic compound ________________________________

Name | Section | Date

Instructor

8 **EXPERIMENT 8**

Pre-Laboratory Questions

1. Complete and balance the following net ionic equations. If no reaction occurs, write NR.

a. $Mg^{2+} + OH^{-}_{limited} \rightarrow$

b. $Cu^{2+} + SCN^{-} \rightarrow$

c. $Co^{2+} + NH_{3(aq)}\ _{limited} + H_2O \rightarrow$

d. $Ba^{2+} + SO_4^{2-} \rightarrow$

e. $Pb^{2+} + NH_{3(aq)}\ _{excess} + H_2O \rightarrow$

f. $Fe^{3+} + SO_4^{2-} \rightarrow$

2. Complete and balance the following net ionic equations. If no reaction occurs, write NR. Tables 4 and 5 of Experiment 7 list amphoteric metal hydroxides that form complex ions in the presence of excess hydroxide ions and aqueous ammonia.

a. $Mg(OH)_2 + OH^{-}_{excess} \rightarrow$

b. $Cu(OH)_2 + OH^{-}_{excess} \rightarrow$

c. $Co(OH)_2 + OH^{-}_{excess} \rightarrow$

d. $Fe(OH)_3 + NH_{3(aq)}\ _{excess} \rightarrow$

e. $Pb(OH)_2 + NH_{3(aq)}\ _{excess} \rightarrow$

3. Write the formula of the precipitate formed in the spaces below when each pair of ions are mixed in *aqueous* solution. Write NP in the spaces if no precipitate forms. General rules for the solubilities of ionic compounds in water are provided in Table 3 of Experiment 7.

	$Pb(NO_3)_2$	$Cu(NO_3)_2$	$(NH_4)_2S$
$KOH_{limited}$			
$FeCl_3$			
$CoSO_4$			

4. Write complete and balanced net ionic equations for each reaction that produced a precipitate in Question 3 above.

Name | Section | Date

Instructor

8 EXPERIMENT 8

Post-Laboratory Questions

1. In Step 4 of the procedure, if ammonium phosphate had been added to the metal ion solutions instead of ammonium sulfate, what substances would have been produced? Write the net ionic equations for these reactions.

2. In Step 34 of the procedure, a student was assigned a colorless unknown. The student added five drops of 1.0 *M* barium chloride solution to 1.0 mL of the unknown solution. A white precipitate formed. In a second experiment, the student added two drops of 0.5 *M* NH_3 solution to 1.0 mL of the unknown solution. A white precipitate formed. The student added five drops of 4.0 *M* NH_3 to the resulting solution in the second experiment. The white precipitate did not dissolve. Identify the colorless unknown solution. Justify your answer with an explanation. Write the net ionic equations for all reactions.

3. Write balanced formula unit and net ionic equations for each of the following chemical reactions in aqueous solution. If no reaction occurs, write NR. Include the states (s, ℓ, g, or aq) of all reactants and products.

A. copper(II) chloride + lead(II) nitrate →

B. zinc bromide + silver nitrate →

C. iron(III) nitrate + ammonia solution →

D. barium chloride + sulfuric acid →

EXPERIMENT 9

Gravimetric Analysis of a Chloride, Sulfate, or Carbonate Compound

OBJECTIVES

Identify an unknown chloride, sulfate, or carbonate compound using gravimetric analysis.

INTRODUCTION

The quantitative determination of a substance by precipitation, followed by isolation and weighing of the precipitate is called **gravimetric analysis**. Because the mass of the precipitate must be accurately determined, this procedure is limited to reactions that produce essentially 100% products. Consequently, gravimetric analysis is limited to precipitation reactions. Given that precipitation reactions can be selective, gravimetric analysis is a useful quantitative procedure.

Quantitative analysis is used to determine the percentage of a particular element or ion in a sample. In a typical gravimetric analysis, the percentage of an ion of interest in a solid compound is determined. This procedure involves dissolving a substance of unknown composition in water, and allowing the ion of interest to react with a counter ion (from a precipitating reagent) to form a precipitate. (Precipitation reactions were introduced in Experiments 7 and 8. Predicting whether a precipitate will form depends on knowledge of the solubilities of the products. General rules for the solubilities of ionic compounds in water are provided in Table 3 of Experiment 7.) The precipitate is then isolated, dried, and weighed. Knowing the mass and the chemical formula of the precipitate, the mass of the ion of interest can be determined from the percent mass composition of the precipitate. From the masses of the ion of interest and the original compound, the percentage of the ion of interest in the original compound can be calculated.

The original sample is dissolved in a *minimum* volume of water. The water must be free of ions that might interfere with the formation of the desired precipitate. The original sample *must be the limiting reagent* to ensure the complete precipitation of the ion of interest. Consequently, the precipitating reagent must be *present in excess*.

Frequently the precipitate forms a colloidal dispersion with particles too small to filter. **Digestion** is the process of heating a solution (*just below the boiling point* to prevent splattering and loss of sample) to encourage aggregation of the colloidal particles into larger, filterable particles. Heating of the solution ceases once the solution clears and the precipitate settles to the bottom of the container.

Following digestion the sample is cooled to room temperature, then filtered using gravity filtration. The precipitate is then washed to remove traces of impurities imparted by the precipitating reagent. **Washing** involves rinsing the precipitate with several small quantities of a volatile electrolyte solution. For example, dilute nitric acid (0.001 *M*) solution is sometimes used to wash colloidal precipitates.

Next, the precipitate is dried by heating it in a drying oven with the temperature set at 50–60 °C (higher temperatures may char the filter paper). After drying, the precipitate is weighed. The percentage of the ion of interest can then be calculated.

Sample Calculation to Determine the Volume of the Precipitating Reagent

A chemist found a bottle of an unknown compound in the stockroom. Preliminary tests indicate that the unknown compound is magnesium chloride hexahydrate ($MgCl_2{\cdot}6H_2O$), magnesium sulfate heptahydrate ($MgSO_4{\cdot}7H_2O$), or magnesium carbonate ($MgCO_3$). A 0.864 gram sample of the unknown magnesium compound is dissolved in approximately 20 milliliters of distilled water. Sodium hydroxide is selected as the precipitating regent because Mg^{2+} ions precipitate as $Mg(OH)_2$ in the presence of OH^- ions. Calculate the volume of 1.00 *M* NaOH required to completely precipitate all of the magnesium ions in the unknown compound according to the following equation.

$$Mg^{2+}{}_{(aq)} + 2\ NaOH_{(aq)} \rightarrow 2\ Na^{+}{}_{(aq)} + Mg(OH)_{2(s)}$$

To ensure that Mg^{2+} in the unknown compound is the limiting reagent, assume the entire mass of the unknown sample is Mg^{2+}.

$$0.864\ g\ Mg^{2+} \times \frac{1\ mol\ Mg^{2+}}{24.30\ g\ Mg^{2+}} \times \frac{2\ mol\ NaOH}{1\ mol\ Mg^{2+}} \times \frac{1000\ mL\ NaOH\ soln}{1.00\ mol\ NaOH} = 71.1\ mL\ NaOH\ soln$$

Sample Calculation for the Mass Composition of the Unknown Sample

A sufficient quantity of 1.00 *M* NaOH is added to the solution from the previous example to precipitate all of the magnesium ions. The solution is digested, filtered, and dried. The mass of $Mg(OH)_2$ collected is 0.607 grams. Identify the unknown compound.

1. Calculate the mass of Mg^{2+} precipitated as $Mg(OH)_2$.

$$0.607\ g\ Mg(OH)_2 \times \frac{1\ mol\ g\ Mg(OH)_2}{58.30\ g\ Mg(OH)_2} \times \frac{1\ mol\ Mg^{2+}}{1\ mol\ g\ Mg(OH)_2} \times \frac{24.30\ g\ Mg^{2+}}{1\ mol\ Mg^{2+}} = 0.253\ g\ Mg^{2+}$$

2. Calculate the percent by mass of Mg^{2+} in the unknown compound.

$$\frac{0.253\text{ g } Mg^{2+}}{0.864\text{ g unknown compound}} \times 100\% = 29.3\% \ \ Mg^{2+}$$

3. Determine the percent by mass of Mg^{2+} in magnesium chloride, magnesium sulfate, or magnesium carbonate. Compare the calculated percent by mass of Mg^{2+} in the unknown compound to the percent mass by mass of Mg^{2+} in magnesium chloride, magnesium sulfate, and magnesium carbonate to determine the identity of the unknown compound.

   The percent error for the experiment is calculated by dividing the absolute value of the difference between the calculated percent mass of Mg^{2+} and the actual percent mass of Mg^{2+} in $MgCO_3$ by the actual percent mass of Mg^{2+} in $MgCO_3$, then multiplying that value by 100%.

PROCEDURE

CAUTION

Students must wear departmentally approved eye protection while performing this experiment. Wash your hands before touching your eyes and after completing the experiment.

Part A – Characterization of the Unknown Compound as a Chloride, Sulfate, or Carbonate Compound

1. Students must determine the formula of a chloride, sulfate, or carbonate compound using gravimetric analysis. The unknown compound is one of the compounds in the following list.

1. Copper(II) chloride dihydrate	$CuCl_2{\cdot}2H_2O$
2. Iron(III) chloride hexahydrate	$FeCl_2{\cdot}6H_2O$
3. Lithium chloride	$LiCl$
4. Copper(II) sulfate pentahydrate	$CuSO_4{\cdot}5H_2O$
5. Iron(III) sulfate	$Fe_2(SO_4)_3$
6. Potassium sulfate	K_2SO_4
7. Potassium carbonate	K_2CO_3
8. Lithium carbonate	Li_2CO_3
9. Ammonium carbonate	$(NH_4)_2CO_3$

2. Obtain ~3 grams of an unknown compound from your laboratory instructor. Should you record the Unknown Number in the Lab Report?
3. The following are available as precipitating reagents: 1.00 *M* aqueous silver nitrate, 1.00 *M* aqueous barium nitrate, and 1.00 *M* aqueous calcium nitrate. The General rules for the solubilities of ionic compounds in water are provided in Table 3 of Experiment 7.

4. Dissolve ~1 gram of the unknown compound in ~20 mL of distilled water. Should the exact mass of the unknown be recorded in the Lab Report, and to how many significant figures?
5. Add a few drops of the unknown solution to each of three separate test tubes.
6. Add 4–5 drops of 1.00 *M* $AgNO_3$ to one of the test tubes in Step 5. Add 4–5 drops of 1.00 *M* $Ba(NO_3)_2$ to another one of the test tubes in Step 5. Add 4–5 drops of 1.00 *M* $Ca(NO_3)_2$ to the remaining test tube in Step 5. Should you record your observations from each test tube in the Lab Report?
7. Discard the solutions into the Waste Container.
8. Based on your observations from Step 6, indicate whether the compound is a chloride, sulfate, or carbonate compound. (*It is possible that more than one precipitating reagent will precipitate Cl^-, SO_4^{2-}, or CO_3^{2-} ions.*) Determine the appropriate precipitating agent to be used in Part B below. Should you record the name of the precipitating agent you choose in the Lab Report?

Part B – Determination of the Percent by Mass of Cl^-, SO_4^{2-}, or CO_3^{2-} in the Unknown Sample

9. Simultaneously perform two duplicate gravimetric analyses using approximately one gram samples of the unknown compound in each analysis. Should the exact mass of each sample be determined? Should the exact mass of each sample be recorded in the Lab Report, and to how many significant figures?
10. What volume of 1.00 *M* aqueous silver nitrate, 1.00 *M* aqueous barium nitrate, or 1.00 *M* aqueous calcium nitrate is required to precipitate all of the Cl^-, SO_4^{2-}, or CO_3^{2-} ions in the unknown compound. Which reactant, the unknown compound or the precipitating reagent, must be in excess? Should your calculations be shown in the Lab Report?
11. Dissolve the unknown samples in a minimum amount of water (add water drop wise until the dissolution is complete). Should the volume of the water used be recorded in the Lab Report, and to how many significant figures.
12. Add the appropriate volume of precipitating reagent to the each unknown solution.
13. Digest the solutions (be sure to heat them *just below their boiling points*). When is the digestion process complete?
14. When digestion is complete, let the solutions cool to room temperature.
15. Fold two pieces of filter paper into the conical shape shown in Figure 1.
16. Place each filter paper on a watch glass. Should the mass of each filter paper and watch glass be determined? Should the mass of each watch glass and filter paper be recorded in the Lab Report, and to how many significant figures?
17. Assemble the gravity filtration apparatus depicted in Figure 2. *Be sure the stem of the funnel touches the wall of the beaker.*
18. Carefully pour each unknown solution into the filter paper in each gravity filtration apparatus. Pour as much of the water into the funnel as is possible before pouring the precipitate into the filter.

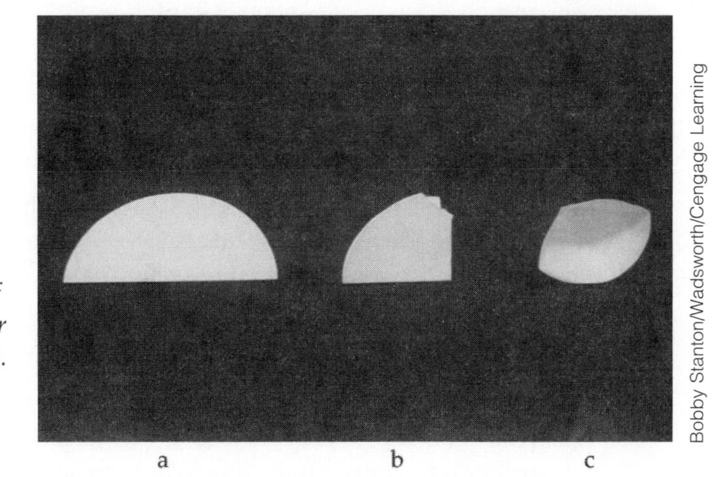

Figure 1
Fold a piece of filter paper in half (1a). Fold the filter paper in quarters and tear off one of the corners (1b). Open the filter paper into a conical shape (1c). Be sure not to open the side with the torn corner when forming the cone

Figure 2
Gravity filtration apparatus

19. Add 4–5 drops of the precipitating agent to the filtrate in the beaker to test for completeness of precipitation. If a precipitate forms, what should be done to complete the precipitation? What should be done to collect the additional precipitate?

20. Should a rubber policeman be used to remove any residual precipitate from the walls of the beaker, or is this mass insignificant enough to be ignored? Should you use a wash bottle filled (with distilled water) to rinse the final portions of the precipitate from the beaker into the filter?

21. If you used 0.1 M $AgNO_3$ as the precipitating reagent, wash the precipitate with 3–5 mL of 0.001 M nitric acid. If you used 0.1 M $Ba(NO_3)_2$ or 0.1 M $Ca(NO_3)_2$ as the precipitating reagent, wash the precipitate with 3–5 mL of distilled water. Repeat the washing process 2–3 times.

22. After all of the liquid has drained from the funnels, *gently* remove the filter and precipitate from the funnel with a spatula. Place each filter with its precipitate on the appropriate watch glass.

23. Place each watch glass and filter in a drying oven set to 50–60 °C. Periodically inspect the filters in the oven. The filter paper *MUST NOT* char.
24. When dry, remove the watch glasses/filter papers from the oven with crucible tongs and let them cool to room temperature.
25. Should the mass of the precipitate on each filter paper/watch glass be determined? Should the masses of the precipitates be recorded in the Lab Report, and to how many significant figures?
26. What is the percent by mass of Cl^-, SO_4^{2-}, or CO_3^{2-} in each sample?
27. What is the average percent by mass of Cl^-, SO_4^{2-}, or CO_3^{2-} in the unknown compound?
28. Knowing the average percent by mass Cl^-, SO_4^{2-}, or CO_3^{2-} in the unknown compound, how do you determine the identify of the unknown compound?
29. What is the average percent error for the calculated percent by mass for the two trials?
30. Discard the filter papers containing the precipitates into the trash can.

Name ______ Section ______ Date ______

Instructor ______

9 EXPERIMENT 9

Lab Report

Part A – Characterization of the Unknown Compound as a Chloride, Sulfate, or Carbonate Compound

Unknown number ________

Observations from the addition of the precipitating reagents to the unknown solution.

Indicate whether the compound is a chloride, sulfate, or carbonate compound.

Precipitating reagent to be used. ____________

Part B – Determination of the Percent by Mass of Cl^-, SO_4^{2-}, or CO_3^{2-} in the Unknown Sample

Experimental data/volume of precipitating reagent calculations - Sample 1

Experimental data/volume of precipitating reagent calculations - Sample 2

Percent by mass of Cl^{-}, SO_4^{2-}, or CO_3^{2-} in the precipitate for each trial.

Average mass percent Cl^-, SO_4^{2-}, or CO_3^{2-} in the unknown compound for the two trials.

What is the identify of the unknown compound?

What is the percent error for each trial and the average percent error for the calculated percent by mass for the two trials.

Name

Section

Date

Instructor

9 EXPERIMENT 9

Pre-Laboratory Questions

1. A chemist found a bottle of an unknown chemical in the stockroom. Preliminary tests indicate that the unknown compound is either zinc nitrate or zinc chloride. A 1.002 gram sample of the unknown zinc compound is dissolved in 20 milliliters of distilled water. Calculate the volume of 1.0 *M* $(NH_4)_2S$ required to completely precipitate all of the zinc ions in the unknown compound. Write a balanced equation for the reaction.

2. A sufficient quantity of 1.00 *M* $(NH_4)_2S$ is added to the solution to precipitate all of the zinc ions. The solution is digested, filtered, and dried. The mass of ZnS collected is 0.717 grams.

   a. Determine the mass of Zn^{2+} precipitated as ZnS.

   b. Determine the percent by mass of Zn^{2+} in the unknown compound.

c. Knowing the percentage of Zn^{2+} ions in the unknown zinc compound, how do you determine its identity? Show all calculations to justify your answer.

Name

Instructor

Section

Date

9 EXPERIMENT 9

Post-Laboratory Questions

1. In Step 24, you were instructed to wash the precipitates on the filter paper.

a. Why was it necessary to wash the precipitate before drying.

b. You were instructed to use dilute nitric acid to wash silver chloride, and to use distilled water to wash barium sulfate and calcium carbonate. Why was water used, in place of nitric acid, to wash barium sulfate and calcium carbonate?

2. Some precipitate is lost when the filter paper and precipitate is removed from the drying oven. The filter paper and precipitate is subsequently weighed. Would the calculated percent by mass of the unknown ion in the compound be higher or lower than the *true* value? Justify your answer with an explanation.

3. If the filter paper containing the precipitate is significantly charred in the drying oven because the oven temperature is to high, would the calculated percent by mass of the unknown ion in the compound be higher or lower than the *true* value? Justify your answer with an explanation.

Emission Analysis of Aqueous Solutions of Groups IA and IIA Metal Salts

PURPOSE

Detection of Group IA and IIA metals ions in aqueous solution from their emission spectra utilizing the MeasureNet spectrophotometer.

INTRODUCTION

Metal ion contamination of water sources is a well known and widely studied environmental problem. From natural sources, metal ions leach into lakes, rivers, and streams as water flows across rock and soil formations containing water-soluble salts. (Remember, **salts** are ionic compounds containing positively and negatively charged ions bonded together in an extended crystal lattice. Frequently, the positive ions in salts are metal ions). Metal ion contamination in human water supplies is especially problematic in areas where metal ore mining has occurred. Rainwater, which is typically acidic, percolates through the mine tailings (scrap ore and waste byproducts) dissolving the metal ions into the water. Various industrial plants release wastewater containing metal ions. Crop fertilizers and salting of roads during winter are other sources of metal ion contaminants. Automobiles also produce metal ions that may end up in rainwater run-off and make their way into drinking water supplies.

Fresh water contamination by metal ions has two significant effects: 1) *salinity* and 2) *toxicity*. **Salinity** is an issue when metal ion concentrations exceed safe limits in water sources (i.e., the salt concentration in the water is too high). Plants and animals (including humans) cannot utilize water containing high metal ion concentrations. High salinity water sources are unusable for irrigation or drinking water, nor can aquatic life survive in these environments. Some metal ions are **toxic** (e.g., Pb^{2+}, Hg^{2+}, Cr^{3+}, Fe^{3+},

Ni^{2+}, Cu^{2+} and others), even when present in extremely low concentrations. These metal ions are commonly called **heavy metals** because they have densities that are five times greater than the density of water. The *d*-transition metals are frequent heavy metal contaminants in water.

Before a potential water supply can be utilized, metal ions present in the contaminated water must be identified and removed if they exceed safe levels. Two common methods for detecting and quantifying metal ion concentrations in water are **absorption** and **emission spectroscopy**. Absorption spectroscopy is the preferred method to detect metal ions that produce colored aqueous solutions. Many transition metal ions produce colored aqueous solutions due to the presence of *3d* or *4d* electrons. Transition metal ions can be identified by the wavelength of light they absorb when light is passed through solutions containing these ions. The amount of light absorbed by an ion is directly proportional to the concentration of the ion in solution. To enhance this effect, frequently a transition metal ion must be mixed with a complexing agent to enhance its ability to absorb light. The complexing agent produces a solution that is more intensely colored than the original ion solution. For example, an aqueous solution of Fe^{3+} is light yellow in color, but when mixed with thiocyanate (SCN^-) ions, the $[Fe(SCN)]^{2+}$ complex produces a blood-red colored solution. This color intensification permits detection of smaller Fe^{3+} concentrations. Absorption analysis will be used in Experiment 11 to detect Cr(VI) (in the form of CrO_4^{2-} ions) in solution.

This experiment uses emission spectroscopy to identify the metal ions present in a water sample. Emission analysis is the preferred method to detect ions that form colorless aqueous solutions such as the Group IA and IIA metal ions. In emission spectroscopy, the intensity and wavelength of light emitted by a metal atom is detected after it has been excited by one of several methods. The concentration of the metal ion is directly proportional to the intensity (brightness) of the emitted light. The wavelength (color) of the emitted light is characteristic of a particular metal.

In this experiment, you will determine if Group IA and IIA metal ions are present in a water sample taken from a local lake. The experiment will test for the presence of ions from Group IA (Li^+, Na^+, and K^+) and Group IIA (Ca^{2+}, Sr^{2+}, and Ba^{2+}). Because most Group IA and IIA metals are present in living organisms, these metals generally are not toxic when present in water supplies, but in elevated concentrations, they can cause salinity problems. (Of the ions used in this experiment, only Ba^{2+} is significantly toxic.)

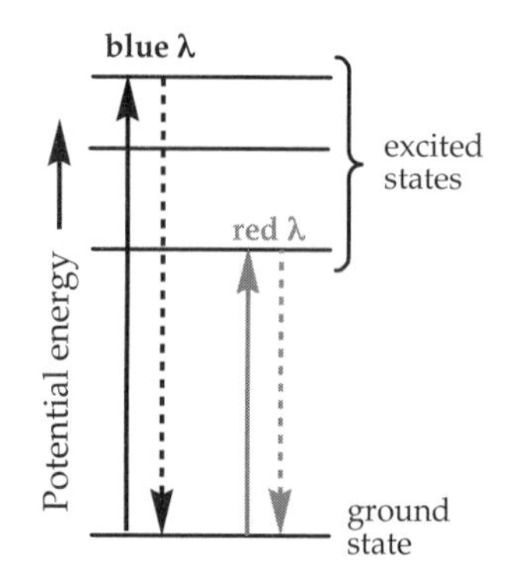

Figure 1
Absorption and emission of energy by electrons. λ is the symbol for the wavelength of light

Atomic emission occurs when an atom's electrons are energetically promoted by the absorption of energy, heat in this particular experiment, from the *ground state* energy level (an electron's normal energy state) to a higher energy level known as an *excited state* (*solid arrows* in Figure 1). Electrons reside in excited states for very brief periods of time, typically 10^{-9} to 10^{-6} seconds. When the electron returns to the ground state, energy is emitted in the form of light (*dashed arrows* in Figure 1).

Using a device called a spectrophotometer, we can detect and record the wavelength and the intensity, or brightness, of the light (wavelengths of 200 to 850 nm) emitted from flames containing Group IA and Group IIA metal atoms. An emission spectrum is a plot of the intensity of light versus the wavelength of the light emitted by the analyte (species of interest in the sample). Metals emit a characteristic pattern of emission wavelengths. By

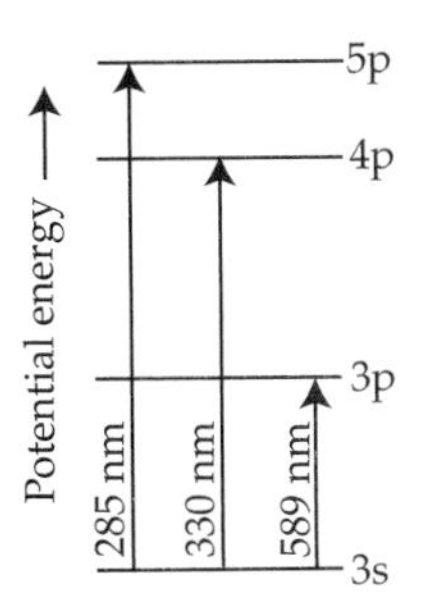

Figure 2
Excited state transitions for a Na atom

comparing the spectrum of an unknown water sample to the spectra of water samples containing known metal ions, it is possible to identify the metal ions present in the unknown water sample.

Consider three of the possible electronic transitions for a sodium atom. The *ground state* electron configuration of Na is [Ne]$3s^1$. The $3s^1$ valence electron can be promoted to any one of several excited states if a solution containing sodium, or sodium ions, is heated in a flame. Commonly, the next lowest energy subshell is the first excited state populated when a metal is heated in a flame. In the case of Na atoms, the next lowest energy subshell is the *3p*. The electron configuration for excitation of a sodium ion's *3s* electron to its *3p* subshell is represented as ***[Ne]**$\mathbf{3p^1}$. (The "*" signifies that the electron configuration is an excited state of Na). Some other possible electronic transitions for Na are $3s \rightarrow 4p$ (*[Ne]$4p^1$) and the $3s \rightarrow 5p$ (*[Ne]$5p^1$). For the *3s* valence electron of Na to be promoted to a *3p*, *4p* or *5p* excited state, the electron must absorb the correct amount (*quantum*) of energy (Figure 2). In this experiment, the Na atom will absorb thermal energy (heat) from a Bunsen burner flame.

The absorbed energy causes electrons in the Na atom to move from the ground state to one of the excited states. After a short period of time, the excited sodium atom loses the energy it absorbed from the flame by emitting that energy in the form of light. An emission spectrum can be recorded for this process. The emitted wavelengths of light are characteristic for a given metal. Different metals emit different wavelengths of light in the emission process. Figure 3 depicts an emission spectrum for sodium atoms in aqueous solution. An intense emission wavelength is observed at 589 nm corresponding to the transition of an electron from the *3p* excited state to the *3s* ground state in Na.

To determine the amount of energy emitted by an electronic transition in an atom, we must know either the wavelength or frequency of the emitted light. The wavelength (λ) of light is related to its frequency (ν) and the speed of light (c) by the following equation. The speed of light, c, is a constant, 3.00×10^8 m/s.

$$c = \lambda \cdot \nu \qquad \text{(Eq. 1)}$$

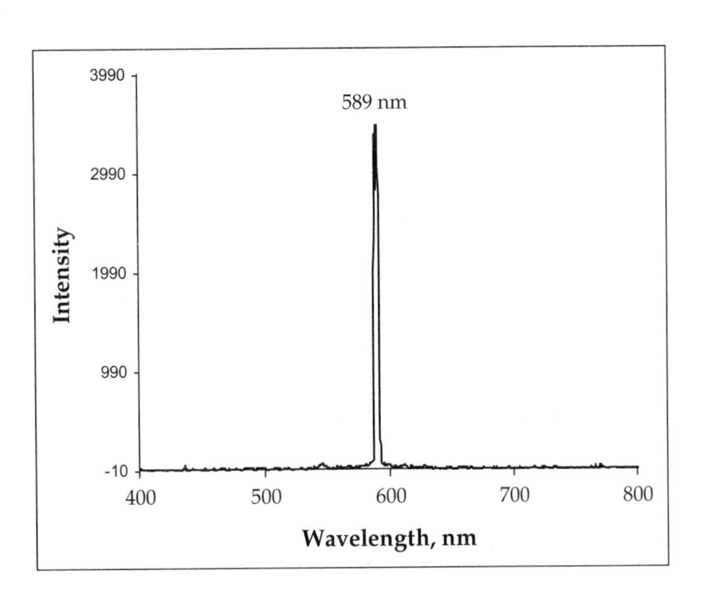

Figure 3
Sodium emission spectrum

The energy, E, of the emitted light is the product of Planck's constant, $h = 6.63 \times 10^{-34}$ J·s, and the frequency of light, ν. This relationship determines the energy emitted by a single atom's electronic transition.

$$E = h\nu \qquad \text{(Eq. 2)}$$

Using Equations 1 and 2, we obtain the relationship between the energy and wavelength of light:

$$\nu = \frac{c}{\lambda} \qquad \text{(Eq. 3)}$$

$$E = \frac{hc}{\lambda} \qquad \text{(Eq. 4)}$$

Using Equation 4, we can calculate the energy of a photon of light emitted by an electron in a sodium atom for the *3p→3s* transition (excited state to ground state transition) which has a wavelength of 589 nm or 5.89×10^{-7} m.

$$E = \frac{hc}{\lambda} = \frac{(6.63 \times 10^{-34} \text{J} \cdot \text{s})\ (3.00 \times 10^{8}\ \text{m/s})}{5.89 \times 10^{-7}\ \text{m}} = 3.38 \times 10^{-19}\ \text{J/atom}$$

We can also calculate the energy emitted by 1 mole of Na atoms in kJ/mole.

$$E = \frac{3.38 \times 10^{-19}\ \text{J}}{\text{atom}} \times \frac{1\ \text{kJ}}{1000\ \text{J}} \times \frac{6.02 \times 10^{23}\ \text{atoms}}{\text{mol}} = 203\ \text{kJ/mol}$$

PROCEDURE

CAUTION

Students must wear departmentally approved eye protection while performing this experiment. Wash your hands before touching your eyes and after completing the experiment. Aqueous chloride solutions can be an irritant and are corrosive. Barium chloride is toxic. Hydrochloric acid is *highly* corrosive.

1. See Appendix E – **Instructions for Recording an Emission Spectrum Using the MeasureNet Spectrophotometer**. Complete Steps 1–7 in Appendix E before proceeding to Step 2 below.
2. Obtain ~1 mL of each of the standard aqueous metal ion solutions (Li^{+}, Na^{+}, K^{+}, Ca^{2+}, Sr^{2+}, and Ba^{2+}), ~1 mL of the unknown water sample, and a nichrome wire. Should you record the Unknown Number in the Lab Report?
3. The nichrome wire must be cleaned before it can be used to introduce samples into the flame. Place 5 mL of 6 *M* HCl in a 50-mL beaker. Dip the nichrome wire into the 6 *M* HCl and swish the wire in the solution

for 30–40 seconds to insure that it is thoroughly clean. Any ions that remain on the wire from a previous lab student's experiment could contaminate your experiment.

4. Rinse the wire with 30 to 50 mL of distilled water. Pour the used acid into the Waste Container.
5. See Appendix E – **Recording an Emission Spectrum**. Complete Steps 8–24 in Appendix E before proceeding to Step 6 below.

   Chemical Warning - Be careful not to get acid on your skin. If acid should contact your skin, immediately wash the affected area with copious quantities of water and inform your laboratory instructor.

Calculations and Experimental Analysis to be Completed Outside of Lab

6. Excel instructions for preparing an XY Scatter Plot are provided in Appendix B-5. Use these instructions to plot emission spectra for each of the standard metal solutions and the unknown water sample.
7. Should you record the two most intense emission lines in the Lab Report for each metal ion in the emission spectra of the known and unknown solutions? What unit should be used to when recording the wavelength of each emission line? Should you record the emission intensities of each emission wavelength?
8. Determine the energy of the light emitted per atom and the kilojoules emitted per mole of atoms for the two most intense emission wavelengths observed in each emission spectrum. Some metals will only exhibit one emission wavelength in their emission spectrum.
9. Determine the metal ions present in the unknown sample and record them in the Lab Report.
10. How many emission spectra should be included in your Lab Report when it is submitted to your instructor?

Name

Instructor

Section

Date

10 EXPERIMENT 10

Lab Report

Unknown number of water sample __________

Flame colors for known metal ions and the unknown water sample.

Two most intense emission lines for each ion in the known and unknown solutions.

Energy of the light emitted per atom and the kilojoules emitted per mole of atoms for for each ion in the known and unknown solutions.

Identity of the metal ions present in the unknown water sample.

____

Name | Section | Date

Instructor

10 EXPERIMENT 10

Pre-Laboratory Questions

1. An emission line of sodium has a wavelength of 330 nm. Calculate the energy of a photon of light emitted in J/atom, and the energy emitted per mole of Na atoms in kJ/mol at this wavelength.

2. Determine the corresponding emission wavelength (in nm) for the emission energies (in kJ/mol) of each Group IA or IIA metal given in the table below.

Element	*Emission Energy (kJ/mol)*	*Wavelength (nm)*
Li	178.2	
Na	203.5	
K	155.3 156.5	
Ca	184.8 198.7 191.4 216.1	
Sr	178.2 197.5	
Ba	144.5 161.9	

Show calculations for completing the table in Question 2.

3. Identify the metal ions present in each of the following water sample emission spectra. Use the data in the Table in Question 2 to assist you in identifying the metal ions present in each spectrum.

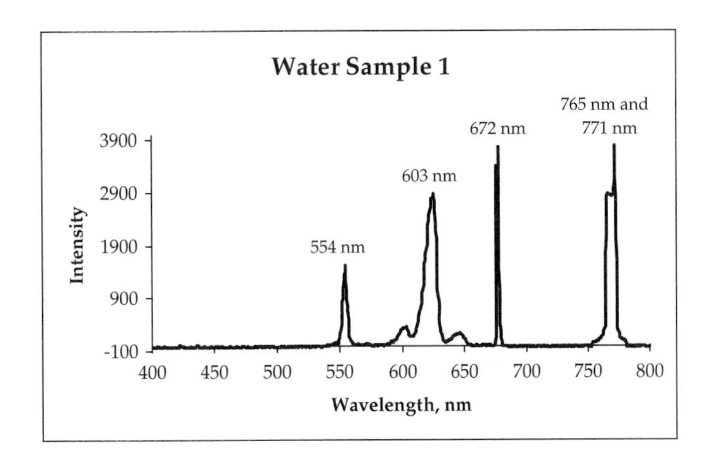

Identity of metal ions in unknown water sample 1. __________

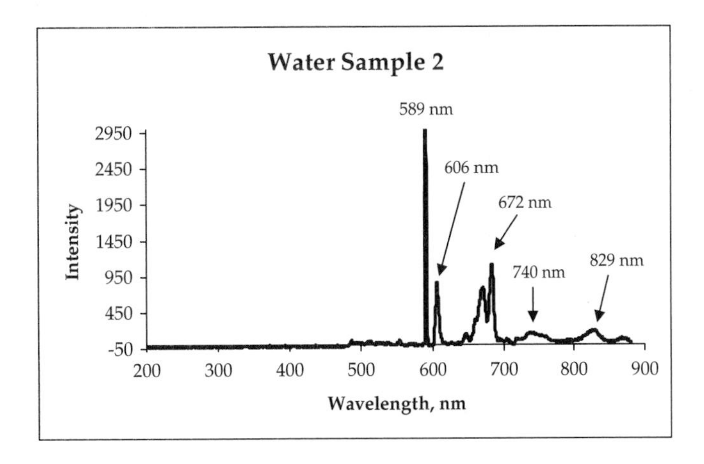

Identity of metal ions in unknown water sample 2. __________

Name

Section

Date

Instructor

10 EXPERIMENT 10

Post-Laboratory Questions

1. A student uses 6 *M* HCl solution contaminated with $Al(NO_3)_3$ to wash the nichrome wire in between recordings of emission spectra for the different known and unknown metal ion solutions. The emission energies for aluminum are 3.90×10^{-19} J/atom, 3.58×10^{-19} J/atom, and 2.97×10^{-19} J/atom. How might this experimental error affect the identification of the metal ions present in the unknown sample? Justify your answer with an explanation.

2. How could you modify the experimental procedure used in this experiment to conduct quantitative analysis (i.e., determining the actual concentrations of the metal ions in the standard and unknown metal ion solutions)?

3. A student records two potassium emission spectra: 1) one using 0.10 *M* KNO_3 solution; 2) one using 0.10 *M* K_2SO_4 solution. Will significant differences be observed in the two emission spectra? Why or why not?

Determination of Chromium(VI) Concentrations Via Absorption Spectroscopy

PURPOSE

To detect Cr(VI) in aqueous solutions using the MeasureNet spectrophotometer and familiarize students with serial dilutions.

INTRODUCTION

In the *Emission Analysis of Aqueous Solutions of Groups IA and IIA Metal Salts* experiment, it was demonstrated that high metal ion concentrations in water supplies might lead to significant salinity problems. It was also pointed out that Emission Spectroscopy is the preferred method of analysis to detect metal ions that produce colorless aqueous solutions (Group IA and IIA metals). Absorption spectroscopy is the preferred method to detect metal ions that produce colored aqueous solutions. Many transition metal ions produce colored aqueous solutions due to the presence of *3d* or *4d* electrons.

NOTE: Cr^{6+} ions do not exist in aqueous solution. Chromium in the 6+ oxidation state, Cr(VI), exists in solution as either chromate ions, CrO_4^{2-}, or dichromate ions, $Cr_2O_7^{2-}$. In this experiment Cr(VI) is present as CrO_4^{2-} ions in solution. Throughout this experiment, CrO_4^{2-} ions will be referred to as Cr(VI).

In this experiment, absorption spectroscopy will be used to determine Cr(VI) concentrations in solution (see Note). Many heavy metals, such as chromium, are toxic at low aqueous solution concentrations. In rivers, lakes, and streams, chromium ions are found in either the trivalent (Cr(III)) or hexavalent (Cr(VI)) states. Oxidizing environments favor Cr(VI) formation, while reducing environments favor Cr(III) formation. Naturally *alkaline* (basic) rivers, lakes, and streams have a much lower risk of heavy metal contamination than naturally acidic water. Rivers that flow over limestone rock formations, which are alkaline, are seldom toxic due to the presence of heavy metals. One example of a naturally alkaline river is the Suwannee River, which flows over a limestone basin in Florida.

One source of chromium contamination in natural waters is automobiles. Chromium is a component in automobile brakes and engine parts. Chromium deposited on highways from brake dust and engine exhaust can seep into lakes and streams via rainwater run-off. Lakes and streams near highly trafficked areas are highly susceptible to chromium contamination. Lake Sidney Lanier (one of Atlanta, Georgia's primary water reservoirs) is a prime example of a popular recreational lake that has several major highways in its vicinity. Lake Lanier is routinely monitored by state and federal environmental agencies for the presence of chromium, and other heavy metal ions. Over the years, these agencies have detected levels of Hg and As in Lake Lanier that exceed state and federal limits.

Chromium(VI) contamination is of particular interest to state and federal environmental agencies because of its known carcinogenic character and toxicity at low concentrations. (Cr(VI) solutions are toxic at significantly lower concentrations than Cr(III) solutions). In this experiment absorption spectroscopy will be used to detect low level concentrations of Cr(VI) in a *simulated* lake water sample.

Colored aqueous solutions contain chemical species that absorb specific wavelengths of light. Transition metals that contain *3d* or *4d* valence electrons produce brightly colored, aqueous solutions. These metals (often referred to as *heavy metals*) can be identified by the wavelengths of light which they absorb. Furthermore, the amount of light absorbed is directly proportional to the concentration of the metal ion in solution. Transition metals are typically reacted with complexing agents to intensify the color of their solutions. By intensifying the color of their solutions, metal ions absorb greater quantities of light, which permits detection of metal ions at very low concentrations in solution. For example, an aqueous solution of Fe^{3+} is light yellow in color. When mixed with SCN^- ions, the $[Fe(SCN)]^{2+}$ complex forms which produces a blood-red colored solution. Aqueous solutions of Cr(VI) are bright yellow in color. Given that Cr(VI) ions readily absorbs ultraviolet and visible light, *it is not necessary* to add a complexing agent to the simulated lake water sample.

Absorption spectroscopy operates on the principle of measuring the amount of light before and after it passes through an aqueous metal solution. The difference in the amount of light before it enters the sample and after it exits the sample is the amount of light absorbed by the chemical species in the sample. For light to be absorbed by a chemical species, the light must have a wavelength, or energy, that exactly matches an energy transition in the absorbing species. Absorption of light by a metal ion promotes an electron from its ground state to an excited state (see Figure 1). Shortly after the electron reaches one of the excited states, typically $10^{-9} - 10^{-6}$ seconds later, the electron will return to the ground state by emitting energy.

Every chemical species has a unique set of excited states, and consequently, absorbs different wavelengths of light. In emission spectroscopy, the wavelengths of light emitted by metal atoms as their electrons return to the ground state are detected and used to identify the metal in solution. In contrast, in absorption spectroscopy, the wavelengths of light absorbed by a metal in solution (that promote an electron to an excited state) are detected and used to identify the metal ion in solution. Because the amount of light absorbed is proportional to the concentration of the metal ion in

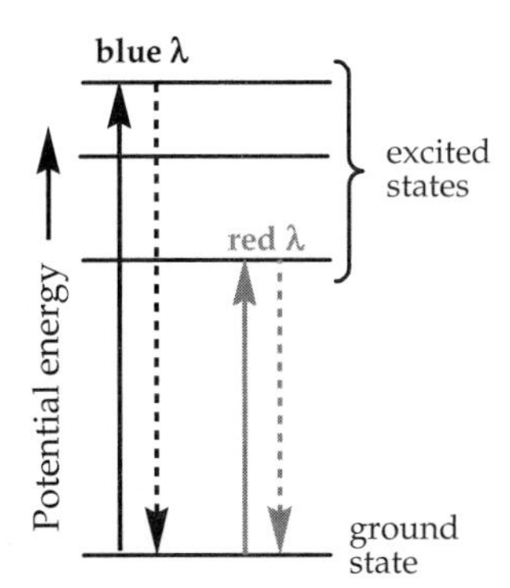

Figure 1
Absorption and emission of energy by electrons. λ is the symbol for the wavelength of light

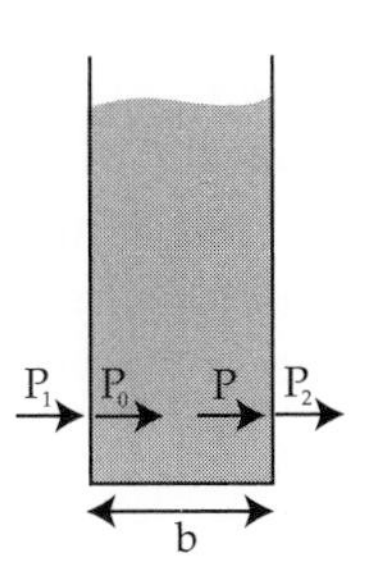

Figure 2
Light passing through a sample

solution, absorption spectroscopy is frequently used to determine very low-level concentrations of dissolved species.

Light is passed through sample solutions contained in an optically transparent cell of known path length. These optically transparent cells are called **cuvettes** (see Figure 2). As light passes through a cuvette containing a sample solution, it can be reflected, refracted, diffracted, or absorbed. Only the absorption of light is directly proportional to the solution's concentration.

Reflection, diffraction, and refraction (essentially scattering of light from the walls of the solution container) can be nullified by the use of a blank solution. A **blank solution** is one that contains all of the species present in the sample solution *except* the absorbing species. The spectrophotometer is designed to subtract the spectrum recorded for the blank solution from the absorbance spectrum of the sample, thus, nullifying the reflection, refraction, and diffraction caused by the walls of the cuvette.

Mathematically, this process can be defined as follows. P_1 is the power of the light before it enters the sample container (Figure 2). P_0 is the power of the light immediately after it passes through the first wall of the sample container, but before it passes through the sample. P is the power of the light after it has passed through the sample. P_2 is the light's power after it exits the second wall of the cuvette. Finally, b is the path length traveled by the light. The difference in power between P_1 and P_0 (or P and P_2) is due to reflection and/or refraction of the light from the cuvette walls. Absorption spectroscopy is only interested in the ratio of P to P_0 (*light absorbed by the sample*). The reflection/refraction effect can be nullified by measuring the power difference between P_1 and P_0, and P and P_2, for a blank solution and subtracting that from the sample spectrum. After the subtraction is performed, the ratio of P divided by P_0 can be determined. This ratio is called the **transmittance** of the solution, T.

$$T = \frac{P}{P_0} \qquad \text{(Eq. 1)}$$

While the spectrophotometer actually measures transmittance, we need to ascertain the amount of light absorbed by the solution to determine its concentration. Absorbance, A, and transmittance, T, are related by the following equation.

$$A = \log\frac{1}{T} \quad \text{(Eq. 2)}$$

Note that as A decreases, T increases. Thus, more transmitted light indicates that less light is being absorbed by the sample, and vice versa.

There are two additional factors besides the solution's concentration that affect the absorbance of the solution. First is the fact that every absorbing species only absorbs a fraction of the light that passes through the solution. The **molar absorptivity coefficient** is a measure of the fraction of light absorbed by a given species. Each absorbing species has its own unique molar absorptivity coefficient. Second, as the path length of the light through the solution is increased (determined by the width of the cuvette), more light is absorbed. The concentration, molar absorptivity coefficient, and the path length of light are directly related to the absorbance of a solution via **Beer-Lambert's Law,**

$$A = \varepsilon bc \quad \text{(Eq. 3)}$$

where A is the absorbance of the solution, ε is the molar absorptivity coefficient of the absorbing species, b is the path length of the light, and c is the concentration of the solution. Beer-Lambert's Law can be simplified if the same absorbing species and same sample container are used in a series of experiments. In that case, ε and b are constant and Beer-Lambert's Law simplifies to

$$A = (\text{constant})c \quad \text{(Eq. 4)}$$

From Eq. 4 we see that the absorbance of a species is directly proportional to its concentration in solution. This convenient, linear relationship between absorbance and concentration makes absorption spectroscopy one of the most popular chemical techniques for measuring concentrations of dissolved species in solution.

Metal Ion Concentrations in Solution

Metal ions dissolved in solution can be identified from their absorbance spectra. Figure 3 is a plot of the absorbance spectra for aqueous solutions of 70 ppm Cr(III) and 70 ppm Cr(VI) ions. Note that absorption bands are typically very broad, covering multiple wavelengths. The wavelength at

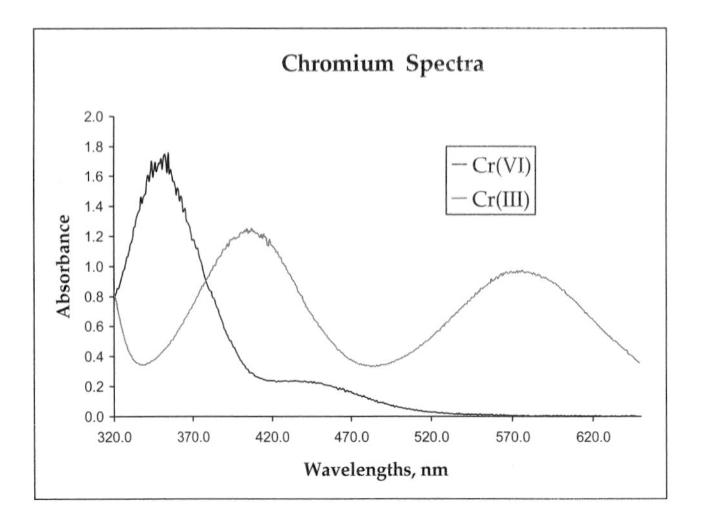

Figure 3
Absorbance spectra of 70 ppm Cr(III) and 70 ppm Cr(VI) aqueous solutions

which maximum absorbance occurs in an absorption band is designated as the **lambda max**, λ_{max}. Chromium(VI) exhibits two absorption bands, one with a λ_{max} of 348 nm, and a second, broader band with a λ_{max} of 435 nm. Chromium(III) also exhibits two absorption bands, one with λ_{max} of 405 nm and one with λ_{max} 576 nm. Chromium(III) and chromium(VI) have different absorption bands because each ionic state has a different electron configuration (Table 1).

Lambda max wavelengths are unique for each metal or ion and the different oxidation states of an ion. Consequently, λ_{max} can be used to identify which metals or ions are present in solution. There are two important considerations in choosing a λ_{max} for a particular metal. *First*, a wavelength is generally chosen in the most intense absorbance band for a given species. In some cases, this is not always feasible. As exhibited in Figure 3, all of the Cr(VI) absorption bands overlap the Cr(III) absorption bands. There are no absorbance bands where only Cr(VI) absorbs. Thus, it is difficult to differentiate aqueous solutions of Cr(VI) from Cr(III) solutions in the 300-700 nm region of the absorption spectrum. However, Cr(III) has a strong absorbance band at wavelengths > 560 nm that does not overlap a Cr(VI) absorbance band. Thus, it is possible to detect Cr(III) in the presence of Cr(VI) using a λ_{max} of 560–580 nm. *Second*, the selected λ_{max} should have an absorbance < 2.0. If the absorbance exceeds 2.0, the spectrophotometer detector saturates, producing a noisy, non-reproducible signal. The onset of detector saturation is exhibited in Figure 4 for Cr(VI) at 360 nm.

Table 1 *Electron configurations*

Cr atom	$[Ar]4s^1 3d^5$
Cr^{3+} ion	$[Ar]3d^3$
Cr^{6+} ion	[Ar]

Once the identity of the metal has been established, the concentration of the metal solution is determined by monitoring changes in its absorbance as a function of concentration. This is accomplished by preparing a series of metal ion **standard solutions**, where the concentration of the species is accurately known, and their absorbance spectra recorded. Typically 4 to 5 standard solutions are prepared that bracket the concentration of the unknown solution. To **bracket the concentration** of the unknown solution, at least one standard solution must have a concentration that is lower than the unknown solution, and one standard solution must have a concentration that is higher than the unknown solution. The absorbance value for each standard solution is determined at λ_{max} of one of the absorbance bands of the metal. If necessary, more than one λ_{max} can be used to determine the absorbance of each standard solution. Typically, the λ_{max} of the most intense absorbance band is used to determine the concentration of the solutions. Figure 4 shows absorbance spectra for five Cr(VI) standard solutions where λ_{max} was selected from the 400-450 nm region. (360 nm, the most intense wavelength in Figure 4, was not used as the λ_{max} due to detector saturation

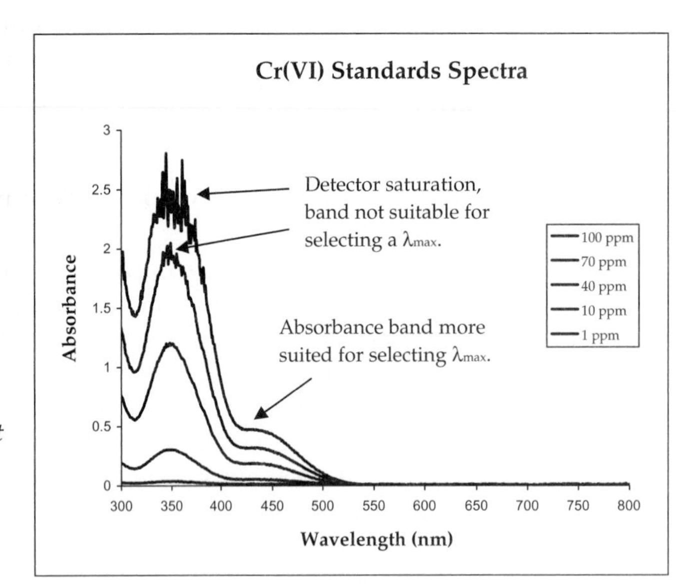

Figure 4
Absorbance spectra of Cr(VI) standard solutions. Note that the concentration of the standard solutions is given in parts per million, ppm. A ***part per million*** *is defined as 1.00 milligram of solute per 1.00 liter of solution, 1.00 mg/L*

(absorbance $\geq$ 2.0) exhibited by the 70 ppm and 100 ppm standard solutions. Note that the absorbance of each solution decreases as the Cr(VI) concentration decreases.

Once the absorbance for each standard solution has been determined, a plot of absorbance (*y*-axis) versus the standard solution concentrations (*x*-axis) is prepared. In accordance with the Beer-Lambert law, the plot should be linear (or very close to linear) and pass through or near the origin. (The latter is true because a solution that has a Cr(VI) concentration of 0 should have an absorbance of 0 at λ_{max}.) The unknown solution's concentration must be within in the concentration range of the standard solutions. Linear regression analysis is performed, using a spreadsheet program such as Excel, to determine the *linear best-fit* for the absorbance versus concentration data (Figure 5). The R^2 value shown in Figure 5 indicates how well the regression analysis fits the absorbance-concentration data (see Figure 5). The closer the R^2 value is to 1.00, the better the linear regression analysis has fit the data.

Finally, a spectrum of the unknown solution is recorded. From the absorbance value of the unknown solution, its concentration can be determined either directly from a plot similar to Figure 5, or more precisely, by using linear regression analysis. The line determined from the regression analysis will be in the form $y = mx + b$, where *y is the absorbance value* and *x is*

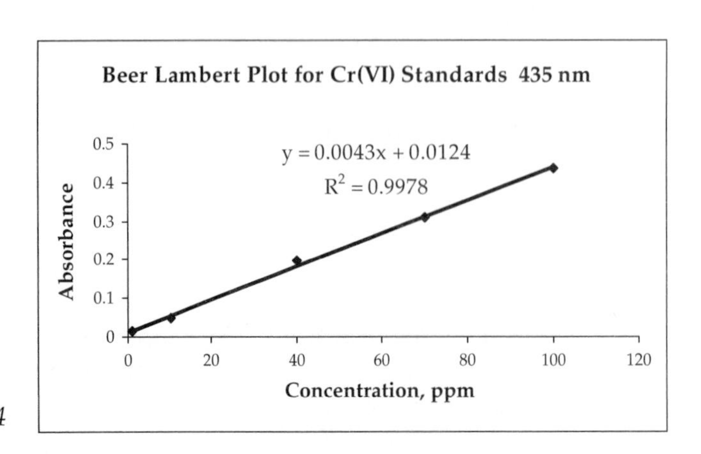

Figure 5
Plot of absorbance versus concentration for the Cr(VI) standard solutions in Figure 4

the concentration of the solution. Algebraic substitution of the absorbance value (y) for the unknown metal solution into the linear regression equation for the line permits the determination of the concentration (x) of the unknown solution.

Standard Solution Preparation

In this experiment standard solutions will be prepared by diluting a 300 ppm (part per million) Cr(VI) standard solution. In natural water supplies, metal or metal ion solution concentrations are typically so small that they are generally reported in the parts per million concentration unit instead of molarity. *Chromium(VI) concentration levels must not exceed 0.1 ppm in drinking water supplies in the United States.*

Students will prepare five standard solutions from a concentrated Cr(VI) standard solution. Dilute solution concentrations can be calculated using the solution dilution formula

$$(M_1)(V_1) = (M_2)(V_2) \qquad \text{(Eq. 5)}$$

where M_1 is the concentration of the concentrated solution, V_1 is the volume of the concentrated solution, M_2 is the concentration of the dilute solution, and V_2 is the volume of the dilute solution. (M usually is expressed in the concentration unit of molarity of the solution, but it can also be expressed in terms of ppm or other concentration units.)

Serial Dilutions

To prepare standard solutions with very low concentrations (< 10 ppm), the method of **serial dilutions** is employed. In serial dilutions, a concentrated standard solution is initially prepared. An **aliquot** (small volume) of the standard solution is subsequently mixed with pure solvent to produce a more dilute solution. An aliquot of the dilute solution can be mixed with additional solvent to produce an even more dilute solution. This process is continued until a solution having the desired concentration is obtained. (We use serial dilutions to prepare very dilute solutions because the balances, available in most introductory laboratories, are not capable of measuring masses small enough to directly prepare dilute solutions.)

Consider the following example. Starting with 25.0 mL of 100.00 ppm Cr(VI) solution and a 10.00-mL volumetric flask, prepare 10.00 mL of 10.00 ppm Cr(VI) solution and 10.00 mL of 1.00 ppm Cr(VI) solution using the serial dilution method.

First, calculate the volume (V_1 from Eq. 5) of 100.00 ppm standard solution that must be diluted with distilled water to prepare 10.00 mL of a 10.00 ppm Cr(VI) solution.

$$(M_1)(V_1) = (M_2)(V_2)$$
$$(100.00\ \text{ppm})(V_1) = (10.00\ \text{ppm})(10.00\ \text{mL})$$
$$(V_1) = 1.00\ \text{mL}$$

This calculation indicates that 1.00 mL of the standard 100.00 ppm solution must be diluted to a volume of 10.00 mL to prepare 10.00 mL of the 10.00 ppm Cr(VI) solution. First, add approximately 5 mL of distilled water to a 10.00-mL volumetric flask. Then add 1.00 mL of the 100.00 ppm standard solution to the flask and swirl to mix the solution. Finally, add sufficient distilled water to the flask until the bottom of the solution's meniscus aligns with the 10.00 mL mark on the neck of the volumetric flask. With the flask's cap securely attached, invert the volumetric flask several times to mix the solution thoroughly. At that point 10.00 mL of a 10.00 ppm Cr(VI) solution has been prepared.

A second dilution is required to prepare 10.00 mL of a 1.00 ppm Cr(VI) solution from the 10.00 ppm Cr(VI) solution. As before, calculate the volume of the 10.00 ppm Cr(VI) standard solution (previously prepared) that must be diluted with distilled water to prepare 10.00 mL of a 1.00 ppm Cr(VI) solution.

$$(M_1)(V_1) = (M_2)(V_2)$$
$$(10.00\ \text{ppm})(V_1) = (1.00\ \text{ppm})(10.00\ \text{mL})$$
$$(V_1) = 1.00\ \text{mL}$$

The calculation indicates that 1.00 mL of the 10.00 ppm Cr(VI) solution must be diluted to a volume of 10.00 mL (using the 10-mL volumetric flask) to prepare 10.00 mL of a 1.00 ppm Cr(VI) solution.

PROCEDURE

CAUTION

Students must wear departmentally approved eye protection while performing this experiment. Wash your hands before touching your eyes and after completing the experiment.

1. Obtain ~ 20 mL of the 300.00 ppm Cr(VI) standard solution.
2. Prepare 10.00 mL each of 5 standard solutions (see Note) having concentrations that bracket the concentration of the 300.00 ppm standard solution and the limit of detection for Cr(VI) ions, 3 ppm, using the MeasureNet spectrophotometer. What should be the concentrations of the most concentrated and the most dilute standard solutions? Should you include all solution dilution calculations in the Lab Report?

NOTE: Ideally, a 10-mL volumetric flask is used to prepare exact quantities of known concentrations of a solution. If a 10-mL volumetric flask is not available, use a 10-mL graduated cylinder.

3. Record absorbance spectra of the 5 standard solutions prepared in Step 2, and an absorbance spectrum of the simulated lake water sample using the MeasureNet spectrophotometer. See **Appendix D – Instructions for Recording an Absorbance Spectrum using the MeasureNet Spectrophotometer**. *Pour all chromium solutions into the Waste Container when the experiment is concluded.*
4. How do you select a λ_{max} for Cr(VI) ions in aqueous solution (see Note). Should you record the selected λ_{max} in the Lab Report?

NOTE: Once a λ_{max} has been selected, that wavelength must be used to determine the absorbance of Cr(VI) in all standard solutions and the simulated lake water sample.

5. Should you prepare a table listing the concentrations of each standard solution and their corresponding absorbencies in the Lab Report?
6. Should you record the absorbance of the simulated lake water sample in the Lab Report?
7. Prepare a Beer-Lambert using the data compiled in Step 5. See **Appendix B-2 - Excel Instructions for Performing Linear Regression Analysis**.
8. How do you determine the concentration of Cr(VI) in the simulated lake water sample?
9. State and federal regulatory agencies consider Cr(VI) to be toxic at concentration levels > 0.100 ppm. Natural waters whose Cr(VI) concentrations exceed 0.100 ppm may not be used for drinking water or for agricultural purposes. Is the simulated lake water sample suitable for drinking water and for agricultural purposes? Explain.

Name

Section

Date

Instructor

11

EXPERIMENT 11

Lab Report

Preparation of five Cr(VI) standard solutions.

Determination of λ_{max} for Cr(VI) ions in aqueous solution.

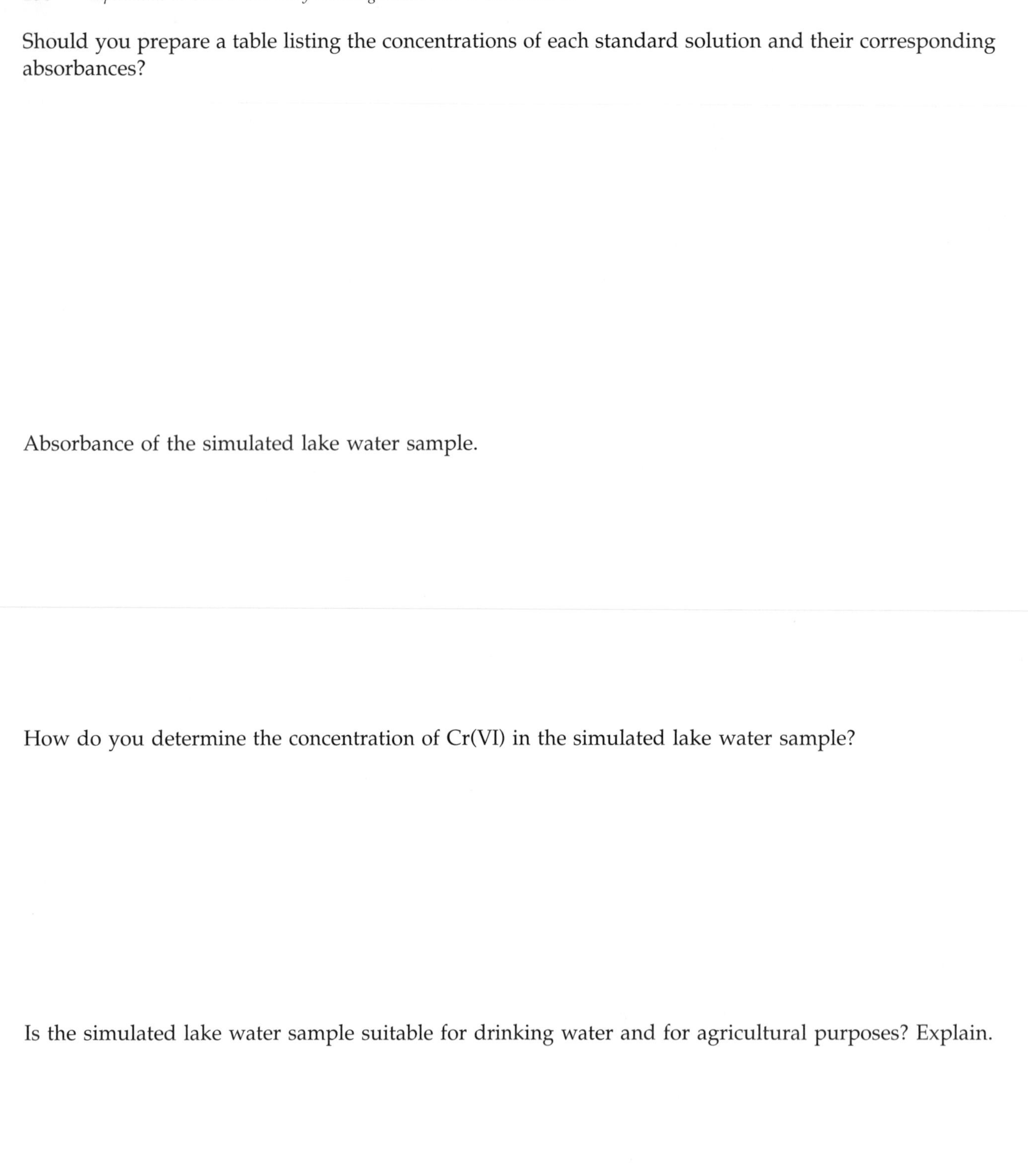

Should you prepare a table listing the concentrations of each standard solution and their corresponding absorbances?

Absorbance of the simulated lake water sample.

How do you determine the concentration of Cr(VI) in the simulated lake water sample?

Is the simulated lake water sample suitable for drinking water and for agricultural purposes? Explain.

Name | Section | Date

Instructor

11 EXPERIMENT 11

Pre-Laboratory Questions

1. A Beer's Law plot will be utilized to determine the concentration of Fe^{3+} in a water sample. The concentration of the standard Fe^{3+} solution is 400. ppm. The limit of detection of Fe^{3+} is 50.0 ppm. Five standard solutions will be used to prepare the Beer's Law plot. An aliquot of the 400. ppm standard solution will serve as one of the five standard solutions. Four additional standard solutions are prepared in the following manner.

a. 7.50 mL of 400. ppm Fe^{3+} solution is diluted to 10.00 mL.

b. 5.00 mL of 400. ppm Fe^{3+} solution is diluted to 10.00 mL.

c. 2.50 mL of 400. ppm Fe^{3+} solution is diluted to 10.00 mL.

d. 5.00 mL of the solution prepared in Step 1c is diluted to 10.00 mL.

Complete the following table by determining the concentrations of each standard solution prepared in Steps 1a – 1d. *Show all calculations*. The absorbance of the solutions prepared in Steps 1a-1d are given in the table.

Concentration, ppm	*Absorbance at 550.8 nm*
400.	2.45
a.	1.79
b.	1.07
c.	0.381
d.	0.124

2. Prepare a Beer-Lambert curve for the data in the table in Question 1. Perform linear regression analysis on this data using Excel. (See **Appendix B-2 –Instructions for Performing Linear Regression Analysis Using Excel.**)

3. A simulated water sample is determined to have an absorbance of 0.169 at 550.8 nm. Determine the concentration of Fe(III) ions in the sample.

____________ ppm

Name Section Date

Instructor

11 EXPERIMENT 11

Post-Laboratory Questions

1. If a contaminated water sample contained only Cr(III) at a concentration less than 70 ppm, what λ_{max} should you choose to determine the Cr(III) concentration in this sample (see Figure 3)? Justify your answer with an explanation.

2. How would each of the following affect the Cr(VI) concentration determined in this experiment? Justify your answers with explanations.

   a. A student adds more standard solution than intended in the initial dilution step when performing serial dilutions.

   b. A student only fills a cuvette ¼ full when taking the absorption spectrum of the simulated lake water sample.

c. A student mistakenly uses the cuvette containing the most dilute standard solution, rather than the cuvette containing deionized water, as the blank solution.

d. A student does not wipe off dirty fingerprints on the cuvette containing the unknown solution before measuring its absorbance.

Determination of the Concentration of Acetic Acid in Vinegar

PURPOSE

Standardize a sodium hydroxide solution using a primary standard acid. Determine the molarity and the percent by mass of acetic acid in vinegar by titration with the standardized sodium hydroxide solution.

INTRODUCTION

The **concentration** of a solution is the amount of solute (species dissolved) in a given amount of solvent (dissolving agent). A concentrated solution contains a relatively large quantity of solute in a given amount of solvent. Dilute solutions contain relatively little solute in a given amount of solvent. Chemists use specific terms to express the concentration of a solution. Two of these terms are molarity and percent by mass. **Molarity** is the number of moles of solute per liter of solution.

$$\text{Molarity } (M) = \frac{\text{moles of solute}}{\text{liter of solution}} \qquad \text{(Eq. 1)}$$

Percent by mass is the mass in grams of solute per 100 grams of solution.

$$\text{Percent solute} = \frac{\text{grams of solute}}{\text{grams of solution}} \times 100\% \qquad \text{(Eq. 2)}$$

The grams of solution are the mass of the solute plus the mass of the solvent. For example, a 5.00 % NaCl solution contains 5.00 grams of NaCl and 95.0 grams of H_2O in 100.0 grams of solution.

Vinegar is a dilute solution of acetic acid (an organic acid). The molecular formula for acetic acid is CH_3COOH. Both the molarity and percent by mass of acetic acid in a vinegar solution can be determined by performing a titration. A **titration** is a process in which small increments of a solution of known concentration are added to a specified volume of a

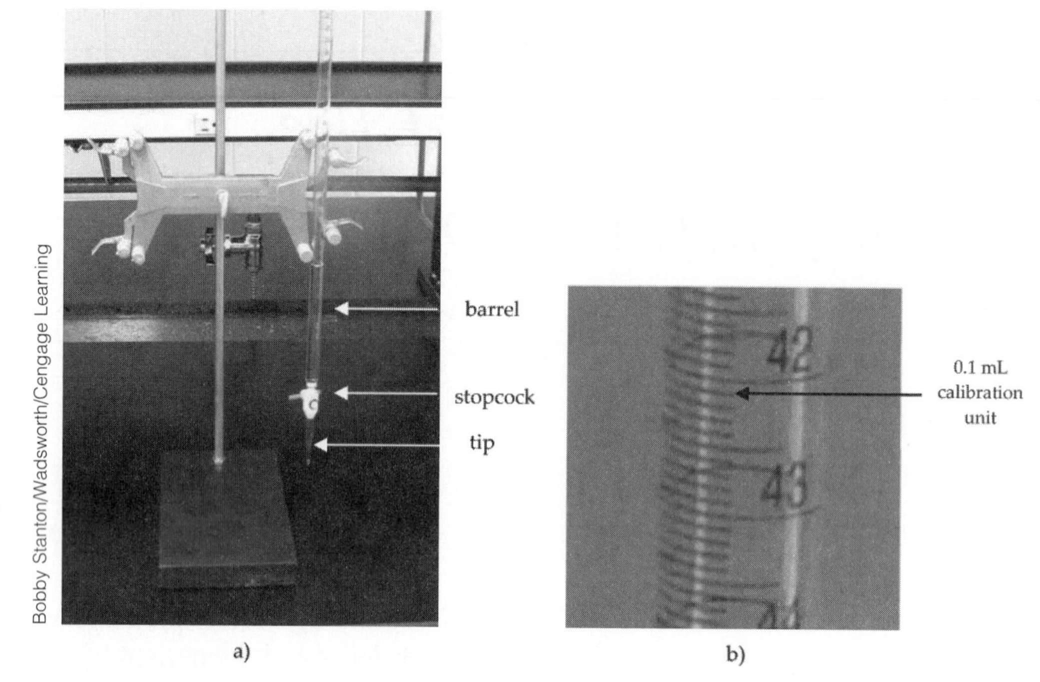

Figure 1
Figure 1a) depicts a typical 50-mL buret. Figure 1b) indicates the smallest calibration unit, 0.1 mL, on a typical 50-mL buret

solution of unknown concentration until the stoichiometry for that reaction is attained. Knowing the quantity of the known solution required to complete the titration, we can calculate the unknown solution's concentration.

In the titration process, a **buret** is used to dispense small, quantifiable increments of a solution of known concentration (Figure 1a). A typical buret has the smallest calibration unit of 0.1 mL (Figure 1b). Therefore, the volume dispensed from the buret should be estimated to the nearest 0.01 mL. (*To be consistent, always read the volume of a liquid inside a buret at the bottom of the liquid's meniscus*).

The purpose of a titration is to determine the equivalence point of the reaction. The **equivalence point** is reached when the added quantity of one reactant is the exact amount necessary for stoichiometric reaction with another reactant. In this experiment, the equivalence point occurs when the moles of acid in the solution equals the moles of base added in the titration. For example, the stoichiometric amount of 1 mole of the strong base, sodium hydroxide (NaOH), is necessary to neutralize 1 mole of the weak acid, acetic acid (CH_3COOH), as indicated in Eq. 3.

$$NaOH_{(aq)} + CH_3COOH_{(aq)} \rightarrow NaCH_3CO_{2(aq)} + H_2O_{(\ell)} \qquad \text{(Eq. 3)}$$

NOTE: Free H^+ ions, produced by an acid when it is dissolved in water, do not actually exist in aqueous solution. Instead, the H^+ ions readily react with water to form H_3O^+ ions. Chemists use H_3O^+ and H^+ interchangeably when referring to hydrogen ions in aqueous solution

One indicator that the titration has reached the equivalence point is a sudden change in the pH of the solution. The pH of an aqueous solution is related to its hydrogen ion concentration. Symbolically, the hydrogen ion concentration is written as $[H_3O^+]$ (see Note). **pH** is defined as the negative of the base 10 logarithm of the hydrogen ion concentration.

$$pH = -\log[H_3O^+] \quad \text{(Eq. 4)}$$

The common base 10 logarithm of a number is the number that 10 must be raised to produce the original number. For example, the logarithm of 1.00×10^{-7} is -7. The pH scale is a convenient, short hand method of expressing the acidity or basicity of a solution. The pH of an aqueous solution is typically a number between 0 and 14. Solutions having a $pH < 7$ are acidic. Solutions with $pH = 7$ are neutral. Solutions having a $pH > 7$ are basic. For example, a solution having $[H_3O^+] = 2.35 \times 10^{-2} M$ would have a pH of 1.629 and is acidic.

In this experiment, we will measure the pH of a solution using the MeasureNet **pH electrode**. The titration is initiated by inserting a pH electrode into a beaker containing the acid solution (pH is in the 3-5 range). As sodium hydroxide, NaOH, is incrementally added to the acid solution, some of the hydrogen ions are neutralized. As the hydrogen ion concentration decreases, the pH of the solution will gradually increase. When sufficient NaOH is added to completely neutralize the acid (most of the H_3O^+ ions are removed from solution), the next drop of NaOH added causes a sudden, sharp increase in pH (Figure 2). The equivalence point is the center of the curve in the region where the pH changes sharply for any acid-base reaction involving 1:1 reaction stoichiometry (Figure 2). The volume of base required to completely neutralize the acid is determined at the equivalence point of the titration.

In this experiment, we will titrate a vinegar sample with a standardized sodium hydroxide solution. To standardize the sodium hydroxide solution, we will initially prepare a primary standard acid solution. In general, **primary standard solutions** are produced by dissolving a weighed quantity of a pure acid or base in a known volume of solution.

Primary standard acids and bases have several common characteristics: 1) they must be available in at least 99.9% purity, 2) they must have a high molar mass to minimize errors in weighing, 3) they must react by one invariable reaction, 4) they must be stable upon heating, and 5) they must be soluble in the solvent of interest. Potassium hydrogen phthalate, $KHC_8H_4O_4$, and oxalic acid, $(COOH)_2$, are commonly used primary

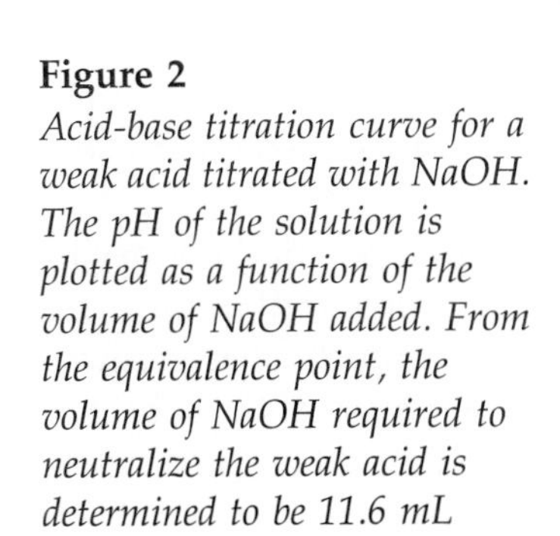

Figure 2
Acid-base titration curve for a weak acid titrated with NaOH. The pH of the solution is plotted as a function of the volume of NaOH added. From the equivalence point, the volume of NaOH required to neutralize the weak acid is determined to be 11.6 mL

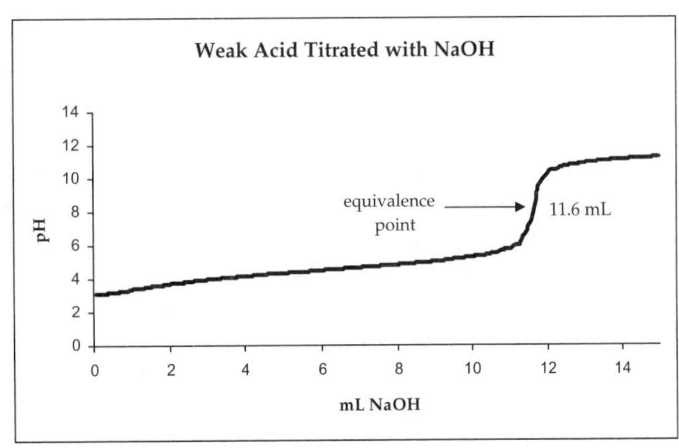

standard acids. Sodium carbonate, Na_2CO_3, is the most commonly used primary standard base.

Most acids and bases (e.g., HCl, CH_3COOH, NaOH and KOH) are not available as primary standards. To standardize one of these acid or base solutions, we titrate the solution with a primary standard. In this experiment, we will titrate a freshly prepared NaOH solution with potassium hydrogen phthalate (often abbreviated as KHP). The equation for the reaction of KHP with NaOH is

$$KHC_8H_4O_{4(aq)} + NaOH_{(aq)} \rightarrow KNaC_8H_4O_{4(aq)} + H_2O_{(\ell)} \qquad \text{(Eq. 5)}$$

Once the sodium hydroxide solution has been standardized, it will be used to titrate 10.00 mL aliquots of vinegar. The equation for the reaction of vinegar with NaOH is

$$CH_3COOH_{(aq)} + NaOH_{(aq)} \rightarrow NaCH_3COO_{(aq)} + H_2O_{(\ell)} \qquad \text{(Eq. 6)}$$

Knowing the standardized NaOH concentration and using Equation 6, we can determine the molarity and percent by mass of acetic acid, CH_3COOH, in the vinegar solution.

Standardizing a Base with KHP ($KHC_8H_4O_4$)

To standardize a solution of NaOH, a known mass of KHP is dissolved in enough water to just dissolve the sample. A buret is used to titrate NaOH solution into the beaker containing the KHP solution. The molarity of the NaOH solution is determined by first calculating the number of moles KHP dissolved in solution. Equation 5 indicates that one mole of NaOH is required to neutralize one mole of KHP. This stoichiometric ratio and the moles of KHP in solution are used to calculate the moles of NaOH required to neutralize the KHP. The volume of NaOH used is the titration is determined at the equivalence point from a plot of pH of the solution versus milliliters of NaOH added. The molarity of the NaOH solution is calculated by dividing the moles of NaOH by the volume of NaOH in liters used to neutralize the KHP solution.

Determining the Acetic Acid Concentration in Vinegar

A 10.00 mL aliquot of vinegar is titrated with the standardized NaOH solution. To determine the molarity of acetic acid in vinegar, we first calculate the moles of NaOH required to neutralize the acetic acid in the sample. The moles of NaOH required is the product of the molarity of the standardized NaOH solution times the volume of NaOH used in the titration. Equation 6 indicates that one mole of NaOH is required to neutralize one mole of acetic acid. This stoichiometric ratio and the moles of NaOH used in the titration are used to calculate the moles of acetic acid in the vinegar sample. The molarity of the acetic acid is obtained by dividing the moles of acetic acid in the sample by the liters of vinegar solution used in the titration.

Dividing the mass of acetic acid in vinegar by the mass of the vinegar solution, then multiplying by 100%, yields the mass percent of acetic acid in vinegar. The mass of acetic acid is the product of the moles of acetic acid in the sample times the molar mass of acetic acid. The mass of the solution is equal to the volume of the vinegar used in the titration times the density of vinegar solution. In this experiment, assume the density of vinegar is the same as that for water, 1.00 g/mL at ambient temperature.

PROCEDURE

CAUTION

Students must wear departmentally approved eye protection while performing this experiment. Wash your hands before touching your eyes and after completing the experiment.

Part A – Standardization of a Sodium Hydroxide Solution

1. Prepare 150 mL of approximately 0.6 *M* sodium hydroxide solution from solid NaOH. Check the calculations with your laboratory instructor prior to preparing the solution. Should your calculations be recorded in the Lab Report?
2. Add approximately 1.5 grams of potassium hydrogen phthalate (KHP) to a 250-beaker. Should the exact mass of the sample be recorded in the Lab Report, and to how many significant figures? Add sufficient distilled water to the beaker to dissolve the KHP. Should the volume of water used be recorded in the Lab Report, and to how many significant figures?
3. Stir to completely dissolve the KHP. Why must the KHP be completely dissolved before beginning the titration? Add enough distilled water to the beaker to ensure that the cut-out notch on the tip of the pH electrode is completely submerged in solution. Why must the cut-out notch be completely submerged in solution?
4. See Appendix F – **Instructions for Recording a Titration Curve Using the MeasureNet pH Probe and Drop Counter.** Complete all steps in Appendix F before proceeding to Step 5 below.
5. Repeat Steps 2–4 to perform a second trial to standardize the NaOH solution.
6. *Steps 7–10 are to be completed after the laboratory period is concluded (outside of lab)*. Proceed to Step 11, *Determination of Acetic Acid Concentration in Vinegar*.
7. From the tab delimited files you saved, prepare plots of the pH versus volume of NaOH added using Excel (or a comparable spreadsheet program). Instructions for plotting pH versus volume curves using Excel are provided in Appendix B-4.
8. How do you determine the volume of NaOH required to neutralize the KHP solution in each titration? Should this volume be recorded in the Lab Report, and to how many significant figures?
9. How do you determine the molarity of the sodium hydroxide solution in Trials 1 and 2?
10. What is the average molarity of the sodium hydroxide solution? The average molarity is the concentration of the standardized NaOH solution.

Part B – Determination of Acetic Acid Concentration in Vinegar

11. Transfer 10.00 mL of vinegar to a clean, dry 250-mL beaker using a 10 mL volumetric pipet. Add sufficient water to the acid so that the cut out notch on the pH electrode tip is completely submerged. Should the volume of water be recorded in the Lab Report, and to how many significant figures?
12. See Appendix F–**Instructions for Recording a Titration Curve Using the MeasureNet pH Probe and Drop Counter.** Complete all steps in Appendix F before proceeding to Step 13 below.
13. Repeat Steps 11 and 12 to perform a second titration of vinegar with the standardized NaOH.
14. From the tab delimited files you saved, prepare plots of the pH versus volume of NaOH added using Excel (or a comparable spreadsheet program). Instructions for plotting pH versus volume curves using Excel are provided in Appendix B-4.
15. How do you determine the volume of NaOH required to neutralize vinegar in Trials 1 and 2? Should this volume be recorded in the Lab Report, and to how many significant figures?
16. How do you determine the molarity of acetic acid in vinegar for Trials 1 and 2?
17. What is the average molarity of acetic acid in the vinegar sample?
18. How do you determine the percent by mass of acetic acid in the vinegar sample?
19. What is the average percent by mass of acetic acid in the vinegar sample?

Name

Instructor

Section

Date

12

EXPERIMENT 12

Lab Report

Part A – Standardization of a Sodium Hydroxide Solution

Preparation of 150 mL of 0.6 *M* sodium hydroxide solution.

Standardization of NaOH with KHP data – Trial 1

Why must the KHP be completely dissolved before beginning the titration?

Standardization of NaOH with KHP data – Trial 2

How do you determine the volume of NaOH required to neutralize the KHP solution in each Trial?

How do you determine the molarity of the sodium hydroxide solution in Trials 1 and 2?

What is the average molarity of the sodium hydroxide solution?

Part B – Determination of the Concentration of Acetic Acid in Vinegar

Titration of acetic acid with standardized NaOH data – Trial 1

Titration of acetic acid with standardized NaOH data – Trial 2

How do you determine the volume of NaOH required to neutralize vinegar in each Trial?

How do you determine the molarity of acetic acid in vinegar for Trials 1 and 2?

What is the average molarity of acetic acid in the vinegar sample?

How do you determine the percent by mass of acetic acid in vinegar for Trials 1 and 2?

What is the average percent by mass of acetic acid in the vinegar sample?

Name

Section

Date

Instructor

12 EXPERIMENT 12

Pre-Laboratory Questions

1. 1.802 grams of KHP is dissolved in 20.0 mL of distilled water. The titration curve below is for the titration of the KHP solution with NaOH. See Eq. 5 for the balanced equation for the reaction.

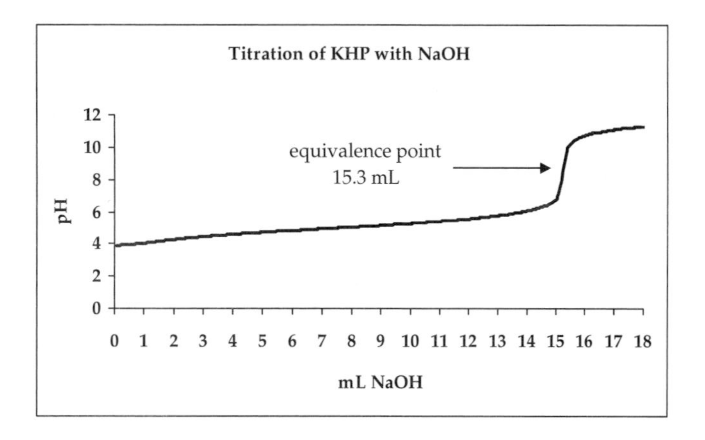

a. Determine the number of moles of KHP in the solution.

b. Determine the number of moles of NaOH required to neutralize the KHP.

c. What is the molarity of the NaOH solution?

2. A 10.00 mL aliquot of vinegar is titrated with the same NaOH solution standardized in Question 1. The titration curve below is for the titration of vinegar with the standardized NaOH solution. See Eq. 6 for the balanced equation for the reaction.

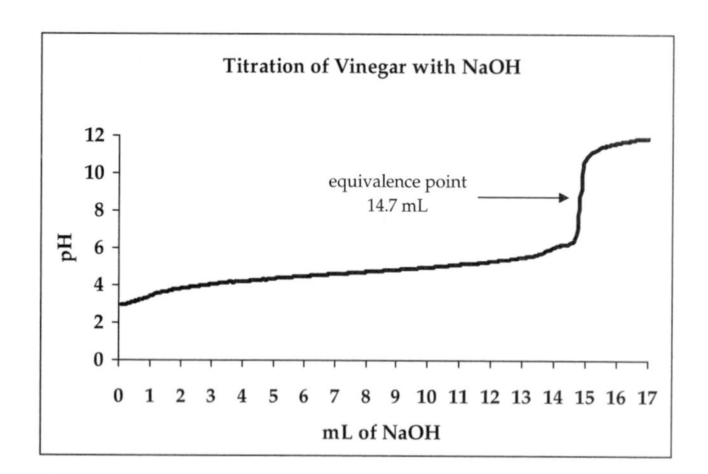

a. Determine the number of moles of NaOH required to neutralize the acetic acid in the vinegar sample.

b. Determine the number of moles of CH_3COOH in the vinegar sample.

c. What is the molarity of CH_3COOH in the vinegar sample?

d. Determine the mass of acetic acid in the vinegar sample.

e. Determine the mass of the 10.00 mL aliquot of vinegar. The density of vinegar is 1.00 g/mL.

f. Determine the percent by mass of acetic acid in the vinegar sample.

Name

Instructor

Section

Date

12

EXPERIMENT 12

Post-Laboratory Questions

1. When performing this experiment, a student mistakenly used impure KHP to standardize the NaOH solution. If the impurity is neither acidic nor basic, will the percent by mass of acetic acid in the vinegar solution determined by the student be too high or too low? Justify your answer with an explanation.

2. When preparing a NaOH solution, a student did not allow the NaOH pellets to completely dissolve before standardizing the solution with KHP. However, by the time the student refilled the buret with NaOH to titrate the acetic acid, the remaining NaOH pellets had completely dissolved. Will the molarity of acetic acid in the vinegar solution, determined by the student, be too high or too low? Justify your answer with an explanation.

3. Distilled water normally contains dissolved CO_2. When preparing NaOH standard solutions, it is important to use CO_2 free distilled water. How does dissolved CO_2 in distilled water affect the accuracy of the determination of a NaOH solution's concentration? (*Hint: Use your textbook, the internet, etc. to research the term acid anhydride.*)

Solubility, Polarity, Electrolytes, and Nonelectrolytes

PURPOSE

Compare the solubilities of some polar and nonpolar species. Observe the solubility of ionic compounds in water. Determine which of several species can be classified as strong, weak, or nonelectrolytes.

INTRODUCTION

An important physical property of substances is their solubility, or ability to dissolve, in a variety of solvents. Water is an excellent solvent with the ability to dissolve a variety of chemical compounds. One of the reasons for water's exceptional dissolving capability is its highly polar nature. The electrons in a water molecule are not distributed equally about the molecule, but are more concentrated near the oxygen atom and less concentrated near the hydrogen atoms. This phenomenon is due to the difference in the electronegativity values of O (3.5) and H (2.1), and the fact that water is a bent (unsymmetrical) molecule (Figure 1).

Consequently, water is a molecular species that can dissolve many ionic and polar covalent compounds. Chemists often refer to the phrase – "*Like Dissolves Like*." In other words, polar compounds tend to be soluble in polar solvents, and nonpolar compounds tend to be soluble in nonpolar solvents.

Water is an effective solvent for many ionic compounds. When an ionic compound such as potassium chloride (KCl) dissolves in water, the three dimensional crystalline lattice is destroyed and K^+ and Cl^- ions are separated from each other. The negative end of a water molecule (the O atom) is attracted to K^+ ions and the positive end of the water molecule (the H atoms) is attracted to Cl^- ions. In solution, all K^+ and Cl^- ions are surrounded by several water molecules in a specific pattern. The process in which an ion is surrounded by water molecules is called **hydration**. The hydration of ions is an energy releasing process. It is the hydration energy

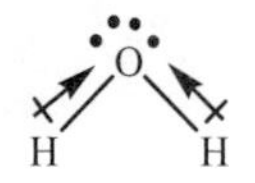

Figure 1
Illustration of the polar nature of water. The electrons in the O—H bond are shared unequally. The O atom, with its higher electronegativity value, attracts the shared pair of electrons much more strongly than the H atom. The shift of electron density produces a "partial negative charge" on the O atom and it leaves H with a "partial positive charge." Arrows are commonly used to indicate the direction of bond polarity. The arrow head points towards the negative end of the bond and the crossed tail indicates the positive end of the bond

that overcomes the electrostatic attraction (ionic bond) between K^+ and Cl^- ions that disrupt the crystal lattice and permits KCl to dissolve in water.

Ionic compounds that have especially strong ionic bonds do not dissolve in water because the water molecules cannot provide sufficient hydration energy to separate the ions. Bond strengths of ionic compounds are related to the charges and sizes of the ions. The lattice energy generally increases with the increasing charge and decreasing size of the ions. This trend is reflected in the *solubility rules* for ionic compounds in water. For example, $Fe(OH)_3$ (containing a trivalent Fe^{3+} ion and a monovalent OH^- ion) is insoluble in water, whereas NaOH (containing monovalent Na^+ and OH^- ions) is soluble in water. The hydration energy supplied by water is insufficient to disrupt the large $Fe(OH)_3$ crystal lattice energy and dissolve the compound.

Many polar covalent compounds are also soluble in water. For a molecular (covalent) compound to be polar, two criteria must be met. The molecule must contain at least one polar covalent bond, and its molecular geometry must be unsymmetrical. For example, methane, CH_4, is a nonpolar compound, whereas methyl chloride, CH_3Cl, and ammonia, NH_3, are both polar covalent compounds (Figure 2).

For a covalent bond to be appreciably polar, the difference in the electronegativity values of the two atoms in the bond must be 0.4 or higher. In a methane molecule, the electronegativity difference for the C—H bond is 0.4 (see Table 1), slightly polar at best.

Since all the bonds in methane are identical (all C—H) and the molecule is symmetrical, the shift in electron density within each bond nullifies or cancels each other, resulting in a nonpolar molecule. In methyl chloride molecule, the C—Cl bond is polar (0.5 electronegativity difference). Since carbon is more electronegative than hydrogen and chlorine is more electronegative than carbon, the molecule is asymmetrical. There is a net shift of electron density towards chlorine (a net dipole moment). As a result, the molecule is polar. The N—H bond in ammonia is polar (0.9 difference in electronegativity values). Due to the presence of the lone pair of electrons

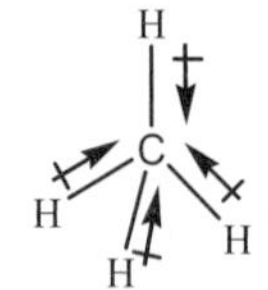

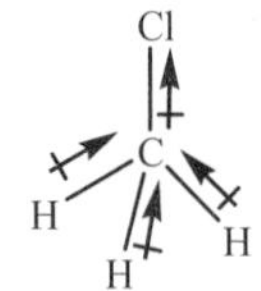

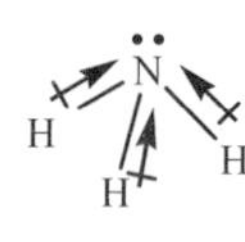

Figure 2
Illustration of the polarity of methane, methyl chloride, and ammonia

Table 1 *Electronegativity values for some elements*

carbon	2.5
chlorine	3.0
fluorine	4.0
hydrogen	2.1
nitrogen	3.0
oxygen	3.5

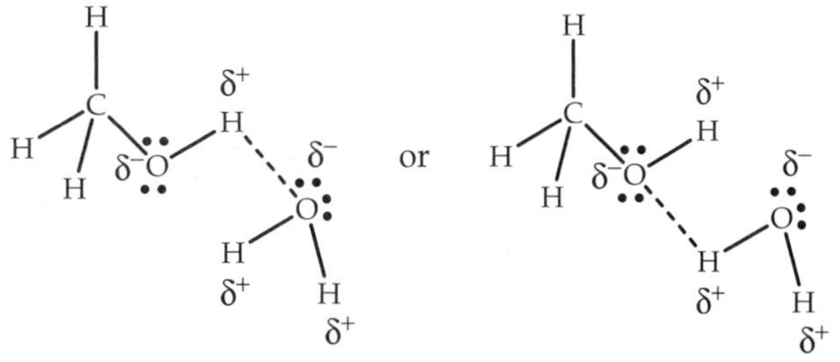

Figure 3
The dashed lines represent hydrogen bonds formed between methanol and water. The δ⁺ over the H atom indicates a partial positive charge and the δ⁻ over the O atom indicates a partial negative charge

on nitrogen, the molecule is asymmetrical, the shifts of electron density within each N—H bond do not cancel. Therefore, the ammonia molecule is polar.

The solubility of a polar compound in water is enhanced if the compound can hydrogen bond with water. Hydrogen bonding occurs between molecules that contain an O—H, N—H, or F—H bond. Oxygen, nitrogen and fluorine are small, highly electronegative elements containing lone pairs of electrons. They can hydrogen bond with hydrogen atoms in adjacent molecules. For example, methanol, CH_3OH, can hydrogen bond with water (Figure 3). The OH group is highly polar; therefore, methanol dissolves in water. When one liquid dissolves in another, the liquids are said to be **miscible**. On the other hand, 1-hexanol, $CH_3CH_2CH_2CH_2CH_2CH_2OH$, is insoluble in water. As the number of carbon atoms increase in an alcohol molecule, the polarity of the OH group is less significant, and the molecule becomes less polar. If the number of hydroxyl groups (OH) in a molecule increases, such as in 1,6-hexanediol, $HOCH_2CH_2CH_2CH_2CH_2CH_2OH$, the more hydrogen bonds it can form with water molecules and the more soluble it is in water.

Nonpolar compounds dissolve in nonpolar solvents because they have similar intermolecular forces. These attractive forces, called **dispersion forces** (or London forces) are temporary forces that result from the attraction of the positive nucleus of one molecule for the electron cloud in a nearby molecule, inducing a temporary dipole in the neighboring molecule. Dispersion forces are significant over very short distances. Two molecules must be very close to one another for them to polarize one another. Dispersion forces increase with increasing molecular weight because the molecules contain larger numbers of electrons. For example, cyclohexane is soluble in pentane because both are nonpolar. In addition, the molecular structures of the two molecules permit them to approach each other very closely (see Figure 4). This *"close fit"* of the molecules favors the distortion of their electron clouds, temporarily polarizing the molecules.

All compounds that dissolve in water are classified as either electrolytes or nonelectrolytes. An **electrolyte** is a substance that conducts electricity when dissolved in water. A substance that does not conduct electricity when dissolved in water is a **nonelectrolyte**. Compounds that produce large quantities of ions in solution (strong acids, strong bases, and water soluble salts) conduct electricity well and are classified as **strong electrolytes**. Compounds that produce small numbers of ions and exist predominantly as molecules in solution (weak acids and bases) are poor

Figure 4
The molecular structures of pentane and cyclohexane permit the molecules to approach each other very closely. (The carbon and hydrogen atoms have been omitted in the structures of pentane and cyclopentane in Figure 4b for purposes of illustrating the "close fit" of the molecules.) The close proximity of the molecules induces temporary dipole moments within each molecule

H_3C $\overset{H_2}{C}$ $\underset{H_2}{C}$ $\overset{H_2}{C}$ CH_3

pentane

$\overset{H_2}{C}$ H_2C CH_2 H_2C CH_2 $\underset{H_2}{C}$

cyclohexane

a)

molecules fit close together

b)

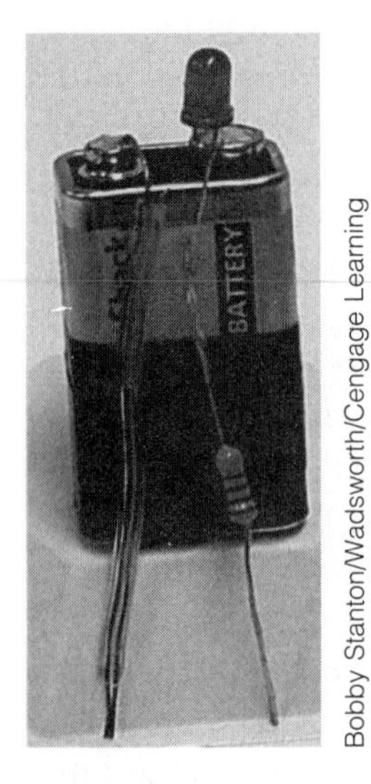

Bobby Stanton/Wadsworth/Cengage Learning

Figure 5
A conductivity detector constructed using a 9V battery, a copper wire electrode, a 1.8V LED, and a 1.8 ohm resistor, connected to the LED that serves as a second electrode

conductors of electricity. These compounds are referred to as **weak electrolytes**. Some compounds, when dissolved in water, produce no ions at all. These species do not conduct electricity and are referred to as nonelectrolytes.

Figure 5 depicts a simple conductivity detector, a device used to distinguish between electrolytes and nonelectrolytes. A pair of electrodes, connected to a source of electric current and a light source (e.g., LED, light emitting diode), is immersed in a solution. To light the LED, an electric current must flow from one electrode to another to complete the circuit. If ions are present in a solution, the movement of ions toward the positive and negative electrodes establishes the electric current needed to light the LED. A strong electrolyte solution contains numerous ions so the LED will light up brightly. A weak electrolyte solution contains a small number of ions and the LED is dimly lit. A nonelectrolyte solution contains no ions, thus, the LED will not light.

In Part A of this experiment, the solubilities of several compounds will be tested in water (a polar solvent) and hexane ($CH_3CH_2CH_2CH_2CH_2CH_3$, a nonpolar solvent) to determine the relative polar/nonpolar nature of the compounds. In Part B, we will examine the relative solubilities of some ionic compounds in water. In Part C of the experiment, we will test the electrical conductivity of several solutions and classify the solutions as strong, weak, or nonelectrolytes.

PROCEDURE

CAUTION

Students must wear departmentally approved eye protection while performing this experiment. Wash your hands before touching your eyes and after completing the experiment.

Several of the molecular compounds used in this experiment are flammable. Open flames are NOT PERMITTED in the laboratory while performing this experiment.

Part A - Determination of the Polar/Nonpolar Nature of Several Compounds

1. Observe the physical appearance of each of the following compounds. Should you record your observations in the Lab Report?

   a. potassium nitrate, KNO_3

   b. ammonium sulfate, $(NH_4)_2SO_4$

   c. calcium chloride, $CaCl_2$

   d. ethanol, CH_3CH_2OH

   e. 1-decanol, $CH_3CH_2CH_2CH_2CH_2CH_2CH_2CH_2CH_2CH_2OH$

   f. glycerin, $HOCH_2CHOHCH_2OH$

   g. urea, NH_2CONH_2

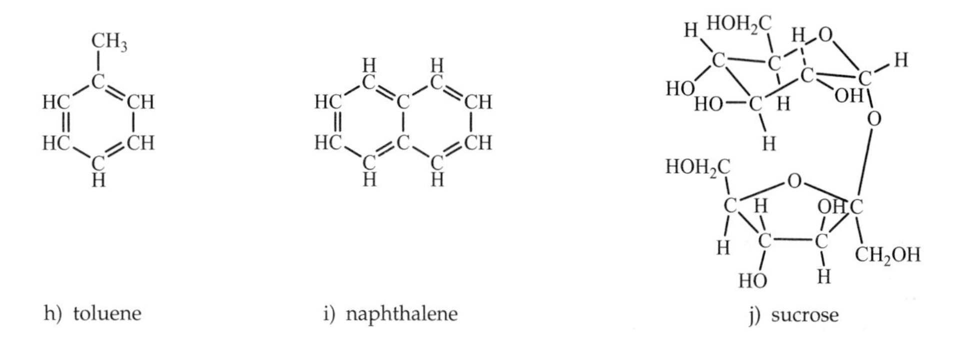

h) toluene i) naphthalene j) sucrose

2. Determine the water solubility of the compounds listed in Step 1. Add a few granules (~ 0.05 g) of each solid compound to ~ 2 mL of water in a small test tube. Shake the test tube to mix the substances. Add 5–6 drops of each liquid compound to ~ 2 mL of water in a small test tube. Shake the test tube to mix the substances. Should you record your water solubility observations in the Lab Report?
3. Empty the contents of each test tube into the Waste Container.
4. Repeat Steps 2 and 3 using hexane as the solvent. Should you record your hexane solubility observations in the Lab Report?
5. Which of the ten compounds listed in Step 1 are polar; which are nonpolar; which are ionic?
6. What structural feature(s) do ethanol and 1-decanol have in common? State a reason for the observed solubility patterns of ethanol and 1-decanol in water and hexane.
7. What structural feature(s) do toluene and naphthalene have in common? State a reason for the observed solubility patterns of toluene and naphthalene in water and hexane.
8. Which of the following compounds can hydrogen bond with water: ethanol, 1-decanol, glycerin, sucrose and urea? State a reason why each compound will or will not hydrogen bond with water. What effect will hydrogen bonding have on the solubility of each compound in water? In hexane?

Part B - Relative Solubilities of Some Ionic Compounds in Water

9. Determine the water solubility of each of the following compounds.
   a. ammonium chloride, NH_4Cl
   b. ammonium nitrate, NH_4NO_3
   c. sodium sulfate, Na_2SO_4
   d. sodium carbonate, Na_2CO_3
   e. calcium chloride, $CaCl_2$
   f. calcium nitrate, $Ca(NO_3)_2$
   g. calcium sulfate, $CaSO_4$
   h. calcium carbonate, $CaCO_3$
10. Should you record your observations from Step 9 in the Lab Report?
11. Empty the contents of each test tube into the Waste Container.
12. Note any trends in the solubilities of the eight compounds determined in Step 10. Provide an explanation for your observations.

Part C - Electrical Conductivity of Several Solutions

13. Clean and dry a watch glass.
14. Add 15 drops of one of the following aqueous solutions or liquids to the watch glass.
    a. distilled water
    b. 0.1 *M* sodium chloride
    c. 0.1 *M* ammonium nitrate
    d. 0.1 *M* calcium sulfate
    e. ethanol
    f. glycerol
    g. 0.1 *M* sucrose solution
    h. glacial acetic acid (CH_3COOH, concentrated, 99.9%)
    i. 0.1 *M* acetic acid
    j. 0.1 *M* ammonia
    k. 0.1 *M* sodium hydroxide
    l. 0.1 *M* hydrochloric acid
15. Simultaneously immerse the conductivity detector electrodes into the solution or liquid on the watch glass. Should you record your observations in the Lab Report?
16. Rinse the electrodes with distilled water. Pour the solution or liquid on the watch glass into the Waste Container. Thoroughly rinse the watch glass with distilled water and dry the watch glass.
17. Repeat Steps 15–16 for each of the remaining solutions in Step 14.
18. Compare the electrical conductivities of sodium chloride ($NaCl$), ammonium nitrate (NH_4NO_3), calcium sulfate ($CaSO_4$) solutions, and distilled water. Account for any differences noted in the electrical conductivities of these compounds.
19. Compare the electrical conductivities of concentrated acetic acid (CH_3COOH, 99.9 %), 0.1 *M* acetic acid (CH_3COOH) and distilled water. Account for any differences noted in the electrical conductivities of these compounds.
20. Compare the electrical conductivities of 0.1 *M* acetic acid, 0.1 *M* HCl, and distilled water. Account for any differences noted in the electrical conductivities of these compounds.
21. Compare the electrical conductivities of 0.1 *M* ammonia (NH_3), 0.1 *M* sodium hydroxide ($NaOH$), and distilled water. Account for any differences noted in the electrical conductivities of these compounds.

Name

Section

Date

Instructor

13 EXPERIMENT 13

Lab Report

Part A – Determination of the Polar/Nonpolar Nature of Several Compounds

Physical appearance of compounds in Step 1.

Solubility of compounds in Step 1 in water.

Solubility of compounds in Step 1 in hexane.

Polarity of compounds in Step 1.

What structural feature(s) do ethanol and 1-decanol have in common? State a reason for the observed solubility patterns of ethanol and 1-decanol in water and hexane.

What structural feature(s) do toluene and naphthalene have in common? State a reason for the observed solubility patterns of toluene and naphthalene in water and hexane.

Which of the following compounds can hydrogen bond with water: ethanol, 1-decanol, glycerin, sucrose and urea? State a reason why each compound will or will not hydrogen bond with water. What effect will hydrogen bonding have on the solubility of each compound in water? in hexane?

Part B – Relative Solubilities of Some Ionic Compounds

Solubility of compounds in Step 9 in water.

Note any trends in the solubilities of the eight compounds tested in Steps 9–10. Provide an explanation as to why some compounds are soluble, some are slightly soluble, and some are insoluble in water.

Part C – Electrical Conductivity of Several Solutions

Electrical conductivity of solutions and liquids in Step 14.

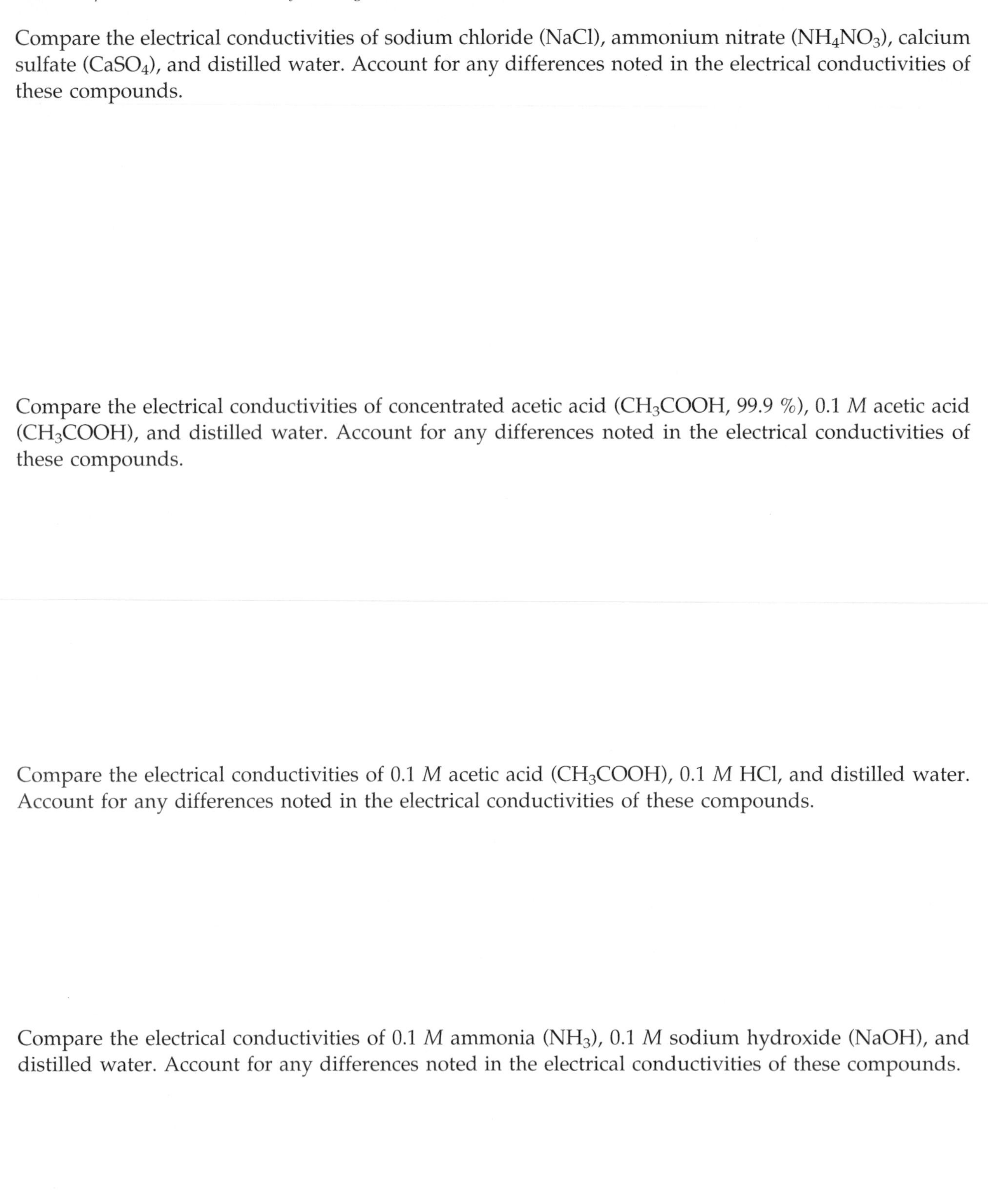

Compare the electrical conductivities of sodium chloride (NaCl), ammonium nitrate (NH_4NO_3), calcium sulfate ($CaSO_4$), and distilled water. Account for any differences noted in the electrical conductivities of these compounds.

Compare the electrical conductivities of concentrated acetic acid (CH_3COOH, 99.9 %), 0.1 *M* acetic acid (CH_3COOH), and distilled water. Account for any differences noted in the electrical conductivities of these compounds.

Compare the electrical conductivities of 0.1 *M* acetic acid (CH_3COOH), 0.1 *M* HCl, and distilled water. Account for any differences noted in the electrical conductivities of these compounds.

Compare the electrical conductivities of 0.1 *M* ammonia (NH_3), 0.1 *M* sodium hydroxide (NaOH), and distilled water. Account for any differences noted in the electrical conductivities of these compounds.

Name ______ Section ______ Date ______

Instructor ______

13 EXPERIMENT 13

Pre-Laboratory Questions

1. The solubilities of four unknown compounds (A, B, C, and D) were tested in water and 1-hexene ($CH_3CH_2CH_2CH{=}CH_2$). The results of the solubility tests are provided below. From the solubility data, determine whether compounds A, B, C, and D are polar or nonpolar.

Compound	*Solubility in water*	*Solubility in 1-hexene*	*Polar or Nonpolar*
A	insoluble	soluble	______
B	soluble	insoluble	______
C	soluble	insoluble	______
D	soluble	insoluble	______

2. The electrical conductivities of the unknown compounds (A, B, C, and D) were determined using a conductivity detector. The results of the conductivity tests are provided below. Determine whether compounds A, B, C, and D are strong, weak, or nonelectrolytes.

Compound	*Conductivity Test*	*Strong/Weak/Nonelectrolyte*
A	LED does not light	______
B	LED does not light	______
C	LED lights brightly	______
D	LED lights dimly	______

3. Based on the data presented in and your responses to Questions 1 and 2, match the letters of unknown compounds (A, B, C, D) to one of the molecular structures or chemical formulas depicted below. (*Write the letter of the unknown compound in the blank space underneath each compound*).

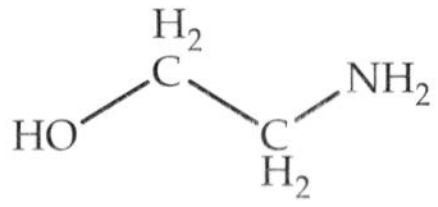

H_2
C
NH_2
HO
C
H_2

unknown compound _______

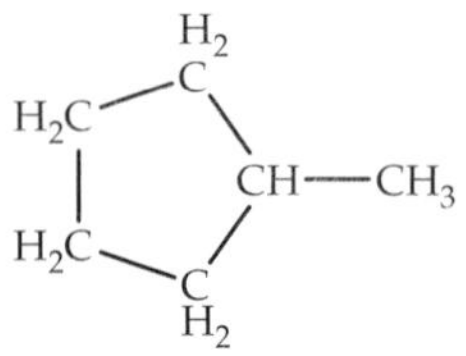

H_2
C
H_2C
CH—CH_3
H_2C
C
H_2

unknown compound _______

$HClO_2$

unknown compound _______

LiCl

unknown compound _______

Name _______________ Section _______________ Date _______________

Instructor _______________

13 **EXPERIMENT 13**

Post-Laboratory Questions

1. Why is the solubility of butane 6.1 mg/100 g water, yet hexane is immiscible in water?

2. Based on their molecular structures, would you expect the compounds depicted below to be soluble in water? Justify your answers with an explanation.

   a. cyclohexylamine

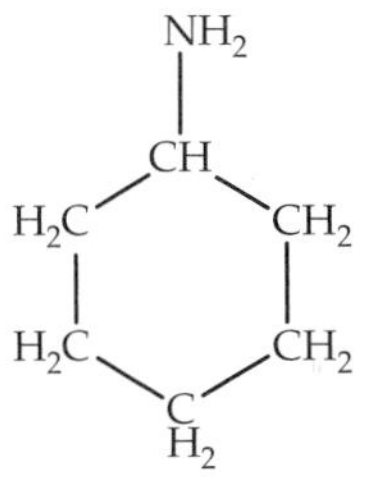

   b. naphthalene

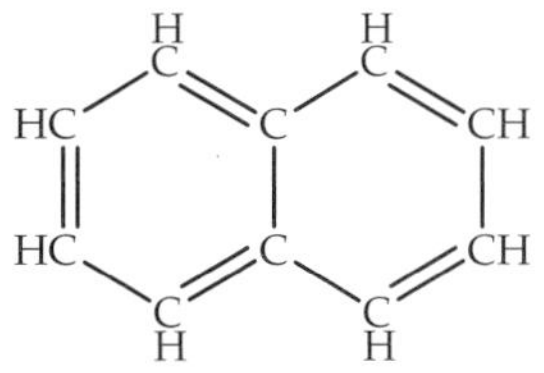

3. Rank the following compounds in order of increasing electrical conductivity: 0.1 *M* formic acid (HCOOH), 0.1 *M* $(NH_4)_3PO_4$, 0.1 *M* KCl, and 0.1 *M* glucose solution ($C_6H_{12}O_6$). Justify your answers with an explanation.

Determination of the Cause of a "Fish-Kill" in the Clark Fork of the Columbia River: A Self-Directed Experiment

PURPOSE

Determine the cause of a "fish-kill" in the Clark Fork of the Columbia River in Montana.

INTRODUCTION

In this experiment, each student team will analyze a *simulated* water sample from the Clark Fork of the Columbia River in Montana where a major fish-kill occurred in 1984. Each team is to analyze their water sample for the presence of Na^+, K^+, Li^+, Ca^{2+}, Ba^{2+}, Sr^{2+}, Cu^{2+}, and Fe^{3+} ions. The presence of Group IA and IIA metal ions contribute to water salinity. Water that is too salty (high salinity) may cause fish-kills. Each team will report if their sample contains toxic levels of Cu^{2+} and Fe^{3+} ions.

This is a *self-directed* experiment wherein teams of students will write the procedure they will use, and rely on lab skills and techniques acquired this semester, to perform this experiment. Each team will submit a **Procedure Proposal** to their lab instructor two weeks before the experiment is to be performed. *The Procedure Proposal must address several important questions, a few examples are listed below.*

1. What is the central question to be answered in this experiment?
2. What experimental technique(s) will be utilized to answer the central question?

3. How will each team determine the λ_{max} for the $[CuSCN]^+$ and for the $[FeSCN]^{2+}$ ions?
4. What calculations should be provided in the Procedure Proposal?
5. What is the concentration range for the standard solutions used to determine the concentrations of Cu^{2+} and Fe^{3+} ions.
6. What safety precautions must be addressed in the Procedure Proposal?
7. What additional questions should be addressed by each team?

Each team may use their lab manual, textbook, reference books, and the internet for assistance in writing the **Procedure Proposal** (the format for writing the Procedure Proposal is provided in Appendix C-1). The graded and annotated Procedure Proposal will be returned to each team one week before the lab period during which the experiment is to be performed.

When the experiment is completed, each team will submit one **Formal Lab Report** (the format for writing the Formal Lab Report is provided in Appendix C-2) addressing the following questions.

1. What conclusions can be drawn from the experimental data collected by each team?
2. Does the experimental data collected answer the central question?
3. Can the team's conclusions be supported by information obtained from reference sources, textbooks, or the internet?
4. What are some possible sources of error in the experimental data?
5. What modifications could be made to the experimental design and procedures to improve the accuracy of the data?

PROCEDURE

Each team will analyze a water sample for the presence of Group IA and IIA metal ions using the MeasureNet Spectrometer. If team members have retained spectra files of solutions containing known Group IA and IIA metal ions from a previous experiment, those spectra can be used in this experiment.

The simulated water sample that each team will analyze is highly concentrated in Cl^- ions, therefore, the Cu^{2+} and Fe^{3+} ions exist in the sample as chloride complexes. The Cu^{2+} and Fe^{3+} chloride complexes weakly absorb visible light. For these ions to absorb visible light *strongly*, a complexing agent must be added to form highly colored complexes with both Cu^{2+} and Fe^{3+} ions. Thiocyanate ions (SCN^-) form complexes with both Cu^{2+} and Fe^{3+} ions. $[CuSCN]^+$ and $[FeSCN]^{2+}$ are highly colored and absorb visible light at *widely separated absorbance bands*.

Teams will be provided with a standard solution containing 400 ppm $[CuSCN]^+$ and 400 ppm $[FeSCN]^{2+}$. Thiocyanate ions have also been added to the Clark Fork of the Columbia River simulated water sample (note the dark color of the water sample). The **limit of detection** (*lowest* concentration that can be detected) for $[CuSCN]^+$ is 50 ppm using this particular analysis method.

LIST OF CHEMICALS

A. Standard 0.5 *M* NaCl solution
B. Standard 0.5 *M* LiCl solution
C. Standard 0.5 *M* KCl solution
D. Standard 0.5 *M* $CaCl_2$ solution
E. Standard 0.5 *M* $BaCl_2$ solution
F. Standard 0.5 *M* $SrCl_2$ solution
G. Standard Fe/Cu solution containing 400 ppm Cu^{2+} and 400 ppm Fe^{3+} in SCN^- solution
H. 0.1 *M* iron(III) nitrate solution (already mixed with SCN^- ions)
I. 0.1 *M* copper(II) nitrate solution (already mixed with SCN^- ions)
J. Simulated Clark Fork of the Columbia River Water Sample

LIST OF SPECIAL EQUIPMENT

A. MeasureNet spectrophotometer
B. Cuvettes
C. Coiled nichrome wires
D. Volumetric flasks
E. Beral pipets
F. Kimwipes®

Quality Control for the Athenium Baking Soda Company: A Self-Directed Experiment

PURPOSE

Determine the purity of baking soda produced by the Athenium Baking Soda Company and identify the chemical nature of the impurities present in the baking soda sample.

INTRODUCTION

Baking soda is the common name for sodium hydrogen carbonate or sodium bicarbonate, $NaHCO_3$. It is an economical, natural compound that can replace many expensive cleaning products that are potentially environmentally harmful. Baking soda has many uses including deodorizer, household cleanser, fire extinguisher, fruit cleaner, hand and face wash, heavy-duty dish cleanser, tooth and denture cleaner, for cooking, and for acid indigestion.

The Athenium Baking Soda Company has recently begun producing baking soda. This company has chosen from one of several methods of producing baking soda to react crystalline ammonium hydrogen carbonate with brine (Eq. 1).

$$NH_4HCO_{3(s)} + NaCl_{(aq)} \rightarrow NaHCO_{3(aq)} + NH_4Cl_{(aq)} \qquad \text{(Eq. 1)}$$

Brine is salt water containing a high concentration of sodium chloride, as well as potassium chloride, lithium chloride, and calcium chloride. When the desired solid product ($NaHCO_3$) is dried and filtered, contaminants (e.g., KCl, LiCl, and $CaCl_2$) may be present in the residue.

Most manufacturing companies employ quality control scientists and technicians to analyze their products for purity, quality of workmanship, structural flaws, composition, and longevity. **Quality control** consists of periodic inspections designed to maintain quality during the manufacturing process.

The Athenium Baking Soda Company is seeking to hire a team of four quality control scientists to analyze their baking soda for composition and purity. The company has prepared a baking soda sample with a known set of contaminants. Applicant teams will compete for the quality control positions by analyzing the composition and purity of the contaminated baking soda samples. The team that uses the most efficient analysis techniques and obtains the most accurate data will be awarded the quality control positions.

This is a *self-directed* experiment wherein teams of students will write their procedure, relying on lab skills and techniques acquired this semester, to perform this experiment. Each team will submit a **Procedure Proposal** to their lab instructor two weeks before the experiment is to be performed. *The Procedure Proposal must address several important questions; a few examples are listed below.*

1. What is the central question to be answered in this experiment?
2. What experimental technique(s) will be utilized to answer the central question?
3. How will the team determine the percentage of baking soda in the sample?
4. If the baking soda sample is impure, how will the team identify the impurities present in the sample?
5. What balanced, chemical equations and calculations should be included in the Procedure Proposal?
6. What safety precautions must be addressed in the Procedure Proposal?
7. What additional questions should be addressed by each team?

Each team may use their lab manual, textbook, reference books, and the internet for assistance in writing the **Procedure Proposal** (the format for writing the Procedure Proposal appears in Appendix C-1). The graded and annotated Procedure Proposal will be returned to each team one week before the lab period during which the experiment is to be performed.

When the experiment is completed, each team will submit one **Formal Lab Report** (the format for writing the Formal Lab Report appears in Appendix C-2) addressing the following questions.

1. What conclusions can be drawn from the experimental data collected by the team?
2. Does the experimental data collected answer the central question?
3. Which technique is more accurate for determining the percentage of baking soda in the sample?
4. Can the team's conclusions be supported by information obtained from reference sources, textbooks, or the internet?
5. What are some possible sources of error in the experimental data?
6. What modifications could be made to the experimental design and procedures to improve the accuracy of the data?

PROCEDURE

Each team will be provided with 6 grams of a baking soda sample produced by the Athenium Baking Soda Company. Each team is to report the percentage of $NaHCO_3$ in the sample. The percentage of $NaHCO_3$ in the sample must be determined by *two different* experimental techniques. If the sample is impure, the team must report whether lithium, potassium, or calcium salts are present as impurities.

Each analysis or test requires no more than 1 gram of solid Na_2CO_3 or $NaHCO_3$. Na_2CO_3 or $NaHCO_3$ solutions should be prepared by dissolving 1 gram of solid in 25 to 30 milliliters of distilled water (see Note).

LIST OF CHEMICALS

A. Sodium carbonate (solid, primary standard base)
B. Baking soda sample (mixture)
C. Approximately 1.0 *M* hydrochloric acid solution
D. 0.1 *M* Calcium chloride solution
E. 0.1 *M* Potassium chloride solution
F. 0.1 *M* Lithium chloride solution

LIST OF SPECIAL EQUIPMENT

A. MeasureNet spectrophotometer, cuvettes and nichrome wires
B. MeasureNet pH probe and drop counter.
C. Volumetric flasks
D. Beral pipets
E. Kimwipes®
F. Crucible and lid

NOTE: In a titration where pH is monitored as a function of the volume of reactant added, the equivalence point is determined as indicated in Experiment 12 for 1:1 reaction stoichiometry (A + B). For a reaction involving 2:1 stoichiometry (e.g., 2A + B) of the reactants, there will be two inflection points on the titration curve. An inflection point will be observed as each mole of A is completely reacted. Either inflection point can be used to determine the equivalence point of the titration. If the first inflection point is used to determine the equivalence point of the titration, 1:1 reaction stoichiometry (A + B) must be used when calculating the volume, moles, or the molarity of a reactant. The second inflection point typically exhibits the greater change in pH, making it easier to determine the equivalence point of the titration. If the second inflection point is used to determine the equivalence point of the titration, 2:1 reaction stoichiometry (2A + B) must be used when calculating the volume, moles, or the molarity of a reactant.

Gas Laws

PURPOSE

Determine the identity of an unknown liquid using the ideal gas law and study the effect of pressure on the volume of a fixed amount of air at constant temperature using the MeasureNet pressure transducer.

INTRODUCTION

Although the chemical properties of different gases are substantially different, their physical properties are quite similar. The physical properties of gases are independent of the identities of the gases at ordinary temperatures and pressures. Observations by scientists of many gases in the 1700's led to the formulation of the gas laws. The behavior of gases can be summarized in the **ideal gas law.** The ideal gas law relates the pressure (P), volume (V), and temperature (T) of a fixed quantity of gas (n) according to Equation 1,

$$PV = nRT \qquad \text{(Eq. 1)}$$

where R is the universal gas constant. (In this experiment, we will use 0.0821 L·atm/mol·K for the value of R. Gas pressures are commonly measured in mmHg (or torr) in the laboratory, where one atmosphere (atm) = 760 mmHg = 760 torr.)

The ideal gas law was formulated from studies in which one quantity (P, V, or T) was varied, and its effect on some other quantity was observed. In the 1600's, Boyle studied the effects of changing pressure on the volume of a gas at constant temperature and moles of gas. In the late 1700's, Charles studied the behavior of the volume of a gas while its temperature is varied at constant pressure and moles of gas. Amontons studied how changing the temperature of a gas effected its pressure when the moles and volume of the gas were held constant. In the 1700's, Dalton established the fact that gases in a mixture behave independently of one another.

Part A – Ideal Gas Law

In Part A of this experiment, we will utilize a boiling water bath to vaporize a small quantity of an unknown, volatile liquid in a pre-weighed flask. This heating process vaporizes the unknown, volatile liquid and drives the air from the flask, leaving it filled with the unknown gas vapor.

Thus, the volume of gas will be equal to the volume of the flask. The temperature of the unknown gas vapor is the same as the boiling water. The pressure inside the flask is equal to atmospheric pressure that is measured with a barometer. The flask is cooled to condense the unknown gas vapor. The mass of the liquid (condensed vapor) is equal to the mass of gas vapor in the hot flask.

The number of moles of gas (n) in the sample is a function of the volume of the gas at a given temperature and pressure. Thus, the number of moles of gas can be determined using Equation 2.

$$n = \frac{PV}{RT} \qquad \text{(Eq. 2)}$$

The number of moles in a sample of gas is equal to the mass of the gas in grams (g) divided by the molar mass (M_m, in grams/mole) of the gas.

$$n = \frac{g}{M_m} \qquad \text{(Eq. 3)}$$

Substituting Equation 3 into Equation 2 for n, and solving for M_m yields Equation 4.

$$M_m = \frac{gRT}{PV} \qquad \text{(Eq. 4)}$$

Assuming the gas behavior approximates that of an ideal gas, and using the values for the volume, pressure, temperature, mass of the gas vapor, and the ideal gas constant R, we can calculate the molar mass of the unknown liquid. The value obtained for the molar mass of the unknown liquid will be compared to the known molar masses for the compounds in Table 1 to identify the unknown liquid.

Part B – Boyle's Law

As a consequence of Dalton's Law, we can study the physical properties of air, even though air is a mixture of gases. In Part B of this experiment we will examine how the pressure of a gas is related to its volume. Air will be trapped in a syringe connected to the MeasureNet pressure transducer via a stopcock. The pressure-volume relationship is determined from a plot of the pressure versus the volume of the air measured with the MeasureNet pressure transducer.

Table 1 *Molar masses in grams per mole of some volatile liquids.*

methanol	32.0
ethanol	46.0
acetone	58.1
1-propanol	60.1
n-hexane	86.2

The pressure transducer measures the pressure of the gas relative to atmospheric pressure. The pressure of the gas (P_{air}) is equal to the pressure transducer value ($P_{transducer}$) plus atmospheric pressure (P_{atm}).

$$P_{air} = P_{transducer} + P_{atm} \quad \text{(Eq. 5)}$$

Atmospheric pressure is independently measured using a barometer.

The volume of the air sample is the volume in the syringe plus the volume of the tubing, V_{tubing}. V_{tubing} includes the tubing that connects the syringe to the pressure transducer and a small volume inside the pressure transducer. There is no convenient method for measuring V_{tubing}. The volume in the syringe changes with the position of the plunger, but V_{tubing} remains constant. Thus, we can still determine whether air behaves according to Boyle's Law. V_{tubing} in this experiment is approximately 5.0 milliliters.

PROCEDURE

CAUTION

Students must wear departmentally approved eye protection while performing this experiment. Wash your hands before touching your eyes and after completing the experiment.

Part A – Determination of the Molar Mass of a Volatile Compound

1. Obtain a clean, dry 125-mL Erlenmeyer flask, a 2-inch by 2-inch piece of aluminum foil, and a rubber band. Using water, devise a method to determine the *exact* volume of the flask.
2. Thoroughly dry the flask with a towel after determining the volume of the flask.
3. Should you determine the mass of the Erlenmeyer flask, the piece of aluminum foil, and the rubber band? If so, to what number of significant figures should the mass be recorded in the Lab Report?
4. Obtain approximately 12 mL of an unknown volatile liquid from your laboratory instructor. Should you record the Unknown Number in the Lab Report?
5. Add approximately 6 mL of the unknown liquid to the Erlenmeyer flask. Should you determine the mass of the unknown liquid? If so, how do you determine the mass, and to what number of significant figures should the mass be recorded in the Lab Report?
6. Crimp the square of aluminum foil tightly over the top of the flask to form a cover. Wrap the rubber band around the flask to secure the aluminum foil. *Be sure there are no loose edges on the sides of the flask where water may get under the foil.*
7. Using a push pin, punch a *small* pinhole in the center of the foil.
8. Setup a water bath as shown in Figure 1. Place the flask containing the volatile liquid in the water bath. Be sure the water level in the beaker is approximately 2 cm below the foil cover on the flask. *Be sure water does not get under the foil cover during the experiment.*

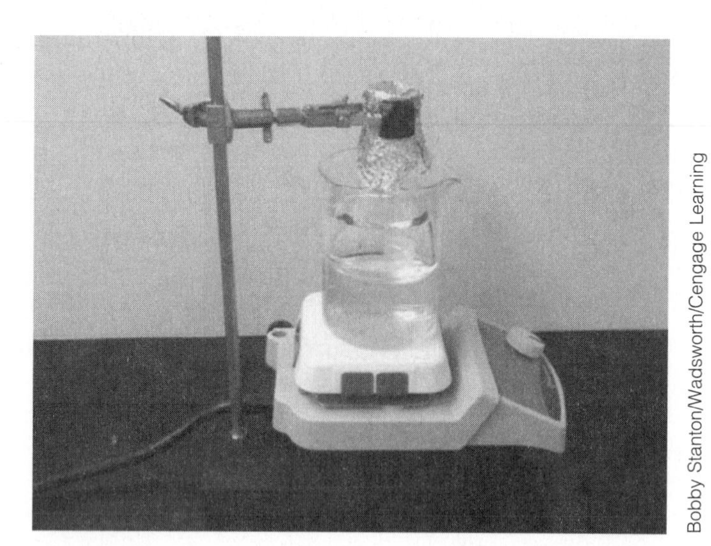

Figure 1
Heating to vaporize the unknown liquid

9. Let the water in the beaker come to a *gentle* boil and continue boiling until the last trace of liquid in the flask has evaporated. Once the liquid has completely evaporated, continue heating for one additional minute.
10. Should you determine the temperature of the gas vapor in the flask, and record it in the Lab Report? If so, how would you determine the gas vapor's temperature? *Do not remove the foil from the flask*. The flask must remain covered until the vapor has completely condensed to a liquid.
11. The laboratory instructor will provide you with the current atmospheric pressure. Should you record the pressure in the Lab Report and in what units?
12. Remove the flask from the beaker and *thoroughly* dry the sides of the flask. Let the flask cool to room temperature.
13. Should you determine the mass of the condensed vapor? If so, how do you determine the mass, and to what number of significant figures should the mass be recorded in the Lab Report?
14. Repeat Steps 1–13 with a second sample of the same unknown liquid.
15. Steps 16–18 are to be completed at the end of the lab period. Proceed to Step 19, *Part B – Pressure-Volume Measurements for an Air Sample.*
16. How do you determine the molar mass of the volatile liquid for Trials 1 and 2?
17. What is the average molar mass of the unknown volatile liquid from the results obtained for the two trials?
18. From the data provided in Table 1, identify the unknown volatile liquid.

Part B – Pressure-Volume Measurements for an Air Sample

19. Turn on the power to the MeasureNet station. Press **Main Menu**. Press **F4 Pressure**. Then press **F2 Pressure and Volume**.
20. Turn the stopcock valve to the setting shown in Figure 2.
21. Press **Calibrate**. Enter the atmospheric pressure in torr. (If atmospheric pressure is not known, estimate atmospheric pressure to be 750 torr). Press **Enter**. Enter 5.0 mL as the volume of the tubing. Press **Enter**.

Syringe Connection Port

Pressure Transducer

Tube connected to unused side

Stopcock Valve Tip

Bobby Stanton/Wadsworth/Cengage Learning

Figure 2
Stopcock position for atmospheric pressure calibration

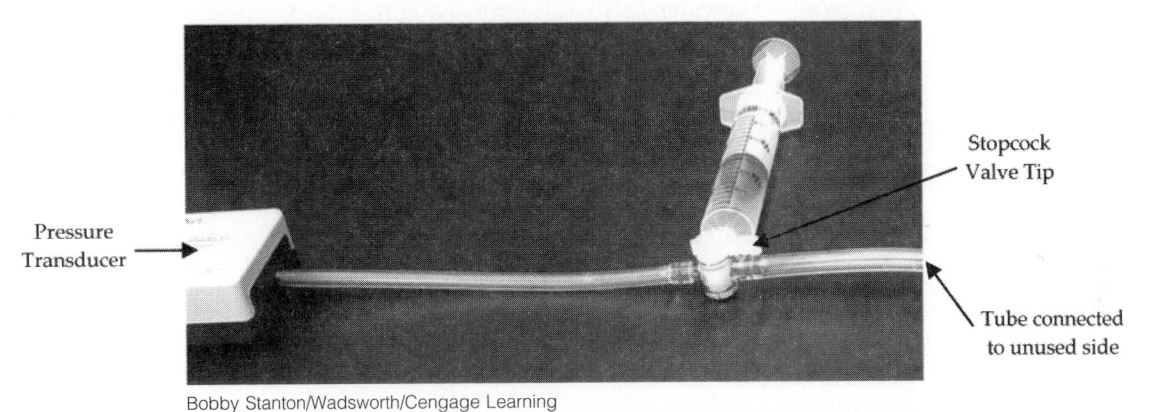

Bobby Stanton/Wadsworth/Cengage Learning

Figure 3
Pressure transducer, syringe assembly for P-V measurements

Bobby Stanton/Wadsworth/Cengage Learning

Figure 4
Unused side of stopcock closed to the atmosphere

22. Press **Display**, the workstation should display a pressure reading near 0.0. If not, repeat the calibration process (Steps 21–22).
23. Pull out the syringe plunger to the 10 mL mark (approximately half-way). *Carefully* screw the syringe into its connection port (*do not over tighten, you may break off the syringe tip*), and turn the valve to the position indicated in Figure 3.
24. While covering the tube on the unused side of the stopcock with your finger, push in the plunger until the pressure reading is approximately 400 torr. With your finger covering the end of the tube on the unused side, turn the stopcock valve to the position indicated in Figure 4.

25. Monitor the pressure reading for 2 minutes. Initially the pressure may drop to 20–40 torr. If the pressure reading does not drop by more than 1 torr in 10 seconds, proceed to Step 26. If the pressure drops at a faster rate, put detergent solution on all tube fittings to check for leaks. Ask your lab instructor for assistance with leak testing using detergent solution.
26. Turn the stopcock so that all three connectors are open to atmosphere (the tip of the stopcock valve is in the same position as shown in Figure 2). The pressure should read approximately 0 torr.
27. Set the bottom ring seal on the syringe plunger (bottom of the black o-ring) to the 4.0 mL mark on the syringe (as precisely on 4.0 as you can), then turn the valve so that the unused connector is closed (see Figure 3).
28. Note the pressure reading on the display. Press **Start/Stop** to record $P_{transducer}$, then enter the volume reading (4.0 mL) when prompted, then press **Enter**. (*Note – Pressure readings will be negative values until we account for atmospheric pressure.*)
29. Holding the flange at the top of the syringe (do not grasp the barrel, you will change the air temperature), pull out the syringe plunger to the 5.0 mL mark. Hold the volume at 5.0 mL for 5–6 seconds. Press **Start/Stop** to record $P_{transducer}$. Enter the volume reading (5.0 mL) when prompted, then press **Enter**. *At no time should the pressure reading exceed ± 750 torr*.
30. Pull out the syringe plunger to the 6.0 mL mark. Hold the volume at 6.0 mL for 5–6 seconds. Press **Start/Stop** to record $P_{transducer}$. Enter the volume reading (6.0 mL) when prompted, then press **Enter**.
31. In a similar fashion to Steps 29–30, increase the volume 1 mL at a time, until a final volume of 20 mL is reached. After each 1 mL increase in volume, press **Start/Stop** to record the new pressure, and enter the new volume to the nearest 0.1 mL, then press **Enter**. This will produce a total of 17 different pressure-volume data points.
32. Press **File Options**. Press **F3** to *save* the scan as a tab delimited file. You will be prompted to enter a 3-digit code (any 3-digit number you choose). The name of the file will be saved as a 4–5 digit number. The first 1–2 numbers represent the workstation number, the last 3 digits are the 3 digit access code you entered. For example, suppose you are working on Station 6 and you select 543 as your code. Your file name will be saved as 6543 on the computer. If you are working at Station 12 and you choose 123 as your save code, the file name will be 12123. Press **Enter** to accept your 3 digit number.
33. Should you record the file name in the Lab Report? Save the file to a flash drive or email the file to yourself via the internet.
34. Press **Display** to clear the previous scan.
35. Repeat Steps 19–32 to record a second trial.
36. Open an Excel spreadsheet (i.e., worksheet).
37. Go to **File Open**, open a MeasureNet tab delimited file containing mL versus pressure data (pressure will be negative numbers). Click **Finish**.
38. Copy the first two columns (containing mL and pressure data) in the tab delimited (*text*) file, and paste it into columns A (mL) and B (pressure) in the Excel worksheet. Close the tab delimited file.

39. The pressures in column B are $P_{transducer}$. We must determine the true pressure (P_{air}) of the air sample in column C of the Excel file. In column C, add the value of atmospheric pressure (750 torr if atmospheric pressure is not known) to each of the pressure readings in column B. You may do this with a calculator or you can do this within Excel. To do this in Excel, click box C1. Then click inside the equation box at the top of the spreadsheet (box beside the $\times$, $\surd$, and $f_{\times}$ symbols). Type = 750 + B1 (this is the equation to calculate P_{air} ($P_{transducer} + P_{atm}$)). If atmospheric pressure is known, use that value instead of 750. Click on the $\surd$ symbol. Double click the black square in the lower right corner of box C1. This should automatically calculate a P_{air} value in column C for each $P_{transducer}$ value in column B. If not, click on the black square in the lower right corner of box C1 and drag down column C until the last cell containing data in column B is reached.

40. Prepare a **XY scatter plot** pressure in torr versus the milliliters of air. You may wish to consult Appendix B1 for additional instructions for preparing an XY scatter plot. The instructions in Appendix B1 are essentially the same for a pressure-volume plot as they are for a temperature-time Plot.

41. From the plot of pressure versus volume of air, describe the relationship between pressure and volume for the sample of air?

42. Did the sample of air behave in accordance to Boyle's Law? You may wish to consult your textbook to see a thorough discussion of Boyle's Law.

43. Include the pressure-volume plot for the sample when submitting your Lab Report.

Name ______ Section ______ Date ______

Instructor ______

16 EXPERIMENT 16

Lab Report

Part A – Determination of the Molar Mass of a Volatile Compound

Method for determining the *exact* volume of the flask.

Unknown number ________

Data and observations for Trial 1.

Data and observations for Trial 2.

How do you determine the molar mass of the volatile liquid for Trials 1 and 2?

What is the average molar mass of the unknown volatile liquid?

From the data provided in Table 1, identify the unknown volatile liquid.

____

From the plot of pressure versus volume of air, describe the relationship between pressure and volume for the sample of air?

Did the sample of air behave in accordance to Boyle's Law?

Name Section Date

Instructor

16 EXPERIMENT 16

Pre-Laboratory Questions

1. A student was assigned the task of determining the identity of an unknown liquid. The student weighed a clean, dry 250-mL Erlenmeyer flask. The student completely filled the flask with water. He then proceeded to weigh the flask containing the water.

mass of 250-mL Erlenmeyer flask	78.639 g
mass of 250-mL Erlenmeyer flask and water	327.039 g

Assuming the density of water is 1.00 g/mL at ambient temperature, what is the exact volume (in mL) of the 250-mL Erlenmeyer flask?

2. Next, the student added 5.2 mL of an unknown liquid to the dry 250-mL Erlenmeyer flask (same flask used in Question 1). He placed the flask in a boiling water bath and left it there until the liquid completely vaporized. The student measured the temperature of the boiling water bath to determine the temperature of the gas inside the flask. The temperature of the water bath was 99.3 °C. Atmospheric pressure was 751 torr when the liquid completely vaporized.

What is the temperature of the gas in Kelvins?

What is the pressure of the gas in atmospheres?

3. The student removed the 250-mL Erlenmeyer flask containing the gas vapor to let it cool to room temperature. When the gas had completely condensed, he weighed the flask and the liquid.

mass of 250-mL Erlenmeyer flask and liquid	79.118 g

What is the mass of the liquid in the flask?

4. What is the molar mass of the unknown liquid in grams per mole?

5. Using Table 1, what is the identity of the unknown liquid? ______________________________

6. Gas laws only apply to ideal gases. Briefly explain why we can use ideal gas law to determine the molar mass of an unknown liquid.

7. The volume of a gas is measured at several temperatures. The moles and pressure of the gas were held constant for each measurement. A plot of the volumes versus the temperatures of the gas sample is given below.

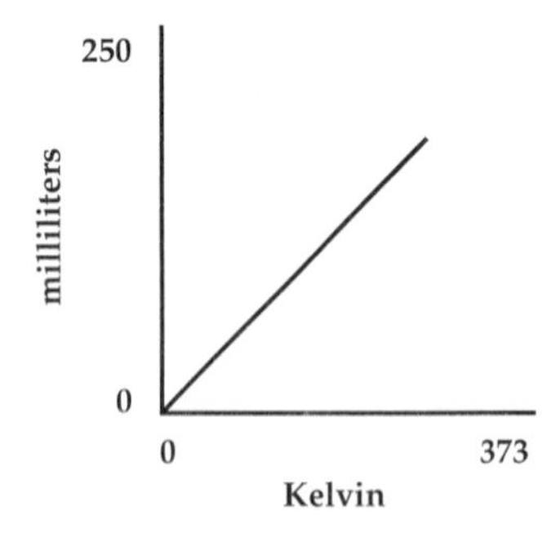

Which gas law governs the gas behavior depicted in the plot above? ______________________________

Name Section Date

Instructor

16 EXPERIMENT 16

Post-Laboratory Questions

1. In Step 12 of the procedure, a student did not dry the flask before weighing it. As a result of this experimental error, would the calculated molar mass be higher or lower than the correct value? Justify your answer with an explanation.

2. In Step 5 of the procedure, a student added 0.1 mL of unknown liquid, instead of 6 mL as stated in the procedure, to a 125 mL flask. What effect would this have on the calculated molar mass of the unknown liquid? Justify your answer with an explanation.

3. Using Excel, prepare a XY plot of pressure versus inverse volume using the data collected in Part B of the experiment. Perform linear regression analysis on this data. (Submit the plot along with your Lab Report.) Do the results support your conclusions in Part B of the experiment? Justify your answer with an explanation.

Colligative Properties

PURPOSE

Identify an unknown, nonelectrolyte compound utilizing freezing point depression and an unknown, electrolyte compound by determining the van't Hoff factor, *i*.

INTRODUCTION

Colligative properties are physical properties of solutions that depend on the total number of solute particles, but not the kind of particles, dissolved in solution. Colligative properties include boiling point elevation, freezing point depression, increasing osmotic pressure, and vapor pressure lowering. A solution (a solute dissolved in a solvent) will always have a higher boiling point, a lower freezing point, a higher osmotic pressure, and a lower vapor pressure than that of the pure solvent.

The degree to which the colligative properties of a solution are affected is directly related to the number (concentration) of solute particles in the solution. Solutions that contain larger numbers of solute particles will have higher boiling points and lower freezing points than solutions that contain smaller numbers of solute particles. In chemistry, the concentration of a solution is generally expressed in terms of molarity or molality to indicate the number of solute particles in solution.

A **nonelectrolyte** is a substance that does not conduct electricity when dissolved in solution. In general, nonelectrolytes are molecular species that produce one mole of solute particles per mole of compound dissolved in solution. For example, a 0.1 *m* aqueous glucose ($C_6H_{12}O_6$) solution has a freezing point depression of 0.186 °C. A 0.2 *m* aqueous glucose solution has a freezing point depression of 0.372 °C because it contains twice as many glucose molecules as a 0.1 *m* solution.

A strong **electrolyte** is a substance that conducts electricity well when dissolved in solution. Strong electrolytes ionize or dissociate into two or more moles of particles (ions) per mole of compound dissolved in solution. Table salt, NaCl, is an example of a strong electrolyte. The freezing point depression of a 0.1 *m* aqueous NaCl solution, 0.348 °C, is approximately twice that of a 0.1 *m* aqueous glucose solution because one mole of NaCl dissociates into two moles of solute particles (Na^+ and Cl^- ions).

As a molecular liquid is cooled, its molecules move slower and approach each other more closely. At the freezing point of a liquid, the attractive forces among its molecules are greater than their kinetic forces (ability to move), causing a change in phase from the liquid to the solid state. The **freezing point** is defined as the temperature at which the liquid state and the solid state co-exist in equilibrium.

The freezing point depression (ΔT_f) of a solution is equal to the freezing point of the pure solvent (T_f^o) minus the freezing point of the solution (T_f).

$$\Delta T_f = T^o_{f\ \text{pure solvent}} - T_f \qquad \text{(Eq. 1)}$$

(ΔT_f *is a positive value because the freezing point of the solvent is always greater than the freezing point of the solution*). The freezing point depression of a solution containing a nonvolatile, nonelectrolyte solute is directly related to the product of the molality (m) of the solution and the molal freezing point depression constant (K_f) for the solvent. K_f has units of °C/m.

$$T_f = K_f\ m \qquad \text{(Eq. 2)}$$

Because the temperature of the solution changes, molality (m) is the concentration unit of choice. Molarity (M,) cannot be used as the unit of concentration because it changes with temperature due to the expansion and contraction of the volume of the solution. **Molality** is defined as the moles of solute dissolved per kilogram of solvent.

$$m = \frac{\text{moles of solute}}{\text{kg of solvent}}$$

Each solvent has a unique K_f value. Table 1 lists K_f values for some common solvents.

In Part A of this experiment, the freezing point of the pure solvent will be determined. In Part B, an unknown, nonelectrolyte compound will be identified by determining its molar mass. In Part C of the experiment, an unknown, electrolyte compound will be identified by determining its van't Hoff factor, *i*.

Table 1 K_f *Values for some common solvents*

acetic acid	3.90 °C/m
benzene	5.12 °C/m
camphor	40.0 °C/m
cyclohexane	20.0 °C/m
ethanol	1.99 °C/m
water	1.86 °C/m

Determination of the Molar Mass of an Unknown, Nonelectrolyte Compound from Freezing Point Depression

The freezing point depression of an unknown solute will be used to calculate the solute's molar mass. There are two factors that must be considered when performing this type of analysis: 1) the solute must be nonvolatile in the temperature range of the freezing point of the solvent, and 2) this type of analysis is generally limited to nonelectrolyte (molecular) solutes.

The first step of this process is the determination of the freezing point of the solvent (water) using a MeasureNet temperature probe. The freezing point of water at standard conditions is 0.00 °C. The observed freezing point of water may be slightly above or slightly below 0.00 °C (± 0.50 °C, see Figure 1). These differences may be attributed to the purity of the water, the calibration of the temperature probe, and atmospheric pressure.

Next, a known mass of the unknown solute is added to a specified mass of water and the freezing point of the solution is determined (see Figure 1). (*To obtain accurate results, the solute must be completely dissolved before the solution is frozen*). The freezing point depression (ΔT_f) of the solution is calculated. Having determined ΔT_f, the molar mass of the unknown solute can be calculated. Table 2 gives the molar masses of some compounds that are nonelectrolytes.

Determination of the van't Hoff Factor of an Unknown, Electrolyte Compound

If the dissociation of NaCl in water were 100%, a 0.100 m aqueous NaCl would have an *effective* molality of 0.2 m (0.100 m Na^+ + 0.100 m Cl^-). The solution would be *expected* to have a freezing point depression of 0.372 °C (0.200 m × 1.86 °C/m). However, the observed freezing point depression is actually 0.348 °C.

How do we account for the difference between the *expected* and the *observed* freezing point depression of a 0.1 m aqueous NaCl solution? Ions in solution are in constant, random motion. Occasionally, ions with opposite charges collide with each other and re-associate to form ion pairs (i.e., Na^+Cl^-). This temporary association reduces the total number of particles in solution, causing a reduction in colligative properties. (During the time the ion pair is associated, two particles are effectively reduced to

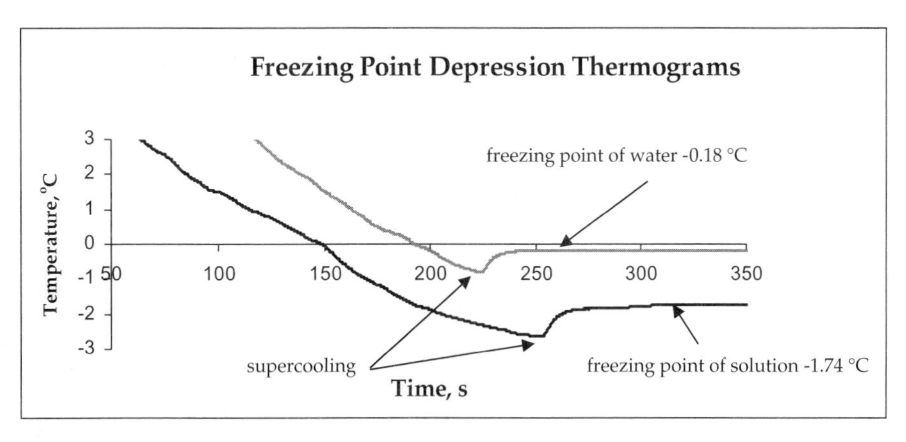

Figure 1
Thermograms for pure water and for an aqueous solution of an unknown, nonelectrolyte compound. Note that pure water and the solution exhibit supercooling. Supercooling results when heat is removed from a solution too rapidly for the molecules to order themselves into a solid. Determine the solution's freezing at the beginning of the flat region on the curve.

Table 2 *List of nonelectrolyte compounds*

urea	$(NH_2)_2CO$
Carbohydrates	
2-deoxy-D-ribose	$C_5H_{10}O_4$
D-ribose	$C_5H_{10}O_5$
glucose	$C_6H_{12}O_6$
sucrose	$C_{12}H_{22}O_{11}$
Amino Acids	
alanine	$NH_2CH(CH_3)CO_2H$
glycine	$NH_2CH_2CO_2H$
phenylalanine	$NH_2CH(CH_2C_6H_5)CO_2H$

one). Ion pair formation is more prevalent in solutions with higher concentrations causing greater deviations from the expected colligative properties. Solutions containing multivalent ions (Ca^{2+}, Fe^{3+}, CO_3^{2-}, etc.) are more likely to form ion pairs than solutions containing monovalent ions (K^+, Br^-, etc.).

The **van't Hoff factor,** ***i***, indicates the extent of dissociation of strong electrolytes or the extent of ionization of weak electrolytes in aqueous solution. The van't Hoff factor is the ratio of the experimentally observed colligative property to the value for a nonelectrolyte solution of the same concentration.

$$i = \frac{\Delta T_{f(observed)}}{\Delta T_{f(predicted\ nonelectrolyte)}} = \frac{T_f\, m_{effective}}{T_f\, m_{stated\ concentration}} = \frac{m_{effective}}{m_{stated\ concentration}} \qquad \text{(Eq. 3)}$$

The effective molality ($m_{effective}$) represents the total number of particles in solution, and $m_{stated\ concentration}$ is the concentration of the compound in solution assuming no dissociation.

For nonelectrolyte solutes (e.g., urea, sucrose, etc.) the value of *i* is 1. In infinitely dilute solutions in which no appreciable ion association occurs, the *i* value for a strong electrolyte such as NaCl should be 2 (NaCl dissociates into two ions, Na^+ and Cl^-), and for a strong electrolye such as $FeCl_3$, *i* should be 4 ($FeCl_3$ dissociates into four ions, 1 Fe^{3+} and 3 Cl^-). The *i* value for a 0.1 *m* aqueous NaCl solution is less than 2 due to ion association in solution. The observed *i* value for a 0.1 *m* aqueous NaCl solution is 1.87 and the observed *i* value for a 0.1 *m* aqueous $FeCl_3$ solution is 3.40. Equation 2 can be modified to calculate the freezing point depression of a strong electrolyte solution by including the van't Hoff factor, *i*.

$$\Delta T_f = i\, K_f\, m \qquad \text{(Eq. 4)}$$

Table 3 *van't Hoff factors for some 0.100 m electrolyte solutions at 25 °C*

$MgCl_2$	2.70
$MgSO_4$	1.21
KBr	1.88
K_2SO_4	2.32
$FeCl_3$	3.40

Equation 4 can be used to calculate $\Delta T_{f\ predicted}$ for the solution using the stated solution concentration. To calculate $\Delta T_{f\ observed}$, subtract the freezing point of the solution from the freezing point of water (pure solvent).

Table 3 list known *i* values for some strong electrolytes. The van't Hoff factor can also be calculated from experimental measurements for weak acid or weak base solutions and used to determine their percent ionization. This method will be utilized in Experiment 28.

PROCEDURE

CAUTION

Students must wear departmentally approved eye protection while performing this experiment. Wash your hands before touching your eyes and after completing the experiment.

Part A – Determination of the Freezing Point of the Pure Solvent

1. See Appendix A-1 – **Instructions for Initializing the MeasureNet Workstation to Record a Thermogram**. Complete all steps in Appendix A-1. Set the *min* temperature to −10 °C and the *max* temperature to 25 °C.
2. Prepare an ice water bath by half filling a 600-mL beaker with ice and water. Add 15 to 20 g of NaCl or $CaCl_2$ to the water and stir the solution to dissolve the salt. Fill the remainder of the beaker with ice. (More salt and ice may be added to the ice bath during the experiment if the ice bath temperature is not sufficiently cold to freeze the solution).
3. Assemble the stopper/wire stirrer/temperature probe assembly shown in Figure 2 on next page. Spread the slit hole in a 2-hole rubber stopper and insert the temperature probe. Insert the wire stirrer (one end has a loop) through the other hole in the stopper. *Make certain that the temperature probe passes through the wire loop.*
4. Secure the rubber stopper/wire stirrer/temperature probe assembly to a ring stand using a utility clamp.
5. Add approximately 15 to 20 mL of distilled water to a 25 × 200 mm test tube.

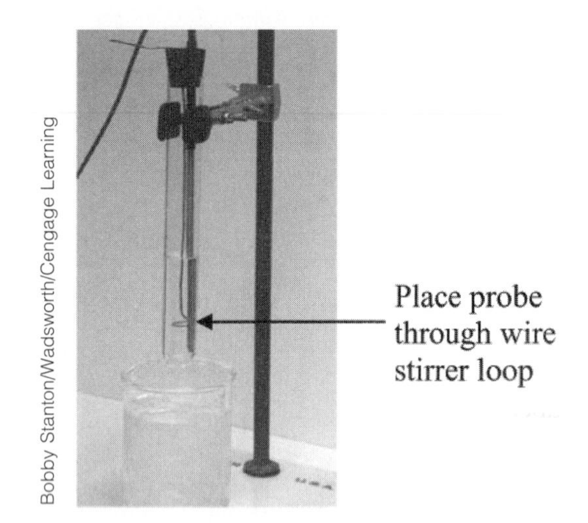

Figure 2
Stirrer/temperature probe assembly

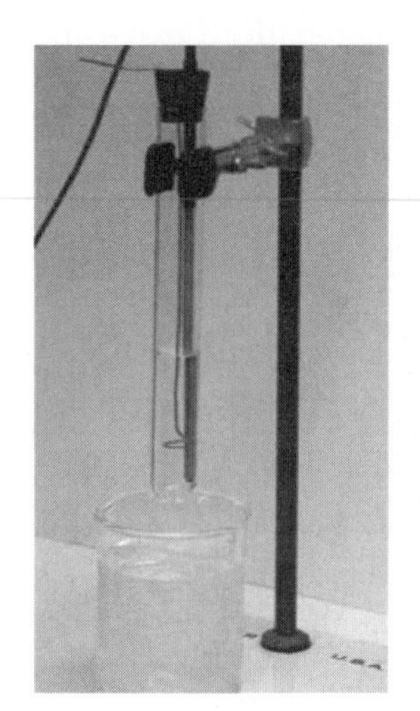
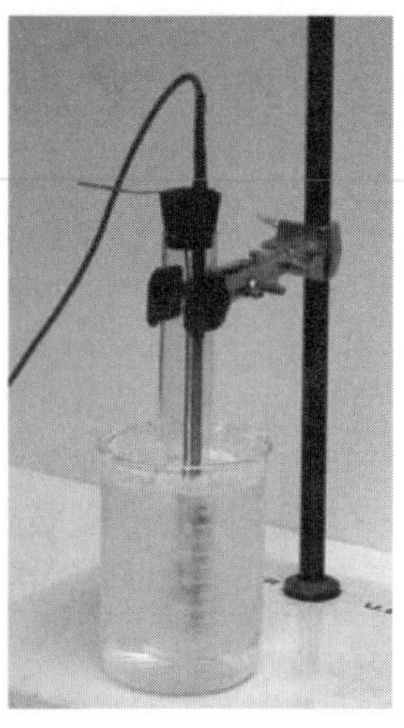

Figure 3
Test tube containing water and the temperature probe/stirrer assembly before and after immersion in an ice bath

6. Insert the temperature probe/wire stirrer assembly into the water in the test tube (Figure 3). Be sure the bottom 1.3 cm–1.9 cm of the temperature probe is covered with water. Snuggly fit the stopper into the test tube.
7. Insert the test tube/temperature probe/stirrer assembly into the ice bath (be sure all of the liquid in the test tube is below the water level in the beaker), as shown in Figure 3. Re-secure the test tube/temperature probe/stirrer assembly to the ring stand. Press **Start** on the Workstation to begin recording the thermogram.

   Should you constantly or occasionally stir the water (by moving the wire stirrer up and down in the test tube) until the water is frozen? When the temperature remains constant for at least 20 to 30 seconds on the Workstation display, the water is frozen.
8. Once the water is frozen, press **Stop** on the MeasureNet Workstation. Press **File Options**, then press **F3**. Enter a 3 digit number to record a

file name for the temperature versus time scan, press **Enter**. Should you record the file name in the Lab Report?

9. Press **Display** to clear the previous scan.
10. Remove the test tube/temperature probe/ stirrer assembly from the ice water bath, and re-secure it to the ring stand. Completely thaw the ice in the test tube. Repeat Steps 7–9 to obtain a second thermogram for determining the freezing point of water.
11. Remove the test tube/temperature probe/ stirrer assembly from the ice water bath, and re-secure it to the ring stand. Completely thaw the ice and decant the water into a laboratory sink. Thoroughly dry the test tube.

Part B – Determination of the Freezing Point of a Nonelectrolyte, Unknown Solution

12. Obtain an unknown, nonelectrolyte (solid) compound from your instructor. Should you record the Unknown Number in the Lab Report?
13. Prepare an aqueous solution of the unknown solid in a 25 × 200 mm test tube. You will need to add approximately 1 gram of sample to approximately 20 grams of water.
14. Should you determine the exact mass of the solid and the exact mass of water in the solution? If so, should you record these masses in the Lab Report and to how many significant figures? (*Use a sheet of weighing paper or a weighing boat to weigh the unknown. DO NOT use a piece of porous filter paper to weigh the unknown*).
15. Stir the solution in the test tube until the unknown solid is *completely* dissolved.
16. Insert the temperature probe/wire stirrer assembly into the solution. Be sure the bottom 1.3 cm–1.9 cm of the temperature probe is covered with solution. Snuggly fit the stopper into the test tube.
17. Secure the test tube/temperature probe/stirrer assembly inside the ice bath. Press **Start** on the Workstation to begin recording the thermogram. Should you constantly or occasionally stir the solution until frozen?
18. Once the solution is frozen, press **Stop** on the MeasureNet Workstation. Press **File Options**, then press **F3**. Enter a 3 digit number to record a file name for the thermogram, press **Enter**.
19. Press **Display** to clear the previous scan.
20. Remove the test tube/temperature probe/ stirrer assembly from the ice water bath, and re-secure it to the ring stand. Remove the frozen solution from the test tube and discard it in the sink.
21. Thoroughly dry the test tube and temperature probe/stirrer with a towel.
22. Perform a second trial by repeating Steps 12–21 to determine the molar mass of the unknown compound.
23. *Steps 24–29 are to be completed after the conclusion of the experiment. Proceed to step 30.*
24. From the tab delimited files you saved, prepare thermograms for pure water and for the solution (both curves must be included in the same plot) for each trial. Instructions for plotting multiple thermograms within the same plot using Excel are provided in **Appendix B-3**.

25. What are the freezing point of the water and the freezing point of the solution for each trial? Should these temperatures be recorded in the Lab Report?
26. Should you determine the freezing point depression, T_f, for the unknown, nonelectrolyte solution, and record it in the Lab Report for each trial?
27. How do you determine the molar mass of the unknown, nonelectrolyte solid for each trial?
28. What is the average molar mass for the unknown solid?
29. Identify the unknown, nonelectrolyte compound (see Table 2).

Part C – Determination of the van't Hoff Factor for a Strong, Electrolyte Unknown

30. Obtain 15 to 20 mL of a 0.200 *m* unknown, ionic compound solution. Should you record the Unknown Number in the Lab Report.
31. Should you experimentally determine the freezing point of water a second time as was done in Steps 5–9? Why or why not?
32. Record a thermogram to determine the freezing point of the ionic compound solution.
33. Once the solution is frozen, be sure to save the thermogram to the MeasureNet computer.
34. Press **Display** to clear the previous scan.
35. Remove the test tube/temperature probe/ stirrer assembly from the ice water bath, and re-secure it to the ring stand. Remove the frozen solution from the test tube. Thaw the solution in the test tube and decant it into a laboratory sink.
36. Perform a second trial to determine the freezing point of the unknown, ionic compound solution.
37. Thaw the frozen solution and discard it into a lab sink. Thoroughly dry the test tube and temperature probe/stirrer with a towel.
38. *Steps 39–44 are to be completed after the conclusion of the experiment. Proceed to Step 45.*
39. From data in the tab delimited files, use Excel to prepare thermograms for the data collected in Steps 30–36.
40. What are the freezing points for water and the unknown ionic solution for each trial? Should you record them in the Lab Report and to how many significant figures?
41. How do you determine freezing point depression for the ionic solution for each trial? Should you record them in the Lab Report and to how many significant figures?
42. How do you determine the van't Hoff factor for each trial?
43. What is the average van't Hoff factor for the unknown, ionic solution?
44. Identify the unknown, ionic compound (see Table 3).
45. When you are finished with the experiment, transfer all files from the computer to a flash drive, or email the files to yourself via the internet.
46. Turn off the MeasureNet workstation before leaving the lab.

Name ____________ Section ______ Date ______

Instructor ____________

17 EXPERIMENT 17

Lab Report

Part B – Determination of the Freezing Point of a Nonelectrolyte, Unknown Solution

Unknown number of nonelectrolyte compound ______________

Data and observations for Trial 1.

Data and observations for Trial 2.

What are the freezing point of the water and the freezing point of the solution for each trial?

Freezing point depression, ΔT_f, for the unknown, nonelectrolyte solution.

How do you determine the molar mass of the unknown, nonelectrolyte compound for each trial?

What is the average molar mass for the unknown compound?

Identity of the unknown, nonelectrolyte compound. ____________________

Part C – Determination of the van't Hoff Factor for a Strong, Electrolyte Unknown

Unknown number of ionic solution ____________________

What are the freezing points for water and the unknown ionic solution for each trial?

How do you determine the freezing point depression for the ionic solution for each trial?

How do you determine the van't Hoff factor for each trial?

43. What is the average van't Hoff factor for the unknown, ionic solution?

Identify the unknown, ionic compound.____________________

Name

Instructor

Section

Date

17

EXPERIMENT 17

Pre-Laboratory Questions

1. The freezing point of 20.024 grams of distilled water was determined by recording a thermogram (see plot below). 1.006 grams of an unknown, nonelectrolyte compound was added to the water, and a second thermogram was recorded to determine the freezing point of the solution.

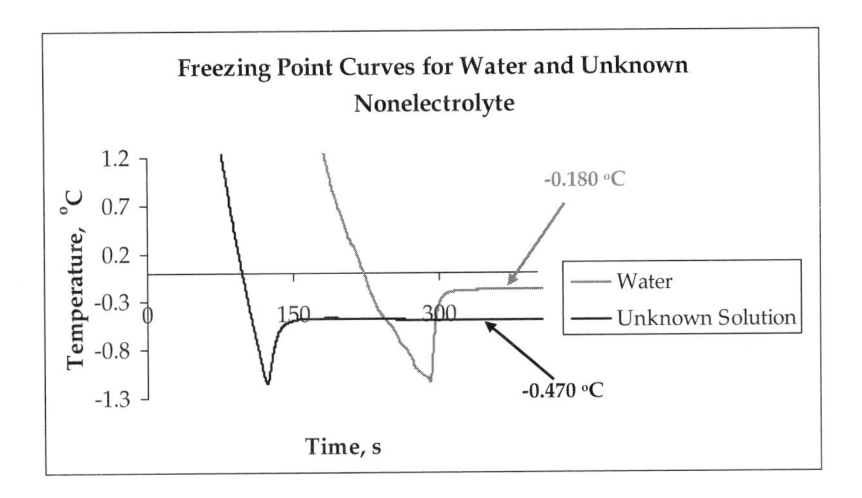

a. Determine the freezing point depression, T_f, for the unknown solution.

b. Determine the molality, *m*, of the unknown solution.

c. Determine the moles of the nonelectrolyte compound in the solution.

d. Calculate the molar mass of the nonelectrolyte compound in grams per mole.

e. Identify the nonelectrolyte compound (see Table 2). ____________________

2. The freezing point of distilled water was determined by recording a thermogram (see plot below). The freezing point of a 0.200 *m* solution containing an unknown, ionic compound was determined by recording a second thermogram.

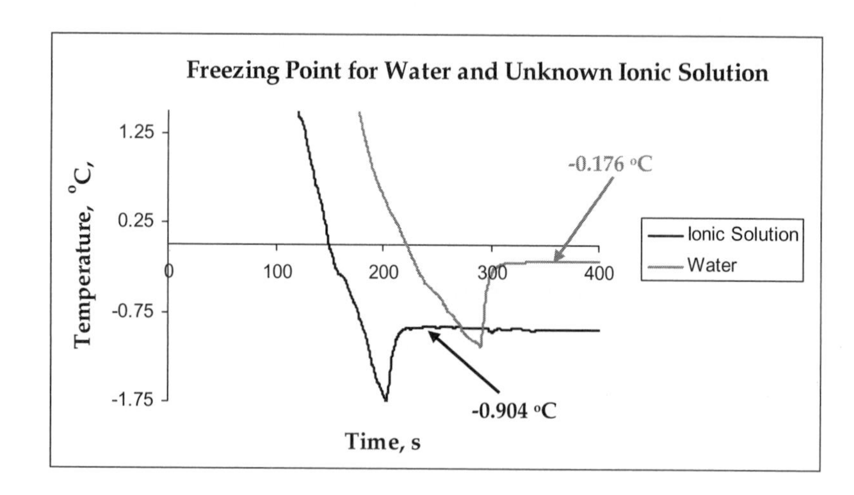

a. Determine the freezing point depression, T_f, for the ionic solution.

b. Determine $m_{effective}$ for the ionic solution.

c. Determine the van't Hoff factor, *i*, for the ionic compound.

Identify the ionic compound (see Table 3). ____________________

Name | Section | Date

Instructor

17 EXPERIMENT 17

Post-Laboratory Questions

1. A student mistakenly used an ethanol solution containing an unknown ionic compound in Part C of the experiment. How would this affect the van't Hoff factor determined experimentally? Justify your answer with an explanation.

2. In Step 5 of the experiment, a student mistakenly used 0.050 *M* HCl solution instead of distilled water. His lab partner performed the remaining parts of the experiment and used distilled water as instructed.

   A. Would the molar mass of the unknown determined by this group of students in Part B be higher or lower than the correct value? Justify your answer with an explanation.

   B. Would the van't Hoff factor determined by this group of students in Part C be higher or lower than the correct value? Justify your answer with an explanation.

3. In a general chemistry lab, a student accidentally added $CaCl_2$ into the carboy containing distilled water to be used by all students.

A. Would the molar mass of the unknown determined in Part B be higher or lower than the correct value? Justify your answer with an explanation.

B. Would the van't Hoff factor determined in Part C be higher or lower than the correct value? Justify your answer with an explanation.

Soaps and Detergents

PURPOSE

Prepare a soap and compare its properties to that of a synthetic detergent. Compare the relative cleansing ability and costs of soap and some detergents.

INTRODUCTION

One of the oldest known chemical reactions is the preparation of soap. The making of soap, or **saponification**, dates back to 600 BC, after humans realized how to control fire and make alcoholic beverages. Soap is a generic term for the sodium or potassium salts of long-chain organic acids (fatty acids) made from naturally occurring esters in animal fats and vegetable oils. All organic acids contain the RCO_2H functional group, where R is a shorthand notation for methyl, CH_3-, ethyl CH_3CH_2-, propyl, $CH_3CH_2CH_2$-, or more complex hydrocarbon chains called **alkyl groups**. Chemists use the R shorthand notation because these groups can be very large and the hydrocarbon chain has little effect on the compound's chemical reactivity. All esters contain the RCO_2R functional group.

The R groups in soaps are hydrocarbon chains that generally contain 12 to 18 carbon atoms. Sodium salts of fatty acids such as lauric (vegetable oil), palmitic (palm oil), and stearic (animal fat) acids are just a few examples of soaps.

$CH_3(CH_2)_{10}COONa$ sodium laurate
$CH_3(CH_2)_{14}COONa$ sodium palmitate
$CH_3(CH_2)_{16}COONa$ sodium stearate

The hydrocarbon chain in soaps may contain saturated (no double bonds) or unsaturated chains (contains double bonds). Sodium salts are usually solids, therefore, most bars of soap are made of sodium salts. Potassium salts are the basis of liquid soaps, shaving creams, and greases. Fats and vegetable oils are triglycerides. A **triglyceride** is an ester derived from glycerol and three fatty acids. A triglyceride made from three lauric acid molecules is shown in Figure 1.

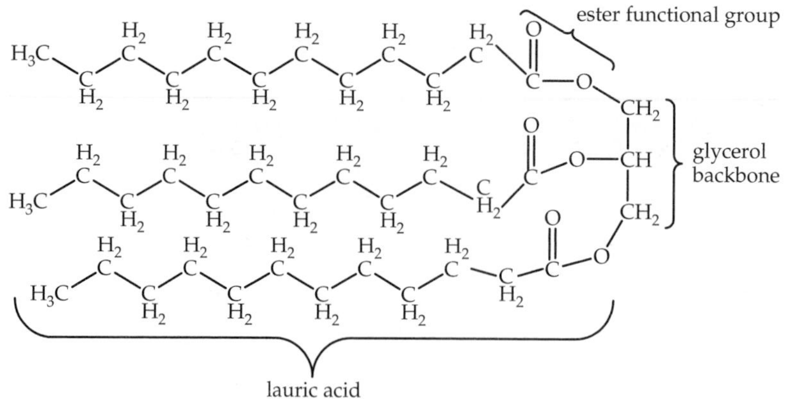

ester functional group

glycerol backbone

lauric acid

Figure 1
A triglyceride molecule made from lauric acid and glycerol.

Saponification is the base hydrolysis of an ester producing a carboxylic acid salt and an alcohol (Eq. 1). A lone pair of electrons on the OH^- ion is attracted to the partially positively charged C atom in the C=O bond in the ester (Eq. 1). The C—OR′ bond breaks generating a carboxylic acid (RCO_2H) and an alcohol (R′OH). In the presence of NaOH, carboxylic acids are converted to their sodium salts ($RCO_2^-Na^+$).

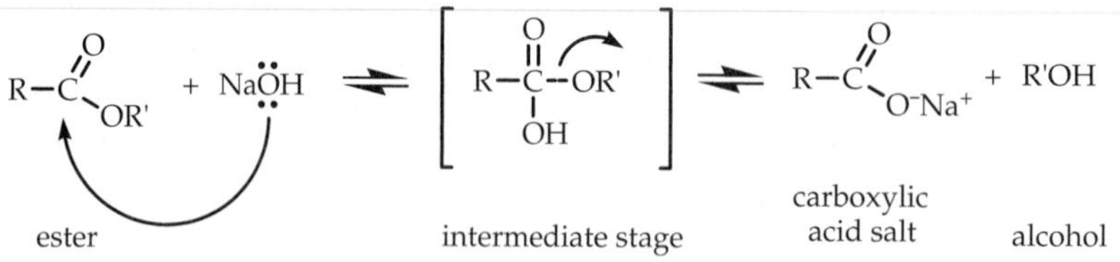

$$\underset{\text{ester}}{R{-}C({=}O){-}OR'} + NaOH \rightleftharpoons \underset{\text{intermediate stage}}{\left[R{-}C(O)(OH){-}OR'\right]} \rightleftharpoons \underset{\text{carboxylic acid salt}}{R{-}C({=}O){-}O^-Na^+} + \underset{\text{alcohol}}{R'OH} \qquad \text{(Eq. 1)}$$

When a triglyceride is saponified, three fatty acid salts (soaps) and glycerol are produced as shown in Equation 2.

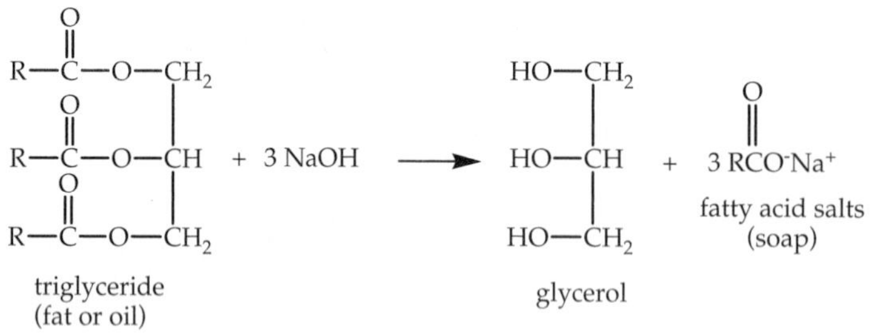

$$\underset{\text{triglyceride (fat or oil)}}{\begin{matrix} R{-}C({=}O){-}O{-}CH_2 \\ | \\ R{-}C({=}O){-}O{-}CH \\ | \\ R{-}C({=}O){-}O{-}CH_2 \end{matrix}} + 3\,NaOH \longrightarrow \underset{\text{glycerol}}{\begin{matrix} HO{-}CH_2 \\ | \\ HO{-}CH \\ | \\ HO{-}CH_2 \end{matrix}} + \underset{\text{fatty acid salts (soap)}}{3\,RC({=}O)O^-Na^+} \qquad \text{(Eq. 2)}$$

The R groups in the triglyceride may or may not have the same chain length (same number of carbons). Thus, different types of soaps may be produced from the saponification of a particular triglyceride. American pioneers made soap from animal fat (triglycerides) from slaughtered cows, pigs, and sheep, and a base consisting of either commercially prepared lye (NaOH or KOH) or sodium carbonate (Na_2CO_3) from the ashes of wood fires.

Soap is the salt of a weak acid. Most organic acids are weak acids. Consequently, hydrolysis occurs to some extent when soap dissolves in water. Soap solutions tend to be slightly alkaline (basic) due to partial hydrolysis of the acid (Eq. 3).

$$R{-}C(=O)O^{-}Na^{+} + H_2O \rightleftharpoons R{-}C(=O)OH + NaOH \quad \text{(Eq. 3)}$$

The cleansing action of soaps results from two effects. Soaps are wetting agents that reduce the surface tension of water, allowing the water molecules to encounter the dirty object. They are also emulsifying agents. "Dirt" frequently consists of a grease or oil along with other organic species. In general, organic compounds are nonpolar. Water is a polar species. These two substances will not dissolve in each other because of their dissimilar characteristics (the "Like Dissolves Like" rule). Soaps cross the boundary between polar and nonpolar because they contain a nonpolar hydrophobic (*water-hating*) end and a polar hydrophilic (*water-loving*) end as shown in Figure 2.

Because soaps have both polar and nonpolar regions in the molecule, they are soluble in both polar and nonpolar species. The hydrophobic (non-polar) portion of soap is soluble in nonpolar compounds like grease and oils. The hydrophilic (polar) end dissolves in water. Soap molecules surround the grease and oils and break them up into microscopic droplets that can remain suspended in the water. These suspended microscopic droplets are called **micelles** (Figure 3). Micelles contain very small amounts of oil or grease in their center. Thus, the oil or grease has been

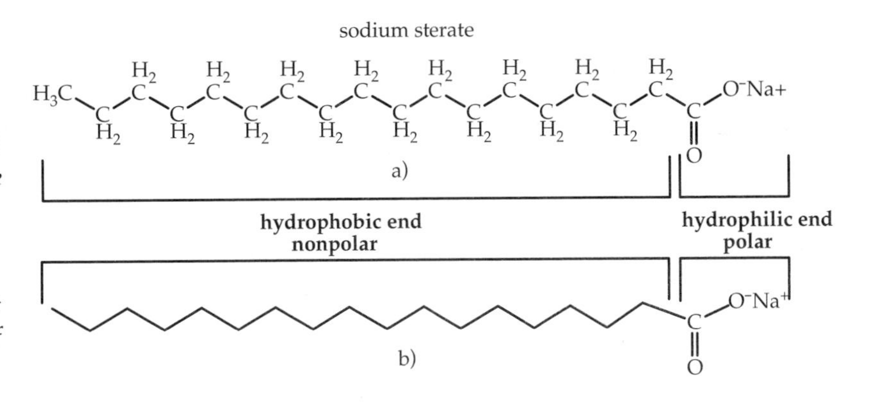

Figure 2
Molecular structure, a), and line drawing, b), of sodium stearate. In a line drawing, all carbon and hydrogen atoms are omitted at the intersection of each line as a shorthand method of drawing the molecule. It is understood that the C and H atoms are part of the molecule

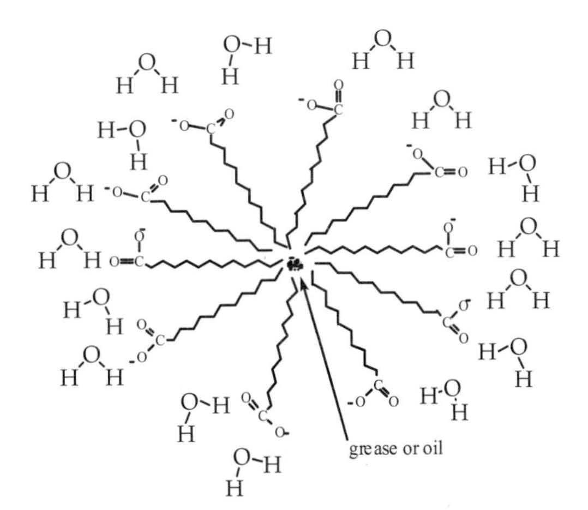

Figure 3
Formation of a micelle

dissolved in water forming an **emulsion**, one form of a colloidal suspension in water.

Water supplies in certain areas are acidic as a result of acid rain or pollution, or "hard" due to the dissolved mineral content. Both acidic and "hard" water reduce the cleansing action of soap. Soap is the salt of a weak acid. In the presence of a stronger acid, the sodium salt is converted to an insoluble organic acid (Eq. 4).

$$\underset{\text{soluble soap}}{CH_3(CH_2)_{16}{-}\overset{O}{\overset{\|}{C}}{-}O^-Na^+} + H^+ \longrightarrow \underset{\text{insoluble acid}}{CH_3(CH_2)_{16}{-}\overset{O}{\overset{\|}{C}}{-}OH} + Na^+ \quad \text{(Eq. 4)}$$

"Hard water" contains dissolved Ca^{2+}, Mg^{2+}, and Fe^{3+} ions from the minerals that the water passes over. Normally, soaps made from sodium and potassium fatty acid salts are soluble in water. However, in the presence of these metal ions, the Na^+ and K^+ soluble salts convert to insoluble Ca^{2+}, Mg^{2+}, or Fe^{3+} salts (Eq. 5).

$$\underset{\text{soluble fatty acid salt (soap)}}{2\,CH_3(CH_2)_{16}{-}\overset{O}{\overset{\|}{C}}{-}O^-Na^+} + Ca^{2+} \longrightarrow \underset{\text{insoluble fatty acid salt}}{(CH_3(CH_2)_{16}{-}\overset{O}{\overset{\|}{C}}{-}O)_2Ca^{2+}} + 2\,Na^+ \quad \text{(Eq. 5)}$$

In either acidic or "hard" water, the soluble soaps form insoluble salts that become a scummy ring on bathtubs and black areas on shirt collars. The cleansing ability of soap is reduced because soap molecules are removed from solution. There are several techniques used to circumvent the problems generated by hard water. Water can be "softened" via removing hard water ions from solution using ion exchange techniques or by adding water-softening agents, such as sodium phosphate (Na_3PO_4) or sodium carbonate (Na_2CO_3). Water-softening agents react with Ca^{2+}, Mg^{2+}, and Fe^{3+} ions, removing them from water (Eqs. 6 and 7) and preventing the reaction of these ions with soap (Eqs. 4 and 5).

$$3\,Ca^{2+}{}_{(aq)} + 2\,PO_4^{3-}{}_{(aq)} \rightarrow Ca_3(PO_4)_{2(s)} \quad \text{(Eq. 6)}$$

$$Mg^{2+}{}_{(aq)} + CO_3^{2-}{}_{(aq)} \rightarrow MgCO_{3(s)} \quad \text{(Eq. 7)}$$

Chemists recognized that it was not always possible to "soften" water and began looking at other methods of improving cleansing properties in "hard" water. In the 1940's and 50's, several chemical companies such as Proctor and Gamble introduced the first synthetic detergents or **syndets**. Syndets differ from soaps in that the nonpolar fatty acids groups are replaced with alkyl or aryl sulfonic acids (RSO_3H). The alkyl or aryl sulfonic acids have long chains of carbon atoms giving the hydrophobic (nonpolar) end. The salt of the sulfonic acid (sulfonate) group forms the hydrophilic end of the molecule. The difference in polar groups is one of the key distinctions between a soap and a synthetic detergent. Syndets form micelles and cleanse in the same manner as soaps. Two examples of synthetic detergents are shown in Figure 4. Synthetic detergents have

$H_3C-(CH_2)_{10}-CH_2-SO_3^-Na^+$

sodium lauryl sulfonate

$H_3C-(CH_2)_{10}-CH_2-C_6H_4-SO_3^-Na^+$

sodium 4-laurylphenylsulfonate

Figure 4
Examples of synthetic detergents

virtually replaced all soaps in homes today. The sole remaining commercially available soap is Ivory®.

Because sulfonic acid is a stronger acid than carboxylic acids, syndets do not precipitate in acidic solutions. Furthermore, alkyl and aryl sulfonates do not form insoluble salts in the presence of the typical hard water ions Ca^{2+}, Mg^{2+}, and Fe^{3+}. Thus, synthetic detergents remain soluble in both acidic and "hard" water.

In the Part A of this experiment, we will prepare a soap from the saponification of a vegetable oil with sodium hydroxide. In Part B of the experiment, we will compare the properties of the prepared soap and commercially available detergents. In Part C, we will compare the cleansing abilities of the prepared soap and two detergents.

PROCEDURE

CAUTION

Students must wear departmentally approved eye protection while performing this experiment. Wash your hands before touching your eyes and after completing the experiment.

Chemical Alert

Concentrated sodium hydroxide (NaOH) solution is extremely caustic and corrosive. NaOH may cause serious burns when it contacts skin, and it can cause irreversible cornea damage to the eye. DO NOT remove your eye protection at ANY TIME during this experiment. If NaOH solution contacts the skin, wash the affected area with copious quantities or water and report the accident to the lab instructor. Ethanol is flammable. No open flames are to be used during this experiment.

Part A – Soap Preparation

1. Obtain a 250-mL Erlenmeyer flask. Add 25 mL of vegetable oil, 20 mL of ethanol (ethyl alcohol), and 25 mL of 6 *M* sodium hydroxide solution to the flask. Stir the mixture with a stirring rod to mix the contents in the flask.

2. Heat the 250-mL flask in a boiling-water bath inside of a 600-mL beaker (see Figure 5).
3. Stir the mixture continuously during the heating process to prevent the mixture from foaming. If the mixture should foam to the point of nearly overflowing, remove the flask from the boiling-water bath until the foaming subsides, then continue heating. Heat the mixture for 20 – 30 minutes, or until the alcohol odor is no longer detectable.
4. Remove the paste-like mixture from the boiling-water bath and cool the flask in an ice bath for 10–15 minutes.
5. While the flask is cooling, assemble the vacuum filtration apparatus shown in Figure 6. Secure the vacuum flask to a ring stand with a utility clamp to prevent the apparatus from toppling over. Your lab instructor will demonstrate the vacuum filtration method.

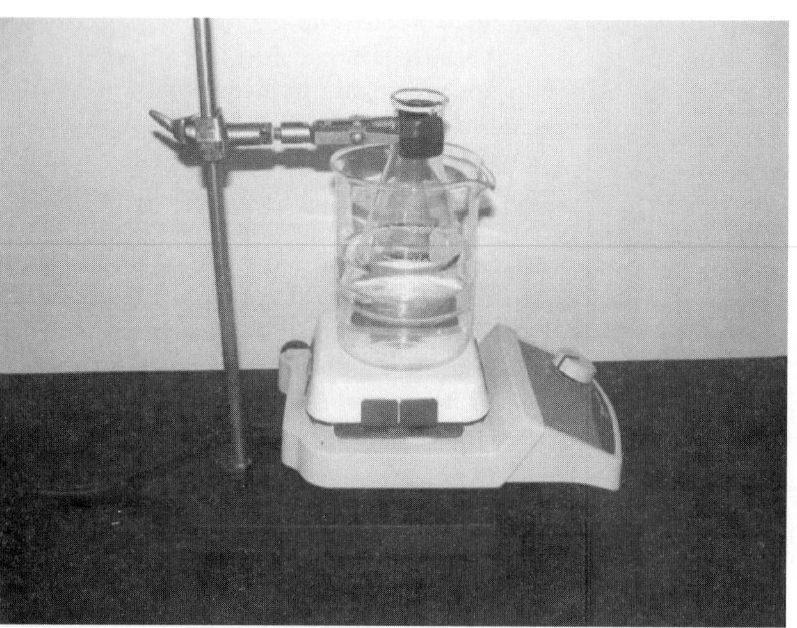

Figure 5
Boiling water bath

Figure 6
Vacuum filtration apparatus

6. Place a piece of filter paper inside the Büchner funnel. Should you determine the mass of the filter paper? If so, to what number of significant figures should the mass be recorded in the Lab Report?
7. Moisten the paper with water so that it fits flush in the bottom of the funnel.
8. Once the flask has cooled, add 150 mL of saturated sodium chloride (NaCl) solution to the flask to "salt out" the soap. What is the purpose of adding excess salt (salting out) to the soap mixture?
9. *Slowly* turn on the water at the aspirator. Pour the mixture from the flask into the Büchner funnel. Once all of the liquid has filtered through the funnel, wash the soap with 10 mL of ice-cold water. Continue the suction filtration until all of the water is removed from the soap. What is the purpose of washing the soap with ice-cold water?
10. Remove the filter paper and soap from the funnel and press it between two paper towels to dry it. Should you determine the mass of the soap recovered? If so, to what number of significant figures should the mass be recorded in the Lab Report?

Part B – Comparison of Soap and Detergent Properties–Precipitation and Emulsification

11. Prepare a stock soap solution by dissolving ~2 g of your prepared soap in 100 mL of boiling, distilled water. Stir the mixture until the soap has dissolved and allow the solution to cool.
12. Repeat Step 11 using ~2 g of synthetic detergent (e.g., Alkanox). Should you record the brand name of the detergent in the Lab Report? When both solutions are cool, determine the pH of each solution. Is using pH paper sufficient for determining the pH of each solution, or should you use the MeasureNet pH probe? Justify your answer with an explanation.
13. Label three test tubes as test tube 1, 2, and 3. Add 4 drops of mineral oil to each test tube. Add ~5 mL of distilled water to test tube 1. Add ~5 mL of stock soap solution to test tube 2. Add ~5 mL of stock synthetic detergent to test tube 3.
14. Mix each solution by shaking and let stand for three to five minutes. Do any of the mixtures emulsify the oil? If an emulsion forms, would you expect to see one layer or two in the solution? Should you record your observations in the Lab Report?
15. Pour the mixtures into the Waste Container. Clean and dry the three test tubes.
16. Label three test tubes as test tube 1, 2, and 3. Place ~2 mL of stock soap solution in each of the three test tubes. Add ~2 mL of 1% $CaCl_2$ solution to test tube 1. Add ~2 mL of 1% $MgCl_2$ solution to test tube 2. Add ~2 mL of 1% $FeCl_3$ solution to test tube 3. Shake each test tube to mix the solutions.
17. Given Ca^{2+}, Mg^{2+}, and Fe^{3+} are "hard water" ions, what should you observe in each test tube prepared in Step 16? Should you record your observations in the Lab Report?
18. Add 4 drops of mineral oil to each of the test tubes in Step 16. Shake each test tube to mix the solutions and let the solutions stand for three to five minutes. Do any of the mixtures emulsify the oil? Should you record your observations in the Lab Report?

19. Repeat steps 16–18 using ~2 mL of stock detergent solution. What should you observe in each test tube? Should you record your observations in the Lab Report?
20. Do any of the mixtures emulsify the oil? Should you record your observations in the Lab Report?
21. Pour the mixtures into the Waste Container.
22. Clean and dry the three test tubes. Place ~5 mL of stock soap solution in one clean test tube and ~5 mL of stock detergent solution in a second test tube. Add 1 *M* HCl one drop at a time to both solutions until the pH in each test tube equals 3. Is pH paper suitable for determining the pH of each solution? Why or why not? Count the number of drops of acid added to each mixture. What should you observe in each test tube? Should you record your observations in the Lab Report?
23. Add 4 drops of mineral oil to each test tube in Step 22. Do any of the mixtures emulsify the oil? Should you record your observations in the Lab Report?

Part C – Comparison of the Cleaning Abilities of a Soap and Two Detergents

24. Clean, dry, and label three beakers. Place ~20 mL of stock soap solution (from Step 11) in the 1st beaker. Place ~20 mL of stock detergent solution (from Step 12) in the 2nd beaker. Place ~20 mL of a commercial liquid detergent (e.g., Joy) in the 3rd beaker.
25. Obtain three cloth test strips that have been soaked in tomato sauce and place one strip in each of the beakers. Place one cloth strip in beaker 1 (from Step 24), one cloth strip in beaker 2, and one cloth strip in beaker 3. Repeatedly stir each solution with a stirring rod for 5 minutes.
26. Remove the cloth strips from the soap and detergent solutions and squeeze out the excess water. Visually compare each cloth strip to determine their relative cleanliness. Should you record your observations in the Lab Report? Indicate which solution cleaned the cloth strip the best.
27. Your lab instructor will inform you of the cost of the detergent solutions. Compare the cleansing abilities of the two detergents and their costs. Which detergent is the best buy?

Name | Section | Date

Instructor

18 EXPERIMENT 18

Lab Report

Part A – Soap Preparation

Should you determine the mass of the filter paper?

What is the purpose of adding excess salt to the soap mixture in Step 3?

Should you determine the mass of the soap recovered?

Part B – Comparison of Soap and Detergent Properties – Precipitation and Emulsification

Should you record the brand name of the detergent in the Lab Report?

Determine the pH of each solution. Is using pH paper sufficient for determining the pH of each solution, or should you use the MeasureNet pH probe? Justify your answer with an explanation.

Part C – Cleansing Comparison of a Soap and Two Detergents

Do any of the mixtures emulsify the oil? If an emulsion forms, would you expect to see one layer or two in the solution?

Given Ca^{2+}, Mg^{2+}, and Fe^{3+} are ''hard water'' ions, what should you observe in each test tube prepared in Step 16?

Do any of the mixtures emulsify the oil?

What should you observe in each test tube containing detergent?

Is pH paper suitable for determining the pH of the soap and detergent solutions with HCl added? Why or why not? What should you observe in each test tube?

Does either the soap or detergent solution emulsify oil?

Part D – Comparison of the Cleaning Abilities of a Soap and Two Detergents

Observations of cloth strips to determine their relative cleanliness. Indicate which solution cleaned the cloth strip the best.

Compare the cleansing abilities of the two detergents and their costs. Which detergent is the best buy?

Name — Section — Date

Instructor

18

EXPERIMENT 18

Pre-Laboratory Questions

4. Complete and balance the following chemical equation for the production of soap. Does this reaction produce a solid or a liquid soap? Justify your answer with an explanation.

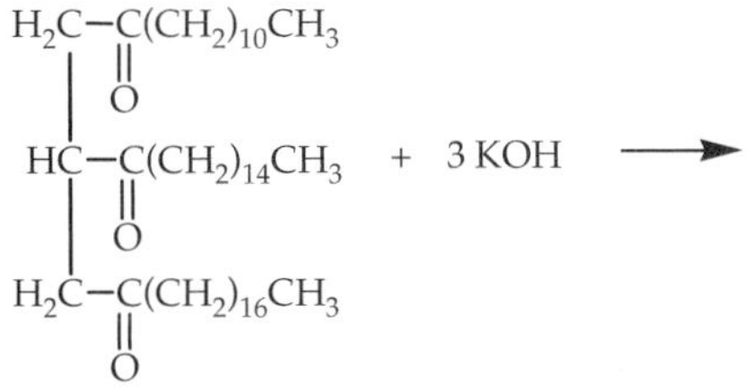

$$\begin{array}{l} H_2C{-}C(=O)(CH_2)_{10}CH_3 \\ \quad | \\ HC{-}C(=O)(CH_2)_{14}CH_3 \\ \quad | \\ H_2C{-}C(=O)(CH_2)_{16}CH_3 \end{array} \; + \; 3\,KOH \longrightarrow$$

4. Complete and balance the following chemical equations. If no reaction occurs, write NR.

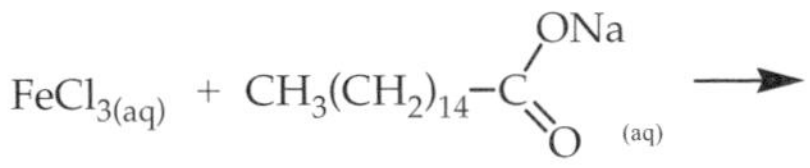

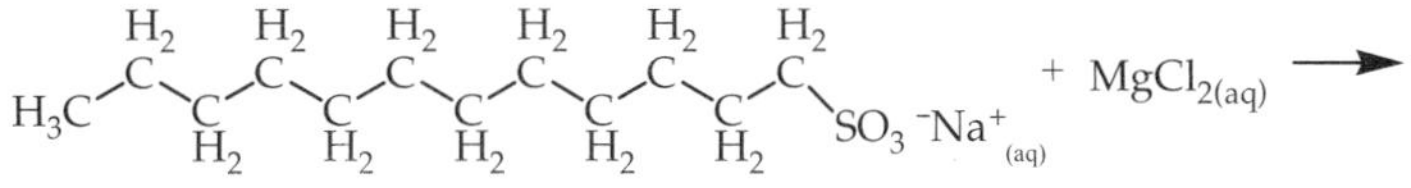

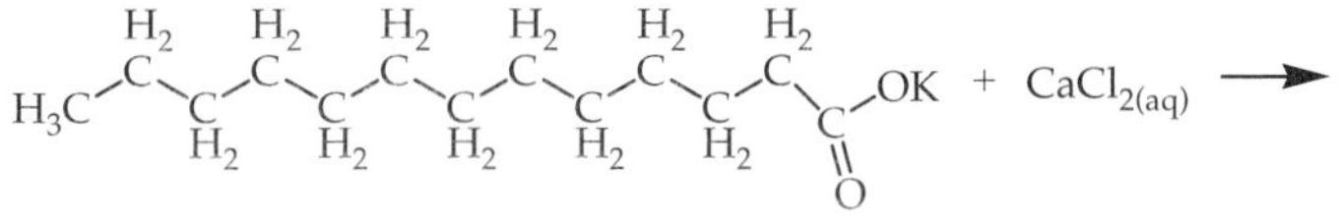

$$FeCl_{3(aq)} + CH_3(CH_2)_{14}{-}C(=O)ONa_{(aq)} \longrightarrow$$

$$H_3C{-}CH_2{-}CH_2{-}CH_2{-}CH_2{-}CH_2{-}CH_2{-}CH_2{-}CH_2{-}CH_2{-}CH_2{-}CH_2{-}SO_3^-Na^+_{(aq)} + MgCl_{2(aq)} \longrightarrow$$

$$H_3C{-}CH_2{-}CH_2{-}CH_2{-}CH_2{-}CH_2{-}CH_2{-}CH_2{-}CH_2{-}CH_2{-}CH_2{-}CH_2{-}C(=O){-}OK + CaCl_{2(aq)} \longrightarrow$$

4. Indicate whether each of the following molecules is a soluble soap, an insoluble soap, or a detergent.

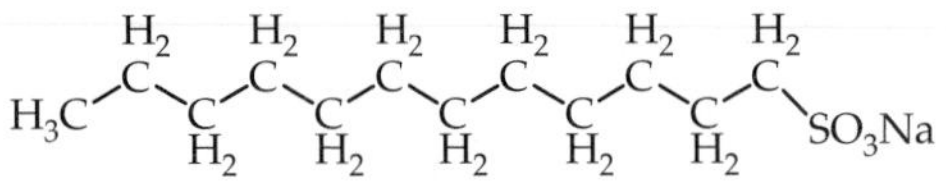

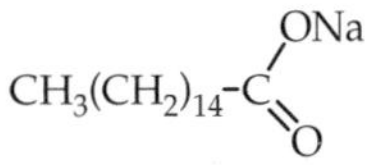

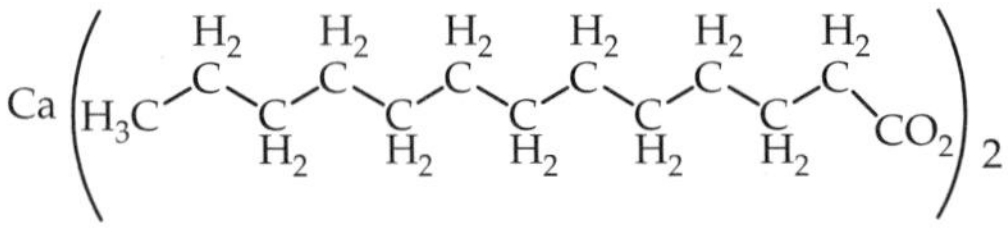

$$H_3C-(CH_2)_{11}-SO_3Na$$

____

$$CH_3(CH_2)_{14}-C(=O)ONa$$

____

$$Ca\left(H_3C-(CH_2)_{11}-CO_2\right)_2$$

____

Name | Section | Date

Instructor

18

EXPERIMENT 18

Post-Laboratory Questions

4. What is the predominant intermolecular force responsible for the dissolution of the hydrophilic end of soap in water? Using dashed lines, illustrate where the intermolecular force occurs between atoms in adjacent molecules? Lone pair electrons have been omitted in the molecules drawn below. Add lone pair electrons to all atoms as needed. Indicate bond polarity, if the molecules contain polar bonds, using δ^+ and δ^- symbols where appropriate.

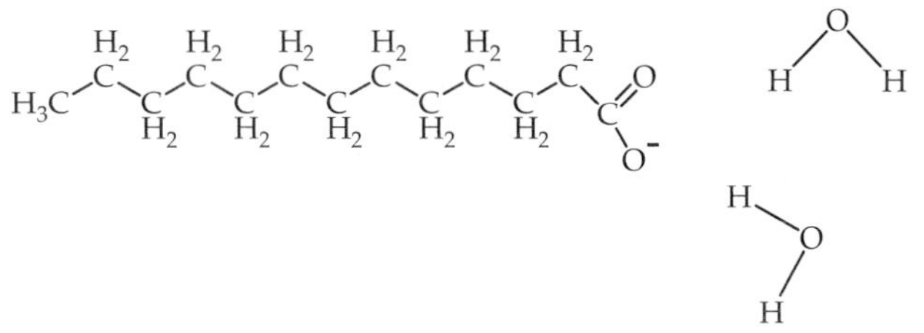

H_3C–CH_2–CH_2–CH_2–CH_2–CH_2–CH_2–CH_2–CH_2–CH_2–CH_2–CH_2–C(=O)O^-

H–O–H

H–O–H

____

2. How does soap acts as an emulsifying agent?

3. A hard water sample contains 121 mg of $CaCO_3$ per liter of water. Assuming the reaction goes to completion, calculate the mass of Na_3PO_4 needed to remove all Ca^{2+} ions from 2.50 L of the water sample.

4. To calculate the percent yield for the soap you produced, what additional information would be required that was not given nor determined in this experiment?

Thermal Energy Associated with Physical and Chemical Changes

PURPOSE

Determine the amount of thermal energy (heat) associated with some physical and chemical changes.

BACKGROUND INFORMATION

Physical and chemical changes are associated with absorption or release of thermal energy (heat). If a process releases heat, it is an **exothermic** process. If a process absorbs heat, it is an **endothermic** process. Endothermic or exothermic processes are associated with many important industrial applications. For example, the conversion of water into steam in thermal power plants produces most of the electricity consumed in the United States. The steam passes through a turbogenerator that generates electricity. This is an example of the First Law of Thermodynamics, which states that energy can neither be created nor destroyed, but it can be converted from one form to another. To design new energy conversion processes, knowledge of the thermodynamics of a variety of chemical and physical processes is required. This experiment demonstrates how to measure the thermal energy involved in chemical or physical processes. In this experiment, the thermodynamics of the dissolution of a salt in water, an acid-base reaction, and a precipitation reaction will be studied.

When a substance dissolves in water, the intermolecular interactions among pure solute particles must be overcome (*step a*). The energy is absorbed and the process is endothermic. Likewise, the intermolecular interactions among pure solvent molecules must also be overcome (*step b*) to separate solvent molecules from one another. This process is also endothermic. As the solute and solvent particles interact in solution (*step c*), energy is released and the process is exothermic. The overall dissolution process is endothermic if the amount of heat absorbed in *steps a and b* is greater than the amount of energy released in *step c*. The overall dissolution process is exothermic if the amount of energy absorbed in *steps a and b* is less than that released in *step c*. The net energy change that accompanies the dissolution of one mole of a substance is called the **molar heat of dissolution** ($\Delta H_{dissolution}$, also called molar enthalpy of dissolution). If the heat absorbed or released when a given amount of substance dissolves in water, $q_{dissolution}$, is measured experimentally, the molar heat of dissolution can be calculated as follows:

$$\Delta H_{dissolution} = \frac{q_{dissolution}}{\text{moles of the substance dissolved}} \qquad \text{(Eq. 1)}$$

Heat transfer also accompanies chemical reactions. For example, when 2 moles of $H_{2(g)}$ reacts with 1 mole of $O_{2(g)}$ to form 2 moles of $H_2O_{(g)}$ at 1 atmosphere and 25 °C, 483.6 kJ of heat is released. This can be expressed using the following thermochemical equation

$$2\ H_{2(g)} + O_{2(g)} \rightarrow 2\ H_2O_{(g)} \qquad \Delta H_{reaction} = -483.6\ \text{kJ/mol} \qquad \text{(Eq. 2)}$$

where $\Delta H_{reaction}$ is called the **molar heat of reaction** (or molar enthalpy of reaction). The negative sign indicates the reaction is exothermic.

The numerical value of $\Delta H_{reaction}$ refers to the amount of heat absorbed or released when the numbers of moles of the reactants react to form the numbers of moles of the products specified by the balanced equation. If different amounts of substances are involved in the reaction, $\Delta H_{reaction}$ must be scaled accordingly. The amount of reaction that corresponds to the number of moles of each substance shown in the balanced equation is sometimes called **one mole of reaction (mol rxn)**.

Changes in thermal energy associated with physical and chemical changes can be measured with a calorimeter. A calorimeter is an insulated container that minimizes heat exchange between the process occurring inside the calorimeter and the outside surroundings. This experiment will employ two styrofoam cups, one nested inside the other, to serve as an effective, but inexpensive calorimeter.

In this experiment, heat released or absorbed by a series of physical or chemical changes will be measured. The process will either release heat to or absorb heat from the reaction mixture and the calorimeter. To measure the heat associated with the process, both the heat absorbed (or released) by the solution and by the calorimeter must be accounted for. The heat absorbed (or released) by the solution can be calculated using the equation

$$q_{solution} = m_{solution} \times C_{solution} \times \Delta T_{solution} \qquad \text{(Eq. 3)}$$

where $q_{solution}$ is the heat absorbed or released by the solution, $m_{solution}$ is the mass of the solution (in grams), $C_{solution}$ is the specific heat of the solution (in J/g°C), and $\Delta T_{solution}$ is the change in the temperature of the solution (in °C). The specific heat of a substance is the amount of energy (in J) required to change the temperature of 1.00 gram of the substance by 1.00 °C. Since all solutions produced in this experiment are dilute aqueous solutions, we can assume that the specific heats and the densities of the solutions are the same as that of pure water, 4.184 J/g°C and 1.00 g/mL, respectively.

The heat absorbed or released by the calorimeter must also be taken into consideration. Therefore, it is necessary to calibrate the calorimeter. The **calorimeter constant** ($C_{calorimeter}$, also called the heat capacity of the calorimeter) is defined as the heat required to raise the temperature of the calorimeter by 1.00 °C. The heat absorbed or released by the calorimeter can be calculated using the equation

$$q_{calorimeter} = C_{calorimeter} \times \Delta T_{calorimeter} \qquad \text{(Eq. 4)}$$

where $q_{calorimeter}$ is the heat absorbed or released by the calorimeter (in J), $C_{calorimeter}$ is the calorimeter constant (in J/°C), and $\Delta T_{calorimeter}$ is the change in temperature (in °C) of the calorimeter (equal to ΔT of the cool water). The calorimeter constant for the calorimeter used in this experiment will be determined experimentally.

Equations 1 and 2 can be combined to calculate the heat released or absorbed by a physical or chemical process. According to the First Law of Thermodynamics, the heat released or absorbed by the process must be *equal in magnitude* to the heat absorbed or released by the solution plus the heat absorbed or released by the calorimeter. Mathematically this can be written as:

$$-q_{process} = q_{solution} + q_{calorimeter} \qquad \text{(Eq. 5)}$$

(The minus sign, preceding $q_{process}$, indicates that heat flows from the substance to the water).

Determination of the Calorimeter Constant

To determine the calorimeter constant, a known amount of warm water is mixed with a known amount of cool water. The hot water loses heat to the cool water until an intermediate, equilibrium temperature is attained (see Figure 1).

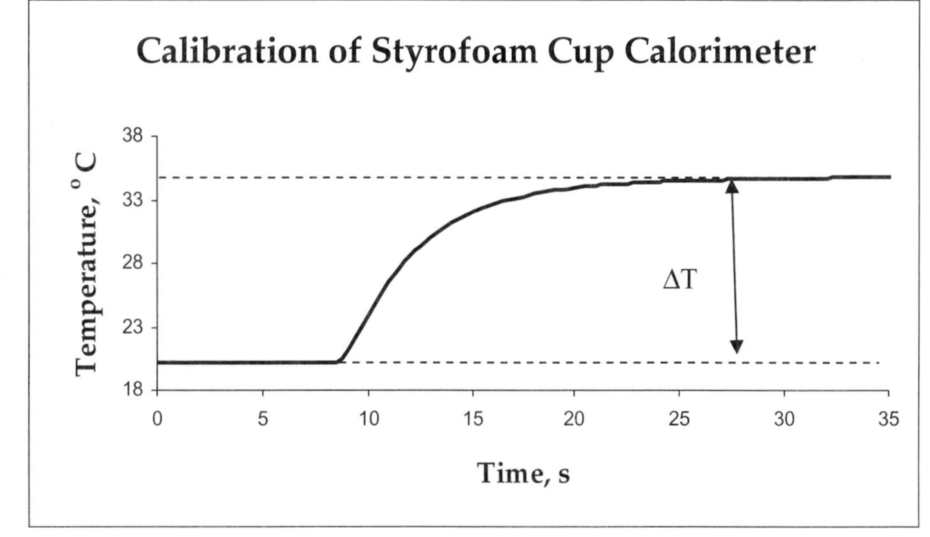

Figure 1
Thermogram for the addition of 31.152 grams of warm water at 51.0 °C to 31.284 grams of cool water at 20.2 °C. The equilibrium temperature of the system is 34.8 °C. ΔT of the warm water is 16.2 °C, and ΔT of the cool water is 14.6 °C

The heat absorbed by the calorimeter ($q_{calorimeter}$) is the difference in the heat lost by the warm water (q_{warm}) and the heat absorbed by the cool water (q_{cool}).

$$q_{calorimeter} = -q_{warm\ water} - q_{cool\ water}$$

$$q_{calorimeter} = -(m_{warm\ water} \times C_{warm\ water} \times \Delta T_{warm\ water}) - (m_{cool\ water} \times C_{cool\ water} \times \Delta T_{cool\ water}) \quad \text{(Eq. 6)}$$

We calculate the calorimeter constant ($C_{calorimeter}$) by rearranging Equation 4.

$$C_{calorimeter} = \frac{q_{calorimeter}}{\Delta T_{cool\ water}}$$

Sample Calculation for Determining the Molar Heat of Reaction

When 20.00 mL of 1.00 *M* $Ba(NO_3)_2$ solution at 21.0 °C is added to 30.00 mL of 1.00 *M* K_2SO_4 solution at 21.0 °C in the calorimeter described above ($C_{calorimeter}$ is 13.7 J/°C), solid $BaSO_4$ forms and the temperature of the mixture increases from 21.0 °C to 25.1 °C. Calculate the molar heat of the reaction.

$$Ba(NO_3)_{2(aq)} + K_2SO_{4(aq)} \rightarrow BaSO_{4(s)} + 2KNO_{3(aq)}$$

$$-q_{reaction} = q_{solution} + q_{calorimeter}$$

$$-q_{reaction} = [m_{solution} \times C_{solution} \times \Delta T_{solution}] + [C_{calorimeter} \times \Delta T_{calorimeter}]$$

$$V_{solution} = 20.00\text{ mL} + 30.00\text{ mL} = 50.00\text{ mL}$$

$$m_{solution} = D_{solution} \times V_{solution}$$

$$m_{solution} = 1.000\text{ g/mL} \times 50.00\text{ mL} = 50.00\text{ g}$$

$$-q_{reaction} = \left[50.00\text{ g} \times \frac{4.18\text{J}}{\text{g °C}} \times (25.1 - 21.0\text{ °C})\right] + \left[\frac{13.7\text{ J}}{\text{°C}} \times (25.1 - 21.0\text{ °C})\right]$$

$$q_{reaction} = -9.1 \times 10^2\text{ J}$$

$$\text{moles of } Ba(NO_3)_2 = 20.00\text{ mL} \times \frac{1.00\text{ mol } Ba(NO_3)_2}{1000\text{ mL}} = 0.0200\text{ moles } Ba(NO_3)_2$$

$$\text{moles of } K_2SO_4 = 30.00\text{ mL} \times \frac{1.00\text{ mol } K_2SO_4}{1000\text{ mL}} = 0.0300\text{ moles } K_2SO_4$$

The number of moles of K_2SO_4 required to react with 0.0200 moles $Ba(NO_3)_2$ is

$$0.0200\text{ moles } Ba(NO_3)_2 \times \frac{1\text{ mol } K_2SO_4}{1\text{ mol } Ba(NO_3)_2} = 0.0200\text{ moles } K_2SO_4$$

Since 0.0300 moles K_2SO_4 is present, which exceeds the required 0.0200 moles, K_2SO_4 is present in excess and $Ba(NO_3)_2$ is the limiting reactant.

$$\Delta H_{reaction} = \frac{-9.1 \times 10^2\text{ J}}{0.0200\text{ moles } Ba(NO_3)_2} \times \frac{1\text{ mol } Ba(NO_3)_2}{1\text{ mol reaction}} = -4.6 \times 10^4\text{ J/mol} = -46\text{ kJ/mol}$$

The reaction is exothermic.

PROCEDURE

CAUTION

Students must wear departmentally approved eye protection while performing this experiment. Wash your hands before touching your eyes and after completing the experiment.

Part A – Determination of the Calorimeter Constant

1. See Appendix A-1 – Instru**ctions for Initializing the MeasureNet Workstation to Record a Temperature versus Time Scan.** Complete all steps in Appendix A-1 before proceeding to Step 2 below.
2. See Appendix A-2 - **Instructions for Recording a Temperature versus Time Thermogram to Determine the Styrofoam Cup Calorimeter Constant.** Complete all steps in Appendix A-2 before proceeding to Step 3.
3. *Steps 4–6 are to be completed after the laboratory period is concluded (outside of lab).* Proceed to Step 8,*Determination the Molar Heat of Dissolution of a Salt.*
4. From the tab delimited files you saved, prepare plots of the temperature versus time data using Excel (or a comparable spreadsheet program). Instructions for plotting temperature versus time thermograms using Excel are provided in Appendix B-1.
5. How do you determine the equilibrium temperature of the hot-cold water mixture in the calorimeter? How do you determine the temperature changes for the warm water and the cold water? Should these temperatures be recorded in the Lab Report and to how many significant figures?
6. How do you determine the calorimeter constant in J/°C? What is the average calorimeter constant?
7. *Use the same two styrofoam cups in Part B of the experiment that were used in Part A. The calorimeter constant determined in Part A will be used in Part B of this experiment to Determine the Molar Heat of Dissolution of a Salt.*

Part B – Determination the Molar Heat of Dissolution of a Salt

8. You will determine the molar heat of dissolution of potassium nitrate. You will dissolve approximately 1.00 g of potassium nitrate in 25 mL of distilled water. Should you determine the exact masses of the potassium nitrate and the water? If so, to what number of significant figures should the mass be recorded in the Lab Report?
9. See Appendix A-4 - **Instructions for Recording a Thermogram to Determine the Enthalpy Change for a Physical or a Chemical Process.** Complete all steps in Appendix A-4 for the dissolution of 1.00 g potassium nitrate (referred to as substance B in Appendix A-4) into 25.00 mL distilled water (referred to as substance A in Appendix A-4) before proceeding to Step 10.
10. *Steps 11–13 are to be completed after the laboratory period is concluded (outside of lab). Proceed to Step 14, Determination of the Molar Heat of Reaction of an Acid-Base Reaction.*

11. From the tab delimited files, prepare thermograms using Excel for each trial. Instructions for plotting thermograms using Excel are provided in Appendix B-1.
12. How do you determine $\Delta H_{dissolution}$ for each trial?
13. What is the average $\Delta H_{dissolution}$ for potassium nitrate? Is the dissolution an endothermic or exothermic process?

Part C – Determination of the Molar Heat of Reaction of an Acid-Base Reaction

14. See Appendix A-4 - **Instructions for Recording a Thermogram to Determine the Enthalpy Change for a Physical or a Chemical Process.** Complete all steps in Appendix A-4 for the neutralization reaction between 20.00 mL of 1.0 *M* HCl (referred to as substance A in Appendix A-4) and 20.00 mL of 1.0 *M* NaOH (referred to as substance B in Appendix A-4) before proceeding to Step 15.
15. *Steps 16–18 are to be completed after the laboratory period is concluded (outside of lab). Proceed to Step 18, Determination of the Molar Heat of Reaction of a Precipitation Reaction.*
16. From the tab delimited files, prepare a thermogram using Excel for each trial. Instructions for plotting thermograms using Excel are provided in Appendix B-1.
17. Write the balanced chemical equation for the reaction. How do you determine $\Delta H_{reaction}$ for each trial?
18. What is the average $\Delta H_{reaction}$ for the neutralization reaction? Is the neutralization reaction endothermic or exothermic?

Part D – Determination of the Molar Heat of Reaction of a Precipitation Reaction

19. See Appendix A-4 - **Instructions for Recording aThermogram to Determine the Enthalpy Change for a Physical or a Chemical Process.** Complete all steps in Appendix A-4 for the reaction between 20.00 mL of 1.0 *M* $Fe(NO_3)_3$ (referred to as substance A in Appendix A-4) and 20.00 mL of 1.0 *M* NaOH (referred to as substance B in Appendix A-4) before proceeding to Step 19.
20. From the tab delimited files, prepare a thermogram using Excel for each trial. Instructions for plotting thermograms using Excel are provided in Appendix B-1.
21. Write the balanced chemical equation for the reaction. How do you determine $\Delta H_{reaction}$ for each trial?
22. What is the average $\Delta H_{reaction}$ for the precipitation reaction? Is the precipitation reaction endothermic or exothermic?

Name

Section

Date

Instructor

19 EXPERIMENT 19

Lab Report

Part A – Determination of the Calorimeter Constant

Experimental data and calculations - Trial 1

Experimental data and calculations - Trial 2

What is the average calorimeter constant?

Part B – Determination of the Heat of Dissolution of a Salt

Experimental data and calculations - Trial 1

Experimental data and calculations - Trial 2

What is the average $\Delta H_{dissolution}$ for potassium nitrate? Is the dissolution an endothermic or exothermic process?

Part C – Determination of the Molar Heat of Reaction of an Acid-Base Reaction

Experimental data and calculations - Trial 1

Experimental data and calculations - Trial 2

What is the average $\Delta H_{reaction}$ for the neutralization reaction? Is the neutralization reaction endothermic or exothermic?

Part D – Determination of the Molar Heat of Reaction of a Precipitation Reaction

Experimental data and calculations - Trial 1

Experimental data and calculations - Trial 2

What is the average $\Delta H_{reaction}$ for the precipitation reaction? Is the precipitation reaction endothermic or exothermic?

Name _______________ Section _______ Date _______

Instructor _______________

19 EXPERIMENT 19

Pre-Laboratory Questions

1. 25.000 grams of water at 45.00 °C was added to 25.00 grams of water at 20.30 °C in a coffee cup calorimeter. Upon mixing, the equilibrium temperature attained by the system was 31.40 °C. The specific heat of water is 4.184 J/g°C. What is the calorimeter constant for the coffee cup calorimeter?

2. 25.00 mL of 1.00 *M* HCl is mixed with 25.00 mL of 1.00 *M* LiOH in the calorimeter (same calorimeter from Question 1). The initial temperature of the HCl and LiOH solutions is 20.50 °C, and the equilibrium temperature of the reaction mixture is 38.50 °C. Assume the density and the specific heat of the solution is the same as for water, 1.00 g/mL and 4.184 J/g°C, respectively.

$$LiOH_{(aq)} + HCl_{(aq)} \rightarrow LiCl_{(aq)} + H_2O_{(l)}$$

a. Determine $q_{solution}$ (heat absorbed or released by the reaction mixture)?

b. Calculate the moles of the limiting reagent in the reaction?

c. What is the $\Delta H_{reaction}$ in kJ/mol for the neutralization reaction?

d. Is the neutralization reaction endothermic or exothermic?

Name

Section

Date

Instructor

19 EXPERIMENT 19

Post-Laboratory Questions

1. A student mistakenly recorded the calorimeter constant as 451.0 J/°C instead of 45.1 J/°C. Assuming no other errors are made in the experiment, will the heat of dissolution of the salt determined by the student be lower than or higher than the correct value? Justify your answer with an explanation.

2. When performing Part D of the experiment, a student spilled three fourth of the $Fe(NO_3)_3$ solution on the lab bench. The student used the remaining fourth of the solution to complete the experiment. Will the heat of reaction determined experimentally be lower than or higher than it would have been if all of the $Fe(NO_3)_3$ solution had been used? Justify your answer with an explanation.

3. Would the reaction between sulfuric acid and calcium hydroxide yield essentially the same heat of neutralization as the reaction between HCl and NaOH did in this experiment? Justify your answer with an explanation.

Hess's Law

PURPOSE

Measure the enthalpies of neutralization for several acid-base reactions, then use that information and Hess's Law to determine the reaction enthalpies for two salts in aqueous solution.

INTRODUCTION

The enthalpies of formation, $\Delta H^0{}_f$, and reaction, ΔH_{rxn}, are important thermodynamic quantities that indicate the amount of thermal energy (heat) that is released or absorbed in a chemical reaction at constant pressure. **Exothermic** reactions are chemical processes that release thermal energy. **Endothermic** reactions absorb thermal energy. Chemical reactions that are exothermic have negative values of ΔH_{rxn} while endothermic reactions have positive values of ΔH_{rxn}.

The thermodynamic state of a system is defined by a set of conditions that specify the properties (e.g., *P*, *V*, *T*, etc.) of the system. **State functions** are properties of a system that depend only on the state of the system and are independent of the pathway by which the system came to be in that state. Enthalpy is an example of a state function. A *change* in a state function describes a difference between two states and it is independent of the process by which the change occurs. For example, if 100 grams of liquid water at 25 °C and 1.00 atmosphere of pressure is heated to 250 °C at 5.00 atmosphere, and then cooled to 15 °C at 0.85 atmosphere of pressure, it would have the same enthalpy change as 100 grams of liquid water that was cooled from 25 °C and 1.00 atmosphere of pressure to –25 °C and 0.50 atmosphere of pressure then heated back to 15 °C at 0.85 atmosphere of pressure. The important point in these changes is that both samples of water began at 25 °C and 1.00 atmosphere of pressure and both processes ended at 15 °C and 0.85 atmosphere of pressure. For state functions, it is the difference (*change*) between the initial and final states of a process that matters, the pathway by which the change occurs is immaterial. This permits chemists to calculate reaction enthalpies using an indirect method known as Hess's law.

Hess's law states that the ΔH_{rxn} for a reaction is the same whether it occurs in one step or a series of steps. For example, consider the decomposition of sulfur trioxide.

$$2\ SO_{3(g)} \rightarrow 2\ SO_{2(g)} + O_{2(g)} \qquad \text{(Eq. 1)}$$

The ΔH_{rxn} for this reaction can be determined indirectly from the ΔH_{rxns} for the series of reactions given below.

$$S_{(s)} + O_{2(g)} \rightarrow SO_{2(g)} \qquad \Delta H_{rxn} = -296.8\ \text{kJ/mol} \qquad \text{(Eq. 2)}$$

$$S_{(s)} + 3/2\ O_{2(g)} \rightarrow SO_{3(g)} \qquad \Delta H_{rxn} = -395.6\ \text{kJ/mol} \qquad \text{(Eq. 3)}$$

We must manipulate Eqs. 2 and 3 in a manner such that the *sum* of the two reactions yields Eq 1. In Eq. 3, note that SO_3 is a product, however, in Eq. 1 SO_3 is a reactant. Therefore, we must reverse Eq. 3 and change the sign of its ΔH_{rxn} from negative to positive. Because Eq. 1 has 2 SO_3 as a reactant, we must multiply Eq. 3 and its ΔH_{rxn} by 2.

$$2\left(SO_{3(g)} \rightarrow S_{(s)} + 3/2\ O_{2(g)}\right) \qquad \Delta H_{rxn} = 2\ (395.6\ \text{kJ/mol}) \qquad \text{(Eq. 4)}$$

Since Eq. 1 has 2 SO_2 as a product, we must multiply Eq. 2 and its ΔH_{rxn} by 2.

$$2\left(S_{(s)} + O_{2(g)} \rightarrow SO_{2(g)}\right) \qquad \Delta H_{rxn} = 2\ (-296.9\ \text{kJ/mol}) \qquad \text{(Eq. 5)}$$

The ΔH_{rxn} for Eq. 1 is the sum of the ΔH_{rxn}'s for Equations 4 and 5.

$$2\ SO_{3(g)} \rightarrow \cancel{2S_{(s)}} + 3\ \cancel{O_{2(g)}} \qquad \Delta H_{rxn} = 791.2\ \text{kJ/mol} \qquad \text{(Eq. 4)}$$

$$\cancel{2S_{(s)}} + 2\ \cancel{O_{2(g)}} \rightarrow 2\ SO_{2(g)} \qquad \Delta H_{rxn} = -593.6\ \text{kJ/mol} \qquad \text{(Eq. 5)}$$

$$2\ SO_{3(g)} \rightarrow 2\ SO_{2(g)} + O_{2(g)} \qquad \Delta H_{rxn} = 197.6\ \text{kJ/mol} \qquad \text{(Eq. 1)}$$

The specific heat (C) of a substance is the quantity of energy (J) required to change the temperature of 1.00 gram of the substance by 1.00 °C. Specific heat typically has units of J/g°C.

$$C = \frac{\text{thermal energy in J}}{\text{(mass)(change in temperature)}} \qquad \text{(Eq. 6)}$$

The thermal energy lost or gained by a substance as its temperature changes can be calculated using the equation

$$q = m \times C \times \Delta T \qquad \text{(Eq. 7)}$$

where q is the thermal energy change of the substance in joules, m is the mass of the substance in grams, and ΔT is the change in the Celsius temperature of the substance.

A **calorimeter** is an insulated container that minimizes thermal energy exchange between the system and the surroundings. Calorimeters are used to accurately measure the thermal energy lost or absorbed by a substance in physical or chemical changes. This experiment will employ two styrofoam cups, one nested inside the other, with a cardboard lid to serve as the calorimeter. Any thermochemical experiment that uses a calorimeter must take into consideration the amount of thermal energy absorbed or released by the calorimeter. Therefore, it is necessary to calibrate the calorimeter. The **calorimeter constant** ($C_{calorimeter}$) is defined as the thermal energy required to raise the temperature of the calorimeter by 1.00 °C. The thermal energy absorbed or released by the calorimeter, $q_{calorimeter}$, can be determined according to the equation

$$q_{calorimeter} = C_{calorimeter} \times \Delta T_{cool\ water} \qquad \text{(Eq. 8)}$$

where $C_{calorimeter}$ is the calorimeter constant in J/°C, and ΔT_{water} is the change in the temperature of the cool water. To determine the calorimeter constant, a known mass of cool water (typically at room temperature) is added to the calorimeter. Next, a known mass of warm water (30–40 °C above room temperature) is added to the calorimeter containing the cool water. From the temperature changes of the two quantities of water, the calorimeter constant is determined.

By combining Eqs. 7 and 8, the thermal energy released by an acid-base reaction in a calorimeter can be calculated using Eq. 9.

$$-(q_{rxn}) = \text{thermal energy absorbed by solution} + \text{thermal energy absorbed by calorimeter}$$

$$-q_{rxn} = (m_{solution} \times C_{solution} \times \Delta T_{solution}) + (C_{calorimeter} \times \Delta T_{solution}) \qquad \text{(Eq. 9)}$$

(The minus sign, preceding q_{rxn}, indicates that thermal energy flows from the reaction to the solution and calorimeter. When determining $\Delta T_{solution}$, subtract the initial temperature of the solution, T_i, from the final temperature of the solution, T_f.) The mass of the solution is equal to the total volume of the acid and the base added to the calorimeter times the density of the solution.

The reaction of potassium hydroxide with nitric acid is an example of a strong base reacting with a strong acid forming potassium nitrate and water. The formula unit (Eq. 10), total ionic (Eq. 11), and net ionic (Eq. 12) equations for this reaction are given below.

$$KOH_{(aq)} + HNO_{3(aq)} \rightarrow KNO_{3(aq)} + H_2O_{(\ell)} \qquad \text{(Eq. 10)}$$

$$K^+{}_{(aq)} + OH^-{}_{(aq)} + H^+{}_{(aq)} + NO_3^-{}_{(aq)} \rightarrow K^+{}_{(aq)} + NO_3^-{}_{(aq)} + H_2O_{(\ell)} \qquad \text{(Eq. 11)}$$

$$OH^-{}_{(aq)} + H^+{}_{(aq)} \rightarrow H_2O_{(\ell)} \qquad \text{(Eq. 12)}$$

The net ionic equation for any strong acid-strong base reaction forming a water-soluble salt is $OH^- + H^+ \rightarrow H_2O$. All strong acid-strong base reactions have essentially the same ΔH_{rxn} because they have the same net ionic equation.

However, less thermal energy is evolved when either the acid or base is weak. Some of the energy released is required to break the stronger bonds in the weak acid or base. Consider the reaction of $HClO_2$ (a weak acid) with KOH (a strong base). The formula unit (Eq. 13), total ionic (Eq. 14) and net ionic (Eq. 15) equations for this reaction are given below.

$$KOH_{(aq)} + HClO_{2\,(aq)} \rightarrow KClO_{2\,(aq)} + H_2O_{(\ell)} \qquad \text{(Eq. 13)}$$

$$K^+{}_{(aq)} + OH^-{}_{(aq)} + HClO_{2\,(aq)} \rightarrow K^+{}_{(aq)} + ClO^-_{2\,(aq)} + H_2O_{(\ell)} \qquad \text{(Eq. 14)}$$

$$OH^-_{(aq)} + HClO_{2\,(aq)} \rightarrow ClO^-_{2\,(aq)} + H_2O_{(\ell)} \qquad \text{(Eq. 15)}$$

Because weak acids are only slightly ionized, the net ionic equation for the neutralization of a weak acid with a strong base is different from the net ionic equation for the neutralization of a strong acid with a strong base (see Eqs. 12 and 15). Consequently, the ΔH_{rxn} for the reaction of $HClO_2$ with KOH has a smaller negative value than the ΔH_{rxn} for the reaction of HNO_3 with KOH.

To employ Hess's law for this experiment, we will measure the heat of neutralization for four different acid-base reactions.

$$HCl_{(aq)} + NaOH_{(aq)} \rightarrow NaCl_{(aq)} + H_2O_{(\ell)} \qquad \text{(Eq. 16)}$$

$$HNO_{3(aq)} + NaOH_{(aq)} \rightarrow NaNO_{3(aq)} + H_2O_{(\ell)} \qquad \text{(Eq. 17)}$$

$$CH_3COOH_{(aq)} + NaOH_{(aq)} \rightarrow NaCH_3COO_{(aq)} + H_2O_{(\ell)} \qquad \text{(Eq. 18)}$$

$$CH_3COOH_{(aq)} + NH_{3(aq)} \rightarrow NH_4CH_3COO_{(aq)} \qquad \text{(Eq. 19)}$$

These equations can be manipulated to yield Eqs. 20 and 21.

$$HCl_{(aq)} + NH_{3(aq)} \rightarrow NH_4Cl_{(aq)} \qquad \text{(Eq. 20)}$$

$$HNO_{3(aq)} + NH_{3(aq)} \rightarrow NH_4NO_{3(aq)} \qquad \text{(Eq. 21)}$$

From the ΔH_{rxn}'s determined for Eqs. 16–19, the ΔH_{rxn}'s for the formation of aqueous ammonium chloride and aqueous ammonium nitrate (Eqs. 20 and 21) can be calculated.

In this experiment, we will determine the ΔH_{rxn}'s for a series of acid-base reactions (Eqs. 16–19) and manipulate those reaction equations, in accordance with Hess's Law, to determine the ΔH_{rxn}'s for the formation of aqueous ammonium chloride and aqueous ammonium nitrate. (**Note:** *When manipulating chemical equations for reactions that occur in aqueous solution, the reactions should be written as total ionic equations*).

Sample calculations for determining the calorimeter constant are provided in Experiment 2. Sample calculations to determine ΔH_{rxn} for a reaction are given in Experiment 19.

PROCEDURE

CAUTION

Students must wear departmentally approved eye protection while performing this experiment. Wash your hands before touching your eyes and after completing the experiment.

Chemical Alert

Acids and bases are corrosive. Be careful not to splatter them on areas of exposed skin. If acid or base does contact exposed skin, wash the affected area with copious quantities of water and inform your laboratory instructor immediately.

NOTE: Contact with nitric acid turns skin yellow.

Part A – Determination of the Calorimeter Constant

1. See Appendix A-1 – **Instructions for Initializing the MeasureNet Workstation to Record a Thermogram**. Complete all steps in Appendix A-1 before proceeding to Step 2 below.

NOTE: Assume the densities of all solutions are 1.00 g/mL. Assume the specific heat values for the aqueous acid and base solutions are the same as water, 4.18 J/g°C.

2. See Appendix A-2 – **Instructions for Recording a Thermogram to Determine the Styrofoam Cup Calorimeter Constant**. Complete all steps in Appendix A-2 before proceeding to Step 3.
3. *Steps 4–6 are to be completed after the laboratory period is concluded (outside of lab). Proceed to Step 7, Determination of the Heat of Neutralization of Several Acid-Base Reactions.*
4. From the tab delimited files you saved, prepare plots of the temperature versus time data using Excel (or a comparable spreadsheet program). Instructions for plotting temperature versus time thermograms using Excel are provided in Appendix B-1.
5. How do you determine the equilibrium temperature of the hot-cold water mixture in the calorimeter? How do you determine the temperature changes for the warm water and the cold water? Should these temperatures be recorded in the Lab Report, and to how many significant figures?
6. How do you determine the calorimeter constant in J/°C? What is the average calorimeter constant?

   This same calorimeter constant must be used in the subsequent steps of this experiment to determine the ΔH_{rxn} for each acid-base neutralization reaction.

Part B – Determination of the Heat of Neutralization for Several Acid-Base Reactions

7. See Appendix A-3 – **Instructions for Recording a Thermogram to Determine the Enthalpy Change for a Neutralization Reaction**. Complete all steps in Appendix A-3 for each acid-base pair (NaOHHCl, NaOH—HNO_3, NaOH—CH_3COOH, and NH_3—CH_3COOH) before proceeding to Step 8.
8. From the tab delimited files you saved for each acid-base reaction, prepare thermograms of the temperature versus time data using Excel (or a comparable spreadsheet program). Instructions for plotting thermograms using Excel are provided in Appendix B-1.
9. How do you determine ΔH_{rxn} for each acid-base reaction (for two trials)?
10. What is the average ΔH_{rxn} for each acid-base pair (NaOH—HCl, NaOH—HNO_3, NaOH—CH_3COOH, and NH_3—CH_3COOH)?
11. What is ΔH_{rxn} for the reaction: $HCl_{(aq)} + NH_{3(aq)} \rightarrow NH_4Cl_{(aq)}$?
12. What is ΔH_{rxn} for the reaction: $HNO_{3(aq)} + NH_{3(aq)} \rightarrow NH_4NO_{3(aq)}$.?

Name Section Date

Instructor

20 EXPERIMENT 20

Lab Report

Part A – Determination of the Calorimeter Constant

Experimental data and calculations – Trial 1

Experimental data and calculations – Trial 2

What is the average calorimeter constant?

Part B – Determination of the Heat of Neutralization for Several Acid-Base Reactions

Experimental data and calculations: HCl + NaOH – Trial 1

Experimental data and calculations: HCl + NaOH – Trial 2

Experimental data and calculations: HNO_3 + NaOH – Trial 1

Experimental data and calculations: HNO_3 + NaOH – Trial 2

Experimental data and calculations: CH_3COOH + NaOH – Trial 1

Experimental data and calculations: CH_3COOH + NaOH – Trial 2

Experimental data and calculations: $CH_3COOH + NH_3$ – Trial 1

Experimental data and calculations: $CH_3COOH + NH_3$ – Trial 2

What is the average ΔH_{rxn} for each acid-base pair (NaOH—HCl, NaOH—HNO_3, NaOH—CH_3COOH, and NH_3—CH_3COOH)?

What is ΔH_{rxn} for the reaction: $HCl_{(aq)} + NH_{3(aq)} \rightarrow NH_4Cl_{(aq)}$?

What is ΔH_{rxn} for the reaction: $HNO_{3(aq)} + NH_{3(aq)} \rightarrow NH_4NO_{3(aq)}$.?

Name Section Date

Instructor

20 **EXPERIMENT 20**

Pre-Laboratory Questions

Assume that the densities of all solutions are 1.00 g/mL and that the specific heat of each reaction mixture is 4.18 J/g°C.

1. 27.0 mL of 1.00 *M* $HClO_4$ are mixed with 27.0 mL of 1.00 *M* LiOH in a coffee-cup calorimeter. The thermogram for the reaction is depicted below. The calorimeter constant is 28.1 J/°C. Both solutions were initially at 21.31 °C. Calculate ΔH_{rxn} (in kJ/mol) for the reaction.

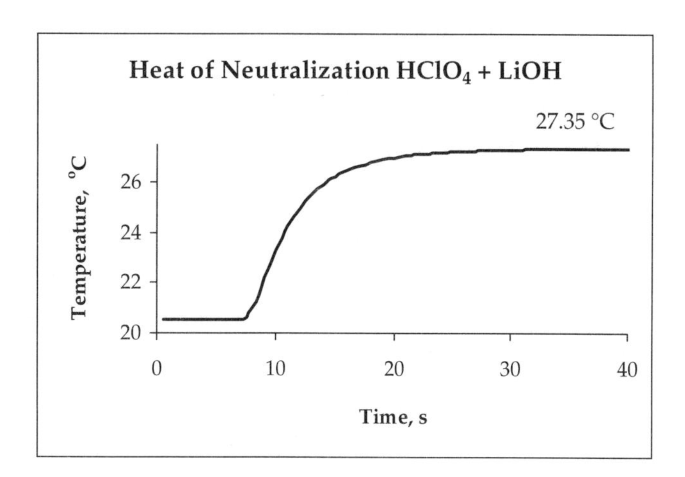

2. 27.0 mL of 1.00 *M* CH_3COOH are mixed with 27.0 mL of 1.00 *M* NH_3 in a coffee-cup calorimeter. The thermogram for the reaction is depicted below. The calorimeter constant is 28.1 J/°C. Both solutions were initially at 20.56 °C. Calculate ΔH_{rxn} (in kJ/mol) for the reaction.

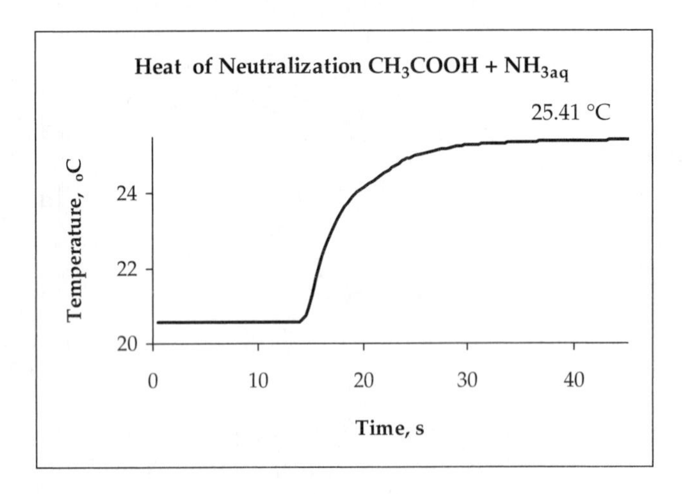

3. 27.0 mL of 1.00 *M* CH_3COOH are mixed with 27.0 mL of 1.00 *M* LiOH in a coffee-cup calorimeter. The thermogram for the reaction is depicted below. The calorimeter constant is 28.1 J/°C. Both solutions were initially at 20.34 °C. Calculate ΔH_{rxn} (in kJ/mol) for the reaction.

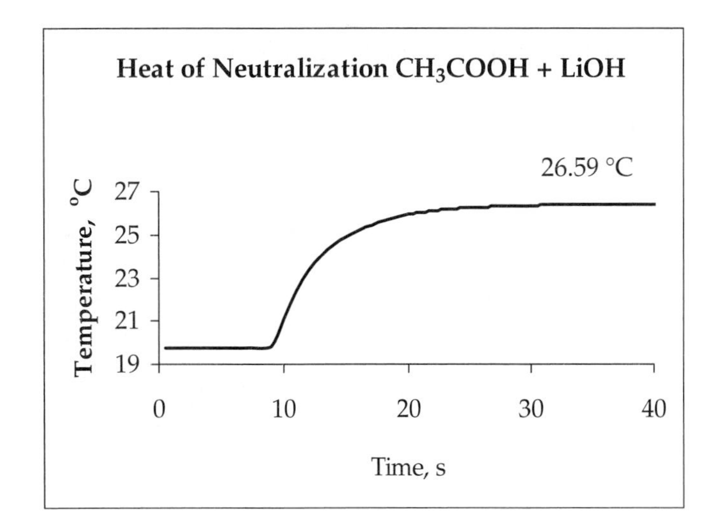

4. In accordance with Hess's Law, manipulate the following equations and their ΔH_{rxn}'s to calculate ΔH_{rxn} for the reaction $HClO_4 + NH_3 \rightarrow NH_4ClO_4$.

$$HClO_{4(aq)} + LiOH_{(aq)} \rightarrow LiClO_{4(aq)} + H_2O_{(\ell)}$$

$$CH_3COOH_{(aq)} + NH_{3(aq)} \rightarrow NH_4CH_3COO_{(aq)}$$

$$CH_3COOH_{(aq)} + LiOH_{(aq)} \rightarrow LiCH_3COO_{(aq)} + H_2O_{(\ell)}$$

______________________________ ΔH_{rxn} = __________ kJ/mol

______________________________ ΔH_{rxn} = __________ kJ/mol

______________________________ ΔH_{rxn} = __________ kJ/mol

$HClO_{4\,(aq)} + NH_{3(aq)} \rightarrow NH_4ClO_{4(aq)}$ ΔH_{rxn} = __________ kJ/mol

Name | Section | Date

Instructor

20

EXPERIMENT 20

Post-Laboratory Questions

1. A student stopped the temperature scan before the last reaction ($CH_3COOH—NH_3$) reached thermal equilibrium in order to leave the lab early. Assuming no other experimental errors were made, would the calculated ΔH_{rxn} for the formation of aqueous ammonium chloride be higher or lower than the correct value? Justify your answer with an explanation.

2. When determining the calorimeter constant for a Styrofoam cup calorimeter, a student splashed some cold water out of the calorimeter before adding the hot water. Would the ΔH_{rxn} for each reaction measured in the experiment be higher or lower than the correct values? Justify your answer with an explanation.

3. Write the balanced formula unit equation for the reaction between hydrochloric acid and silver nitrate. To determine the ΔH_{rxn} for this reaction using Hess's law and the reactions given in Eqs. 16–19, the ΔH_{rxn}'s for what additional reaction(s) would have to be determined? Justify your answer with an explanation.

4. Referring to Question 3, write the formula unit and net ionic equations for the additional reaction(s).

Determination of the Heat of Neutralization of a Variety of Strong Acids and Bases: A Self-directed Experiment

PURPOSE

Determine the heats of neutralization of four acidic or basic compounds.

INTRODUCTION

This is a *self-directed* experiment in which teams of students will rely on chemical knowledge learned, laboratory skills and techniques acquired so far in this course to design and perform the experiment. Your lab instructor will specify the number of students per team. Each team will submit a **Procedure Proposal** to their laboratory instructor two weeks before the experiment is performed. The Procedure Proposal must address several important questions:

1. What is the central question to be answered in this experiment?
2. What experimental techniques will be utilized to answer the central question?
3. What calculations should be provided in the Procedure Proposal?
4. What safety precautions must be addressed in the Procedure Proposal?
5. What additional questions should be addressed by each team?

Each team may use their lab manual, textbook, reference books, and reliable internet resources for assistance in writing the Procedure Proposal. The format for writing the Procedure Proposal is provided in Appendix C-1. The graded and annotated Procedure Proposal will be returned to each team one week before the lab period during which the experiment is to be performed.

When the experiment is completed, each team will submit one **Formal Lab Report** addressing the following questions:

1. What conclusions can be drawn from the experimental data colleted?
2. Does the data answer the central question?
3. Are the team's conclusions in agreement with the information obtained from reference resources, textbook, or the internet?
4. What are some possible sources of error in the experimental data?
5. What modifications could be made to the experimental design and procedures to improve the accuracy of the data?

The format for writing the Formal Lab Report is provided in Appendix C-2. Your lab instructor will inform you of the submission date for the Formal Lab Report.

PROCEDURE

In this experiment, each team of students will be given 50.0 mL of an unknown solution. The unknown is a 1.0 *M* solution of one of the following acids or bases: HCl, H_3PO_4, NaOH, or Na_2CO_3.

First, each team must determine whether the solution is acidic or basic, and the number of ionizable species per formula unit of the compound. Second, each team must determine the heat of neutralization for the unknown solution by reacting it with either 1.0 *M* HCl or 1.0 *M* NaOH. Once the heat of neutralization for each compound has been determined, each team will compare their heat of neutralization with the other students in their lab section to determine any periodic trends that are evident from the experiment. The heat of neutralization data will be used to identify the unknown solution.

Each team must choose a second experimental technique to identify the unknown solution. For example, each team can choose from the following set of experiments: freezing point depression, chemical reactions, titrations, emission and absorption spectroscopy, or other techniques. The data obtained from the second experimental procedure will be used to identify the unknown solution.

Equipment that will be provided for the experiment includes the MeasureNet temperature probe, pH meter, drop counter, the MeasureNet spectrophotometer, and a coffee-cup calorimeter. The normal laboratory equipment will also be available.

LIST OF CHEMICALS

A. 1.0 *M* Hydrochloric acid

B. 1.0 *M* Phosphoric acid

C. 1.0 *M* Sodium hydroxide

D. 1.0 *M* Sodium carbonate

LIST OF SPECIAL EQUIPMENT

A. MeasureNet temperature probe

B. MeasureNet pH probe and drop counter

C. Styrofoam Cups and Lids

D. Magnetic stirrer and stir bar

Dystan Medical Supply Company – Cold Packs and Hot Packs: A Self-Directed Experiment

PURPOSE

Determine which ionic compounds are best suited for the production of cold packs and hot packs. Determine the unit cost of production per cold pack and hot pack.

INTRODUCTION

When ionic compounds dissolve in water, they either absorb energy from or release energy to the surroundings. A chemical reaction that absorbs heat from the surroundings is an **endothermic process**. A chemical reaction that releases heat to the surroundings is an **exothermic process**. The enthalpy of dissolution for a salt ultimately determines whether it can be used as an ingredient to prepare a cold or a hot pack.

Instant cold and hot packs are frequently used by athletes to treat injuries. For example, cold packs are routinely applied to sprained ankles. Heat flows from the ankle to the pack, reducing the temperature of the injured area. Lowering the injured ankle's temperature produces vasoconstriction of the blood vessels that reduces blood flow, which in turn reduces inflammation. Hot packs can be applied to reduce muscle spasms, muscle soreness, inflammation, and relieve pain. Heat flows from the pack to the affected area, increasing its temperature. This produces vasodilation, which increases blood flow into the target tissue. Increased blood flow brings needed oxygen and nutrients to the injured area, aiding the healing process.

A typical cold or hot pack consists of a plastic bag with two compartments, one containing water and the other a salt. When the bag is broken, the contents of the two compartments mix, dissolving the salt. How cold or hot the pack gets depends on the salt's concentration in the water. Cold packs can reach temperatures as low as 0 °C. Hot packs can reach temperatures as high as 90 °C. The duration of cold or heat therapy for commercially available packs is 15–20 minutes.

The Dystan Medical Supply Company has recently hired a team of chemists for their Research and Development Division. The team has been assigned the task of determining which salts should be used to produce efficient and economical cold and hot packs. The company has large quantities of four different salts available for testing: ammonium nitrate, calcium chloride, lithium chloride, and potassium chloride. In addition, the team must determine the mass of salt to be used to produce a 100-mL cold that can attain a tempera-ture of 0 °C and a 100-mL hot pack that can attain a temperature of 65 °C. Both types of packs are to be designed for storage at an ambient temperature of 25 °C. Finally, the team must determine the production cost per unit for a cold pack and a hot pack. The Chief Executive Officer (CEO) of the Dystan Company informed the Research and Development Division that the production cost per pack must be less than $5.00 for the company to return a profit.

This is a *self-directed* experiment wherein teams of students will write the procedure they will use, and rely on lab skills and techniques acquired this semester, to perform this experiment. Each team will submit a **Procedure Proposal** to their lab instructor two weeks before the experiment is performed. *The Procedure Proposal must address several important questions, a few examples are listed below.*

1. What is the central question to be answered in this experiment?
2. What experimental technique(s) will be utilized to answer the central question?
3. How will each team determine which ionic compound to use to prepare a cold pack and which ionic compound to use to prepare a hot pack?
4. When comparing enthalpies of dissolution for different salts, how critical is the concentration of each salt solution?
5. What safety precautions must be addressed in the Procedure Proposal?
6. What additional questions should be addressed by each team?

Each team may use their lab manual, textbook, reference books, and the internet for assistance in writing the **Procedure Proposal** (the format for writing the Procedure Proposal is provided in Appendix C-1). The graded and annotated Procedure Proposal will be returned to each team one week before the experiment is to be performed.

When the experiment is completed, each team will submit one **Formal Lab Report** (the format for writing the Formal Lab Report is provided in Appendix C-2) addressing the following questions.

1. What conclusions can be drawn from the experimental data collected by each team?
2. Does the experimental data collected answer the central question?
3. Can the team's conclusions be supported by information obtained from reference sources, textbooks, or the internet?
4. What are some possible sources of error in the experimental data?
5. What modifications could be made to the experimental design and procedures to improve the accuracy of the data?

PROCEDURE

Each team will measure the enthalpy of dissolution for ammonium nitrate, calcium chloride, lithium chloride, and potassium chloride. Teams will be provided with 3 grams of each salt. A minimum of two determinations of the enthalpy of dissolution for each salt are required. Use sufficient distilled water to dissolve the salt and cover the temperature probe tip. From the enthalpy of dissolution data, determine which salt should be used for cold packs and which salt should be used for hot packs.

Each cold pack and hot pack will contain 100 mL of water. Estimate the mass of salt that must be added to the cold pack to attain a temperature of 0 °C. Estimate the mass of salt that must be added to the hot pack to reach a temperature of 65 °C. From the information (current market prices) given below, estimate the unit production cost per cold pack and per hot pack. Based on the estimated unit production costs, will the Dystan Company return a profit?

Ammonium nitrate	\$26.20 per 500 g
Calcium chloride	\$31.70 per 500 g
Lithium chloride	\$65.00 per 500 g
Potassium chloride	\$28.19 per 500 g
Labor cost per unit	\$0.73
Capital/overhead cost per unit	\$0.36
Plastic bag cost per unit	\$0.19

LIST OF CHEMICALS

A. Ammonium nitrate

B. Calcium chloride

C. Lithium chloride

D. Potassium chloride

LIST OF SPECIAL EQUIPMENT

A. MeasureNet temperature probe

B. Styrofoam Cups and Lids

C. Magnetic stirrer and stir bar

Chemical Kinetics

PURPOSE

Determine the rate law expressions for the iodination of acetone (2-propanone) and for the reaction of crystal violet with sodium hydroxide.

INTRODUCTION

Chemical kinetics is the study of reaction rates and reaction mechanisms. Some reactions, such as acid-base or precipitation reactions, occur instantaneously. Other reactions, such as the rusting of iron, occur very slowly. Understanding the factors that affect reaction rates has important biological, industrial, and environmental significance.

The rate of a reaction depends on the nature and concentrations of the reactants, the temperature at which the reaction occurs, and the presence of a catalyst. For a given chemical reaction,

$$aA + bB \rightarrow \text{product(s)} \qquad \text{(Eq. 1)}$$

the reaction rate depends only on the concentrations of the reactants under certain experimental conditions (temperature, presence of catalysts, etc.). The **rate law** describes how the reaction rate depends on the concentrations of the reactants (Eq 2).

$$\text{rate} = k[A]^m[B]^n \qquad \text{(Eq. 2)}$$

The constant k is the **specific rate constant** for the reaction at given temperature. The exponents m and n are called **reaction orders**. Typically, they are small, whole number integers. For example, a reaction order of *zero* indicates that the reaction rate does not depend on the concentration of that reactant (as long as the reactant is present). A reaction order of *one* indicates that the reaction rate is directly proportional to the concentration of that reactant. A reaction order of *two* indicates that the reaction rate is directly proportional to the square of the concentration of that reactant. (Except for elementary reactions, the exponents m and n are *not* related to the coefficients in the balanced chemical equation for the overall reaction. They must be determined experimentally.) The **overall reaction order** is the sum of m and n.

Method of Initial Rates for Determination of Reaction Orders

The **method of initial rates** is a commonly used method to determine the reaction orders (m and n) for a rate law equation. The initial rate of a reaction is the instantaneous rate at the moment the reaction begins. To determine the reaction orders, we conduct a series of experiments in which the concentration of only one reactant at a time is changed. The initial rate is measured for each of these experiments. The results from these experiments are compared to see how the initial rate depends on the initial reactant concentrations. The order with respect to each reactant is deduced from the experimental data.

For example, the initial rate of the reaction described in Equation 1 is determined in a series of experiments. The results of these experiments are presented in Table 1.

Comparing Experiments 1 and 2, we see that the initial concentrations of A remain the same, but the initial concentration of B doubles from 0.00250 M to 0.00500 M. We also note that the *Initial Rate* doubles from $1.23 \times 10^{-3} M/s$ to $2.46 \times 10^{-3} M/s$. Mathematically, we solve for the ratio of the rate laws for Experiments 1 and 2 to determine n. Substituting the experimental data for Experiment 1 into Eq. 2 gives

$$1.23 \times 10^{-3}\ M/\mathrm{s} = k(0.100\ M)^m(0.00250\ M)^n$$

Similarly for Experiment 2,

$$2.46 \times 10^{-3}\ M/\mathrm{s} = k(0.100\ M)^m(0.00500\ M)^n$$

The ratio for Experiment 2 to Experiment 1 is

$$\frac{\text{rate}_{\text{experiment 2}}}{\text{rate}_{\text{experiment 1}}} = \frac{2.46 \times 10^{-3}\ \cancel{M/s}}{1.23 \times 10^{-3}\ \cancel{M/s}} = \frac{\cancel{k(0.100\ M)^m}(0.00500\ M)^n}{\cancel{k(0.100\ M)^m}(0.00250\ M)^n}$$

$$\frac{\text{rate}_{\text{experiment 2}}}{\text{rate}_{\text{experiment 1}}} = 2 = \left(\frac{0.00500}{0.00250}\right)^n$$

$$2 = 2^n$$

$$n = 1$$

Because $n = 1$, the reaction is 1st order with respect B. Therefore, the reaction rate is directly proportional to the concentration of B. For example, if we double the concentration of B, the reaction rate doubles. Mathematically, we write the rate law expression as

$$\text{rate} = k[\text{A}]^m[\text{B}]^1$$

The same method can be used to determine the reaction order, m, with respect to A. In Experiments 2 and 3, the initial concentrations of B remain

Table 1 *Experimental data for initial rate method*

Experiment	*Initial [A], M*	*Initial [B], M*	*Initial Rate, M/s*
1	0.100	0.00250	1.23×10^{-3}
2	0.100	0.00500	2.46×10^{-3}
3	0.200	0.00500	9.84×10^{-3}

the same, but the initial concentration of A doubles from 0.100 M to 0.200 M. The observed initial rate increases by a factor of 4 from $2.46 \times 10^{-3} M/s$ to $9.84 \times 10^{-3} M/s$. Mathematically, we solve for the ratio of the rate laws for Experiments 2 and 3 to determine m.

$$\frac{\text{rate } 3}{\text{rate } 2} = \frac{9.84 \times 10^{-3} \cancel{M/s}}{2.46 \times 10^{-3} \cancel{M/s}} = \frac{\cancel{k}(0.200\ M)^m \cancel{(0.00500\ M)^1}}{\cancel{k}(0.100\ M)^m \cancel{(0.00500\ M)^1}}$$

$$\frac{\text{rate } 3}{\text{rate } 2} = 4 = \left(\frac{0.200}{0.100}\right)^m$$

$$4 = 2^m$$

$$m = 2$$

Because $m = 2$, the reaction is 2nd order with respect to A. The reaction rate is directly proportional to the square of the initial concentration of A. For example, if the concentration of A is reduced by one half, the reaction rate is reduced by one fourth. The rate law expression is written as

$$\text{rate} = k[\text{A}]^2[\text{B}]^1$$

The reaction is second order with respect to A ($m = 2$), first order with respect to B ($n = 1$), and third order overall ($m + n = 3$).

The specific rate constant, k, is calculated by substituting any of the three sets of experimental data into the rate law expression. Using the data from Experiment 1 gives

$$\text{rate} = k[\text{A}]^2[\text{B}]^1$$

$$1.23 \times 10^{-3}\ M/s = k(0.100\ M)^2(0.00250\ M)^1$$

$$k = \frac{1.23 \times 10^{-3}\ M/s}{(0.0100\ M^2)(0.00250\ M)} = 49.2\ M^{-2}s^{-1}$$

Integrated Rate Laws

The rate equation for any reaction can be integrated using calculus. The simplest type of reaction is

$$a\text{A} \rightarrow \text{products} \qquad \text{(Eq. 3)}$$

where a is the stoichiometric coefficient of A, the lone reactant in this reaction. There are three possible rate laws for this type of reaction: 1) the reaction is zero order with respect to A, 2) the reaction is first order with respect to A, or 3) the reaction is second order with respect to A.

$$\text{rate} = k[\text{A}]^0$$
$$\text{rate} = k[\text{A}]^1$$
$$\text{rate} = k[\text{A}]^2$$

Integrated rate laws express reactant concentrations as a function of time (t). Table 2 summarizes the integrated rate laws for the different reaction orders for the reaction in Eq. 3.

Table 2 *Summary of the integrated rate laws for different reaction orders for a A → products.*

	Reaction Order		
	zero	*first*	*second*
Rate law	*rate* = *k*	*rate* = *k*[*A*]	*rate* = *k*[*A*]^2
Integrated Rate Law	$[A] = -akt + [A]_0$	$\ln[A] = -akt + \ln[A]_0$	$\frac{1}{[A]} = akt + \frac{1}{[A]_0}$
Plot that gives straight line	[A] versus *t*	ln[A] versus *t*	$\frac{1}{[A]}$ versus *t*
Slope	−*ak*	−*ak*	*ak*
Half-life	$t_{1/2} = \frac{[A]_0}{2k}$	$t_{1/2} = \frac{\ln 2}{ak}$	$t_{1/2} = \frac{1}{k[A]_0}$

Integrated rate laws can also be used to determine the reaction order. This method employs graphical analysis of experimental data using various forms of the integrated rate equations. The concentration of a reactant is measured as a function of time (*t*). The concentration and time data are plotted in different ways. If the reaction obeys *zero* order kinetics, a plot of [A] vs. *t* should be linear. If the reaction obeys *first* order kinetics, a plot of ln[A] vs. *t* should be linear. If the reaction obeys *second* order kinetics, a plot of 1/[A] vs. *t* plot should be linear. Only one of these three plots will appear linear (assuming the reaction times are not too short) for a given reaction. Once the reaction order is determined, the rate constant, *k*, can be calculated from the slope of the line from the linear plot (see Table 2).

The **half-life** of a reactant is the time required for one half of a reactant to be converted to a product or products. When the time for a reaction, *t*, equals the half-life, $t_{1/2}$, the concentration of A is equal to one half of the initial concentration of A ($[A] = \frac{1}{2}[A_0]$). Substituting $\frac{1}{2}[A]_0$ for [A] into the integrated rate equations for a zero, first, or second order equation, we can derive equations to calculate the half-life for each reaction order (see Table 2).

Alternatively, the half-life can be determined by monitoring changes in the concentration of a colored reactant as a function of time. The time for the concentration of a reactant to decrease by one half is the half-life of the reaction. For example, assume the concentration of A in Eq. 3 is 0.30 *M* at the beginning of the reaction. If the time required for the concentration of A to decrease to 0.15 *M* is 75 seconds, the half-life of the reaction is 75 seconds.

Using the MeasureNet Colorimeter to Monitor Changes in the Concentration of a Colored Reactant

In this experiment, the MeasureNet colorimeter will be used to monitor changes in the concentrations of colored reactants. A **colorimeter** is an instrument that measures a solution's absorbance as a function of time. The absorbance of a solution is related to its concentration according to Beer-Lambert's Law,

$$A = \varepsilon bc \quad \text{(Eq. 4)}$$

where A is the absorbance of the solution, ε is the molar absorptivity coefficient of the absorbing species, b is the path length of the light, and c is the concentration of the solution. Beer-Lambert's Law can be simplified if the same absorbing species and same sample container are used in a series of experiments. In that case, ε and b are constant and Beer-Lambert's Law simplifies to

$$A = (\text{constant})c \quad \text{(Eq. 5)}$$

In this experiment, we will monitor changes in the concentration of a colored reactant as a function of time. Equation 5 indicates that the absorbance of the reactant is directly proportional to its concentration in solution. *Therefore, we can substitute absorbance values for the actual reactant concentrations into the integrated rate equations.* Thus, we will prepare plots of absorbance versus time, ln(absorbance) versus time, and 1/absorbance versus time to determine the order of the reaction.

Determination of the Rate Law for the Iodination of Acetone in Acidic Solution

In Part A of this experiment, we will determine the rate law for the iodination of acetone (CH_3COCH_3, also called 2-propanone) in the presence of an acid catalyst (HCl). The overall equation for the reaction is

$$CH_3COCH_{3(aq)} + I_{2(aq)} \xrightarrow{H^+} CH_3COCH_2I_{(aq)} + HI_{(aq)} \quad \text{(Eq. 6)}$$

The rate of this reaction should depend on the concentrations of acetone, iodine, and H^+ as reflected in Eq. 7,

$$\text{rate} = k[CH_3COCH_3]^m[I_2]^n[H^+]^p \quad \text{(Eq. 7)}$$

where k is the specific rate constant, and m, n, and p are the orders of the reaction with respect to acetone, I_2, and H^+, respectively.

Based on previous experimental studies, the order with respect to I_2, n, is assumed to be zero. (This assumption will be verified in this experiment.) Thus, $[I_2]^n = [I_2]^0 = 1$, and the rate law expression in Eq. 7 can be simplified as

$$\text{rate} = k[\text{acetone}]^m[H^+]^p \quad \text{(Eq. 8)}$$

To monitor reaction rates visually or using the MeasureNet colorimeter, one of the species in the reaction must have a detectable color. Acetone and HCl form clear, colorless solutions. However, aqueous iodine (I_2) solutions at low concentrations have a distinct yellow color. The rate of disappearance of I_2 (yellow color) will be used to monitor the rate of the reaction.

In this experiment, I_2 is the limiting reactant in the presence of a large excess of acetone and HCl. We will measure the time required for the known initial concentration of I_2 to be completely consumed. Given that the concentrations of acetone and HCl are large, the change in their concentrations is negligible in the course of an individual experiment. The reaction rate remains essentially constant until I_2 is completely consumed, at which time the reaction stops.

The overall reaction rate can be expressed as the change in concentration of I_2, $\Delta[I_2]$, divided by the time required for the change to occur, Δt:

$$\text{rate} = -\frac{\Delta[I_2]}{\Delta t} \qquad \text{(Eq. 9)}$$

Since $\Delta[I_2]$ is negative ($[I_2]$ decreases as the reaction proceeds), a negative sign is added to make the reaction rate positive. If it takes t seconds for the yellow color of iodine to disappear, the *rate of the reaction* becomes

$$\text{rate} = -\frac{\Delta[I_2]}{\Delta t} = -\frac{0 - [I]_{\text{initial}}}{t - 0} = \frac{[I]_{\text{initial}}}{t} \qquad \text{(Eq. 10)}$$

Using the method of initial rates, we will determine the reaction orders with respect to acetone and hydrochloric acid, and verify that the order of reaction is zero with respect to iodine. In a series of 4 experiments, the concentrations of acetone, HCl, and I_2 will be systematically varied. The reaction rate for each experiment will be determined by monitoring the time required for the color of I_2 to disappear, $[I]_{\text{initial}}/t$. From the experimental data, we can deduce the reaction order with respect to each reactant.

Also in Part A of the experiment, we will verify the value of n (order with respect to I_2) colorimetrically. We will monitor the disappearance of I_2 as a function of time. Since the initial concentrations of acetone and HCl are much larger than that of I_2, they remain essentially constant during the course of the reaction. The rate law in Equation 7 becomes

$$\text{rate} = k'[I_2]^n \qquad \text{(Eq. 11)}$$

where the pseudo rate constant $k' = k[\text{acetone}]^m[H^+]^p$. From the MeasureNet absorbance data, we will prepare plots of absorbance vs. time, ln(absorbance) vs. time, and 1/(absorbance) vs. time to determine the reaction order with respect to iodine.

Determination of the Rate Law for the Reaction of Crystal Violet with Sodium Hydroxide

In Part B of this experiment, we will determine the rate law for the reaction of crystal violet (abbreviated as CV^+) with hydroxide ions (OH^-). Crystal violet is a brightly colored organic dye and a pH indicator. The molecular structure of crystal violet is shown in Figure 1. The multiple double bonds in the structure permit π electrons in the benzene rings to be delocalized throughout the molecule, forming a large conjugated system. This highly conjugated system is responsible for the violet color of the CV^+ ion in

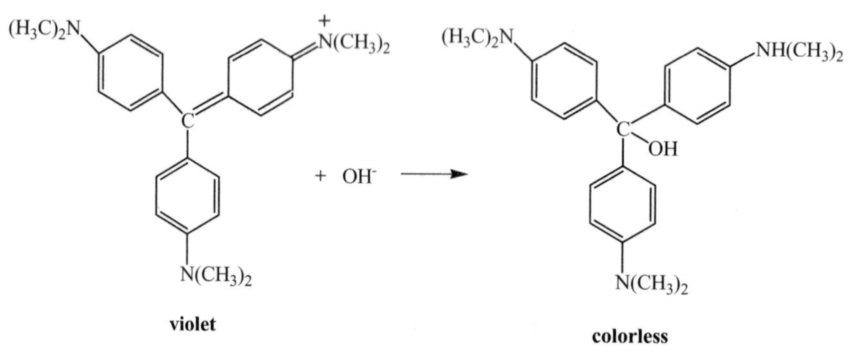

(H₃C)₂N — N⁺(CH₃)₂ — C — N(CH₃)₂ + OH⁻ ⟶ (H₃C)₂N — NH(CH₃)₂ — C — OH — N(CH₃)₂

violet — colorless

Figure 1
Reaction of crystal violet with sodium hydroxide

solution. When CV^+ reacts with OH^-, the addition of OH^- to the ''central'' carbon atom disrupts the conjugation of the system, and the solution becomes colorless.

In a simplified form the chemical equation for the reaction can be written as

$$CV^+ + OH^- \rightarrow CVOH \qquad \text{(Eq. 12)}$$

The rate law for the reaction is

$$\text{rate} = k[CV^+]^q[OH^-]^r \qquad \text{(Eq. 13)}$$

where k is the rate constant, $[CV^+]$ and $[OH^-]$ are the concentrations of crystal violet and of OH^- in the reaction mixture, respectively; q and r are orders with respect to crystal violet and the hydroxide ion, respectively.

In Part B of the experiment, we will determine the reaction order, q, with respect to crystal violet. Crystal violet will be the limiting reactant in this experiment. Since the initial concentration of OH^- is much larger than that of crystal violet, the concentration of OH^- remains constant during the course of the reaction. The rate law becomes

$$\text{rate} = k'[CV^+]^q \qquad \text{(Eq. 14)}$$

where the pseudo rate constant $k' = k[OH^-]^r$.

We will monitor the change in the concentration of crystal violet as it reacts with OH^- using the MeasureNet colorimeter. Plots of absorbance vs. time, ln(absorbance) vs. time, and 1/(absorbance) vs. time will be prepared. Linear regression analysis will be performed for each plot. The plot that yields the straightest line is used to determine the order, q, with respect to crystal violet. The pseudo rate constant, k', is determined from the slope of the line from the linear plot.

The reaction order, r, with respect to OH^- will be determined. We will perform two experiments using different OH^- concentrations (referred to as $[OH^-]_1$ and $[OH^-]_2$), while the concentration of CV^+ remains constant. The pseudo rate constants for the two experiments, k'_1 and k'_2, will be calculated from the slopes of the straight lines from the linear plots. Substituting the values of $[OH^-]_1$, $[OH^-]_2$, k'_1 and k'_2 into Equation 15, the ratio of k'_1 to k'_2 can be solved to determine the reaction order, r, with respect to OH^-.

$$\frac{k'_1}{k'_2} = \frac{k[OH^-]_1^{\,r}}{k[OH^-]_2^{\,r}}$$

$$\frac{k'_1}{k'_2} = \left(\frac{[OH^-]_1}{[OH^-]_2}\right)^r \qquad \text{(Eq. 15)}$$

Substituting the values of k', $[OH^-]$, and r (from either experiment) into Equation 16 yields the specific rate constant for the reaction.

$$k = \frac{k'}{[OH^-]^r} \qquad \text{(Eq. 16)}$$

PROCEDURE

CAUTION

Students must wear departmentally approved eye protection while performing this experiment. Wash your hands before touching your eyes and after completing the experiment.

Chemical Alert

Crystal violet and iodine will stain clothing and skin. Be careful not to spill crystal violet and iodine solution on your skin or clothes.

Gloves should be worn when performing this experiment.

Part A – Determination of the Rate Law for the Iodination of Acetone in Acidic Solution

1. Add approximately 60 mL of 4.0 *M* aqueous acetone, 60 mL of 1.0 *M* HCl, and 60 mL of 0.0050 *M* I_2/0.050 *M* KI solution to three separate beakers (one solution per beaker).
2. Obtain two clean, dry 125-mL Erlenmeyer flasks.
3. Add 50.0 mL of distilled water to one of the Erlenmeyer flasks. This will be used as a reference flask for color comparison purposes in subsequent steps.
4. In the first of four experiments (see Table 3), add 10.0 mL of 4.0 *M* acetone, 10.0 mL of 1.0 *M* HCl, and 20.0 mL of distilled water to the second 125-mL Erlenmeyer flask from Step 2.
5. Place the two Erlenmeyer flasks side by side on a sheet of white paper.
6. Add 10.0 mL of 0.0050 *M* I_2/0.050 *M* KI solution to the Erlenmeyer flask containing the acetone/HCl mixture and **immediately begin timing the reaction once you add the I_2/KI solution to the acetone/HCl mixture**. Continually swirl the flask to mix the solution during the experiment.
7. Initially, the solution will appear yellow due to the presence of I_2. The yellow color will fade slowly as I_2 reacts with the acetone. When the yellow color just disappears (it is colorless like the distilled water in the reference flask), stop timing. Should you record the time in the Lab Report (see Note)?

Table 3 *Experiments to determine reaction orders for the iodination of acetone*

Experiment	*mL of 4.0 M acetone*	*mL of 1. 0 M HCl*	*mL of 0.0050 M I_2/0.050 M KI*	*mL of distilled H_2O*
1	10	10	10	20
2	—	—	—	—
3	—	—	—	—
4	—	—	—	—

NOTE: In this experiment a significant quantity of data will be collected and presented. The use of tables is an efficient way to report large quantities of data. It is recommended that the data collected in Steps 4–18 be presented in the Lab Report in tabular form (similar to Table 3).

---

8. You must determine the reaction temperature. How do you determine the reaction temperature, and should you record it in the Lab Report?
9. Pour the reaction mixture into the Waste Container.
10. Repeat Steps 4–9 to perform a second trial. The amount of time required for the two trials should agree within 20 seconds of each other. The reaction temperatures should agree within ±2.0 °C. If either or both conditions are not met, perform additional trials.
11. Repeat Steps 4–10 for Experiments 2, 3, and 4 in Table 3. In Experiments 2–4, double the concentration of either acetone, HCl, or I_2 one at a time (the concentration of the other two will remain the same as in Experiment 1). How do you double the concentration of one of the reactant solutions in each experiment? The total volume of the reactants in Experiments 2–4 must be 50 mL (the same total volume used in Experiment 1). What reagent should you use to control the total volume of each Experiment? Record the volume of each reagent used in tabular form in the Lab Report.

    Could you reduce the concentration of a reactant by one half in Experiments 2–4 and achieve the same purpose as doubling their concentration? Why or why not?
12. *Steps 13–18 are to be performed after the experiment is concluded (outside of the lab). Proceed to Step 19.*
13. What is the average time required for the disappearance of the yellow color for Experiments 1–4? Should the average time be recorded in a table in the Lab Report?
14. What are the initial concentrations of the reactants in each experiment? Should the initial concentrations of the reactants be recorded in a table in the Lab Report?
15. What is the reaction rate for each experiment? Should the reaction rate for each experiment be recorded in a table in the Lab Report?
16. Determine the reaction orders with respect to each reactant: acetone, H^+, and I_2. Determine the overall order for the reaction.
17. What is the specific rate constant for each Experiment?
18. What is the average rate constant for the reaction? Write the rate law expression for the reaction.
19. Press **On/Off** switch to turn on the power to the MeasureNet workstation.
20. Press **Main Menu**, then press **F6 Colorimetry**, then press **F1 Colorimetry**.
21. Press **F1 Red LED**, then press **F1 Kinetics**.
22. Fill two cuvettes with solvent (0.050 *M* KI solution). Place one cuvette in the reference cell (labeled **R**) and the second cuvette in the solvent cell (labeled **S**) inside the colorimeter.
23. Close the colorimeter lid, then press **Enter**. Wait while the colorimeter adjusts 0% T and 100% T. After the tone (a beep), press **Display**, then **Press Start**.

24. Remove the cuvette from the sample cell (**S**). Leave the other cuvette in the reference cell (**R**).
25. Add 1.0 mL of 4.0 *M* acetone and 1.0 mL of 1.0 *M* HCl to a clean, dry cuvette.

NOTE: Steps 26–28 must be carried out quickly, but as carefully as possible.

---

26. See the Note in the margin. Add 2.0 mL 0.0050 *M* I_2/0.050 *M* KI solution to the cuvette containing the acetone/HCl mixture. **At the moment you add the I_2/KI solution to the cuvette**, press **Start** on the workstation to start timing the reaction.
27. Quickly place the cap on the cuvette. Securing the cap with your fingers, invert the cuvette 2–3 times to mix the solution.
28. Place the cuvette in the sample cell (**S**) inside the colorimeter. Close the lid. Press **Start** on the workstation to start monitoring the absorbance change. After 400 seconds have elapsed, press **Stop**. If the absorbance value reaches as low as 0.01 before 400 seconds elapse, press **Stop**.
29. Press **File Options**. Press **F3** to save the scan as a tab delimited file. You will be prompted to enter a 3 digit code (any 3 digit number you choose). Press **Enter**.
30. *Steps31–49 are to be performed after the experiment is concluded (outside the lab). Proceed to Step 50.*
31. Open an Excel spreadsheet (i.e., worksheet).
32. Go to **File Open**, open a MeasureNet tab delimited file containing time and absorbance data. Click **Finish**.
33. Copy the first two columns (containing time and absorbance data) in the tab delimited (*text*) file, and paste it into columns A (seconds) and B (absorbance) in the Excel worksheet. Close the tab delimited file.
34. In column C, calculate the natural logarithms (ln) of each absorbance value in column B. Click box C1. Then click inside of the equation box at the top of the page (box beside the X, √, and f_x symbols). Type = ln(B1). Double click the black square in the lower right corner of box C1. This should automatically calculate a ln(absorbance) value in column C for each absorbance value in column B.
35. In column D, calculate the inverse of each absorbance value in column B. Click box D1. Then click inside of the equation box at the top of the page (box beside the X, √, and f_x symbols). Type =1/(B1). Double click the black square in the lower right corner of box D1. This should automatically calculate a 1/(absorbance) value in column D for each absorbance value in column B.
36. Click on the "**Chart Wizard**" Icon on the Toolbar in the Excel program (looks like a bar graph).
37. Click on **XY scatter plot**. Click on the "scatter with data points connected by smooth line with no markers."
38. Click **Next**. Click **Series**, then click **Add**. Click in the **x values box**. Select all data in column A (time in seconds). Click in the **y values box**. Select all data in column B (absorbance).

39. Click **Next**. Click **Titles**. Enter a name for your plot (e.g., Absorbance vs. Time for the Iodination of Acetone). The **x-value box** is for labeling the x-axis (time, s), and the **y-value box** is for labeling the Y axis (absorbance) on your plot.
40. If you wish to remove the gridlines, Click on **Gridlines** and click on the axes that are checked to turn off the gridlines. This is a suggested cosmetic function, it is not necessary.
41. Click **Legend**. Click on **Show Legend** to remove the legend, it is not required for a single curve.
42. Click **Next**. You can save the plot as a separate sheet which can then be printed. Alternatively, you can save it as an object in the current worksheet. The plot will appear in the "new" Excel worksheet beside columns A and B. If you save the file in this manner, you will have to select and highlight a copy area box (i.e., highlight the plot area) under **Page Set-Up** to print the plot.
43. Alternatively, you can select the plot, copy, and paste it into a "**Word**" document, and print the plot from within the word document.
44. There are numerous other functions and options available to you to enhance the appearance of your plot. Trial and error is the best way for you to become proficient with Excel.
45. **Scaling a Plot** – You may scale the plot to exhibit only those portions of a scan that are significant to your experiment.
46. Repeat Steps 36–45 to prepare a ln(absorbance) vs. time plot. Select all data in column C, ln(Absorbance), for the **y values box** in Step 38, instead of Column B. Change the title of the plot accordingly.
47. Repeat Steps 36–45 to prepare 1/(absorbance) vs. time plot. Select all data in column D, 1/(Absorbance), for the **y values box** in Step 38, instead of Column B. Change the title of the plot accordingly.
48. Perform linear regression analysis by following Steps 10–13 of **Appendix B-2**. Based on the R^2 values, which of the three plots prepared in Steps 36–47 is most linear? Determine the reaction order with respect to I_2. Does the order, n, with respect to I_2 determined from the method of initial rates agree with the value obtained from the plots of the integrated rate laws?
49. Submit the Excel plots along with your Lab Report.

Part B – Determination of the Rate Law for the Reaction of Crystal Violet with Sodium Hydroxide

50. Press **Main Menu**, then press **F6 Colorimetry**, then press **F1 Colorimetry**.
51. Press **F2Green LED**, then press **F1 Kinetics**.
52. Fill two cuvettes with solvent (2% ethanol/98% water). Place one cuvette in the reference cell (labeled **R**) and the second cuvette in the solvent cell (labeled **S**) inside the colorimeter.
53. Close the colorimeter lid, press **Enter**. Wait while the colorimeter adjusts 0% T and 100% T. After the tone (a beep), press **Display**, then press **Start**.
54. Remove the cuvette from the sample cell (**S**). Leave the other cuvette in the reference cell (**R**).

55. Add 3.0 mL of $3.0 \times 10^{-5} M$ crystal violet solution to a clean, dry cuvette.

NOTE: Steps 56–58 must be carried out as quickly as possible.

---

56. See the Note in the margin. Add 1.0 mL 0.10 *M* NaOH solution to the cuvette containing the crystal violet solution. Press **Start** on the workstation **at the moment you add the NaOH solution to the cuvette**.
57. Quickly place the cap on the cuvette. Securing the cap with your fingers, invert the cuvette 2–3 times to mix the solution.
58. Place the cuvette in the sample cell (**S**) inside the colorimeter. Close the lid. Press **Start** on the workstation. After 400 seconds have elapsed, press **Stop**.
59. Press **File Options**. Press **F3** to save the scan as a tab delimited file. You will be prompted to enter a 3 digit code (any 3 digit number you choose). Press **Enter**. *Each tab-delimited file must be assigned a unique 3 digit code.*
60. Repeat Steps 55–59 using 0.050 *M* NaOH solution (trial 2).
61. Repeat Steps 31–47 to plot absorbance vs. time, ln(absorbance) vs. time, and 1/(absorbance) vs. time for each trial. In Step 39, the title should be Absorbance vs. Time for the Reaction of CV^+ with NaOH.
62. Perform linear regression analysis by following Steps 10–13 of **Appendix B-2**. Based on the R^2 values, which of the three plots prepared in Step 61 is most linear for each trial? Determine the reaction order with respect to crystal violet.
63. Submit the Excel plots along with your Lab Report.
64. Determine the slope of the straight line from the most linear of the three plots.
65. What is the pseudo rate constant, k', for each trial from the slope of the line.
66. What is the initial concentration of OH^- in the reaction mixture for each trial.
67. What is the reaction order, r, with respect to OH^-.
68. Determine the specific rate constant, k, for the reaction from the pseudo rate constant, the initial concentration of OH^- in the reaction mixture, and the reaction order with respect to OH^- for each trial.
69. What is the average rate constant?
70. Write the rate law expression for the reaction of CV^+ with OH^-.

Name Section Date

Instructor

23 EXPERIMENT 23

Lab Report

Part A – Determination of the Rate Law for the Iodination of Acetone in Acidic Solution

Data and observations for Experiments 1–4 to determine initial concentrations of reactants, average reaction times, and reaction orders for the iodination of acetone. *The data (from Steps 4–10 and 13–15) should be presented in tabular form.*

How do you double the concentration of one of the reactant solutions in each experiment? What reagent should you use to control the total volume of each Experiment?

Could you reduce the concentration of a reactant by one half in Experiments 2–4 and achieve the same purpose as doubling their concentration? Why or why not?

Determine the reaction orders with respect to each reactant: acetone, H^+, and I_2.

Determine the overall order for the reaction.

What is the specific rate constant for each Experiment?

What is the average rate constant for the reaction?

Write the rate law expression for the reaction.

Which of the three plots is the most linear?

Determine the reaction order with respect to I_2.

Does the order, n, with respect to I_2 determined from the method of initial rates agree with the value obtained from the plots of the integrated rate laws?

Which of the three plots is the most linear?

Determine the reaction order with respect to crystal violet.

Determine the slope of the straight line.

What is the pseudo rate constant, k', from the slope of the straight line.

What is the initial concentration of OH^- in the reaction mixture?

What is the reaction order, *r*, with respect to OH^-?

Determine the specific rate constant, *k*, for the reaction from the pseudo rate constant, the initial concentration of OH^- in the reaction mixture, and the reaction order with respect to OH^- for each trial.

What is the average rate constant?

Write the rate law expression for the reaction of CV^+ with OH^-.

Name | Section | Date

Instructor

23 EXPERIMENT 23

Pre-Laboratory Questions

1. The kinetics for the reaction 2 A + 3 B → C + 2 D were studied at 20 °C, and the following data were obtained by monitoring the rate of consumption of A. If it takes *t* seconds for A to be consumed, the initial rate for the consumption of A is equal to the initial concentration of A divided by the product of the time required for A to be consumed and the stoichiometric coefficient of A.

$$\text{rate} = -\frac{1}{a}\left(\frac{\Delta[A]}{\Delta t}\right) = \frac{1}{2}\left(\frac{[A]_{initial}}{t_{consuumption\ of\ A}}\right)$$

Experiment	*Volume in mL* 0.0080 M *A*	*Volume in mL* 0.20 M *B*	*Volume in mL distilled water*	*Time required for consuming A in seconds*
1	10.0	10.0	30.0	82
2	20.0	10.0	20.0	41
3	10.0	30.0	10.0	82

A. Calculate the initial concentrations of A and B, and the initial rate for each Experiment. Enter the results in the table below. *Show all calculations.*

Experiment	*Initial [A] in* M	*Initial [B] in* M	*Initial Rate for Consuming A in* M/s
1			
2			
3			

B. Determine the reaction orders with respect to A and B.

C. Determine the overall order of the reaction.

D. Calculate the specific rate constant for each reaction mixture.

E. Calculate the average rate constant for the reaction.

F. Write the rate law expression for the reaction.

2. The decomposition of A is studied and the following experimental data were obtained.

$$A \rightarrow \text{products}$$

Time, s	*[A]*	*ln[A]*	*1/[A]*
10	0.234		
50	0.193		
100	0.155		
150	0.124		
200	0.100		
250	0.082		
275	0.075		

A. Complete the following. Prepare Excel plots of [A] vs. time, ln[A] vs. time, and 1/[A] vs. time for the data in the table above (see Steps 34–45 in the Procedure for Excel instructions). Submit the plots along with your Pre-Lab Questions to your instructor.

B. Determine the reaction order for the conversion of A to the products. Justify your answer with an explanation.

C. Calculate the specific rate constant for the reaction.

D. Is it possible to estimate the half-life for the conversion of A to the products from the *Time, s* and *[A]* data in the table? If so, what is the estimated half-life of the reaction?

Name | Section | Date

Instructor

23 EXPERIMENT 23

Post-Laboratory Questions

1. The reaction A + 3B → 2C is second order with respect to A and first order with respect to B. How will the initial reaction rate change if the initial concentration of A is halved and the initial concentration of B is doubled? Justify your answer with an explanation.

2. If 0.035 *M* NaOH is used instead of 0.050 *M* NaOH in Step 60, how would the experimentally determined reaction order with respect to OH^- be affected? Justify your answer with an explanation.

3. A student performed a series of experiments to study the rate law of a chemical reaction, $A + 2B \rightarrow C + 2D$. The initial concentration of B is held constant at 0.010 *M* in all trials. The half-life of the reaction is 98 seconds if the initial concentration of A is 0.020 *M*. The half-life of the reaction is 65 seconds if the initial concentration of A is 0.030 *M*. Which graph, [A] vs. time, ln[A] vs. time, or 1/[A] vs. time, should be linear for the reaction? Write the rate law, including the value of the rate constant, for the reaction.

EXPERIMENT 24

Le Châtelier's Principle

PURPOSE

Observe Le Châtelier's principle in action as chemical systems at equilibrium respond to different stresses.

INTRODUCTION

Chemical reactions attain a reaction rate that depends upon the nature and concentration of the reactants and the reaction temperature. For a given reaction performed at a constant temperature, the reaction rate depends solely on the concentrations of the species. To understand chemical equilibrium, we must realize that a chemical reaction involves two opposing processes: the reaction in the forward direction in which the reactants react to form the products, and the reaction in the reverse direction in which the products react to form reactants. For example, consider the hypothetical reaction

$$aA \rightleftharpoons bB \qquad \text{(Eq. 1)}$$

where a and b represent the stoichiometric coefficients and A and B represent the reactants and products involved in the reaction. If we assume that the reaction is an elementary reaction, the forward reaction rate (which describes how quickly A forms B) has the mathematical form

$$\text{rate}_{\text{forward}} = k_f[A]^a \qquad \text{(Eq. 2)}$$

The reverse reaction rate (which describes how quickly B reforms A) has the mathematical form

$$\text{rate}_{\text{reverse}} = k_r[B]^b \qquad \text{(Eq. 3)}$$

Notice that the reaction rates depend on the concentrations of each species. Thus, if the concentrations are changed, the rates of formation of the products and reactants also change.

At equilibrium, the forward reaction rate equals the reverse reaction rate. Externally, it appears that nothing is happening in chemical reactions at equilibrium. However, if we could see the atoms, ions, or molecules involved in a reaction at equilibrium, they are far from static. Reactants are forming products and products are forming reactants at the same rate.

It should be noted that all chemical reactions, even those that *''go to completion''*, attain equilibrium. In those cases, the product equilibrium concentrations are very large compared to the reactant equilibrium concentrations.

Because the forward and reverse reaction rates are equal, we can set Eq. 2 equal to Eq. 3 and derive the equilibrium constant expression.

$$\text{rate}_{\text{forward}} = \text{rate}_{\text{reverse}}$$
$$k_f[A]^a = k_r[B]^b$$
$$\frac{k_f}{k_r} = \frac{[B]^b}{[A]^a} \quad \text{(Eq. 4)}$$
$$K_c = \frac{[B]^b}{[A]^a}$$

Because k_f and k_r (reaction rate constants) are constant at a given temperature, their ratio, k_f/k_r, is also a constant. This constant, K_c, is the called the **equilibrium constant**. Notice that K_c is a ratio of the product concentrations, raised to their stoichiometric powers, divided by the reactant concentrations raised to their stoichiometric powers.

For a more complex reaction, such as the hypothetical reaction given in Eq. 5

$$aA + bB \rightleftharpoons cC + dD \quad \text{(Eq. 5)}$$

the equilibrium constant expression is written as

$$K_c = \frac{[C]^c[D]^d}{[A]^a[B]^b} \quad \text{(Eq. 6)}$$

The magnitude of the value of K_c is a measure of the extent to which a reaction occurs.

If K_c > 10, equilibrium product concentrations >> reactant concentrations

If K_c < 0.1, equilibrium reactant concentrations >> product concentrations

If 0.1 < K_c < 10, neither equilibrium product or reactant concentrations predominate

Changes (stresses) that affect a reaction rate will also affect reactant and product equilibrium concentrations. **Le Châtelier's principle** states that a system at equilibrium changes in a manner that tends to relieve the stress placed on the system. Stresses that disturb a reaction at equilibrium include changes in concentration, changes in the reaction temperature, or changes in the pressure or volume (for gaseous reactions). These stresses preferentially affect the rate of either the forward or the reverse reaction. The forward and reverse reaction rates are unequal until the reaction can reestablish equilibrium.

For example, if the reactant concentrations are increased, the forward reaction rate exceeds the reverse reaction rate and the equilibrium shifts to the right (product side). If the product concentrations are increased, the reverse reaction rate exceeds the forward reaction rate and the equilibrium shifts to the left (reactant side).

Effect of Concentration Changes on Systems at Equilibrium

Assume that the reaction shown below is at equilibrium in a closed reaction vessel.

$$N_{2(g)} + O_{2(g)} \rightleftharpoons 2\ NO_{(g)} \qquad \text{(Eq. 7)}$$

What happens to the equilibrium if more N_2 is added to the vessel? In this case, the stress applied to the equilibrium initially increases the concentration of N_2. To offset this stress, some O_2 reacts with the N_2, producing more NO and the equilibrium shifts to the right (favors the forward reaction). The N_2 and O_2 concentrations decrease while the NO concentration increases until a new equilibrium is established.

What happens to the equilibrium if more NO is added to the reaction vessel? The stress applied to the equilibrium is an increase in the concentration of NO. Some NO decomposes producing more N_2 and O_2, the equilibrium shifts to the left (the reverse reaction is favored). The NO concentration decreases and the N_2 and O_2 concentrations increase to reestablish equilibrium.

What happens to the equilibrium if some NO is removed from the equilibrium system? The stress applied to the equilibrium is a decrease in the concentration of NO. In that case, N_2 reacts with O_2 to replenish the NO that was removed from the system. The equilibrium shifts to the right, favoring the forward reaction.

In Part A of this experiment, we will study the effects of changing reactant and product concentrations in an aqueous chemical system at equilibrium. One reaction that visually illustrates Le Châtelier's principle is the reaction of solid antimony trichloride ($SbCl_3$) with water. When solid antimony trichloride ($SbCl_3$) is dissolved in water, antimonyl chloride (SbOCl) precipitates according to Equation 8.

$$SbCl_{3(s)} + H_2O_{(\ell)} \rightleftharpoons \underset{\text{white precipitate}}{SbOCl_{(s)}} + 2\ HCl_{(aq)} \qquad \text{(Eq. 8)}$$

By adding either distilled water or hydrochloric acid and monitoring the presence or absence of the precipitate, we can illustrate the effects of changing the reactant and product concentrations on an equilibrium system.

Effect of Changing pH on a Complex Ion Equilibrium

Most *d*-transition metals form complex ions in aqueous solution. These complexes tend to be brightly colored. The dissolution of cobalt(II) nitrate in water produces a pink colored solution from the formation of the hexaaquacobalt(II) ion, $Co(OH_2)_6^{2+}$. In the presence of concentrated HCl, the hexaaquacobalt(II) ions form tetrachlorocobalt(II) ions, $CoCl_4^{2-}$, that are blue colored in solution. We will use color changes (pink to blue and vice versa) to study the effects of changing the pH of the equilibrium mixture shown in Equation 9.

$$\underset{\text{pink}}{Co(OH_2)_6{}^{2+}{}_{(aq)}} + 4\ Cl^-{}_{(aq)} \rightleftharpoons \underset{\text{blue}}{Co(Cl)_4{}^{2-}{}_{(aq)}} + 6H_2O_{(\ell)} \qquad \text{(Eq. 9)}$$

Effect of Changing Reaction Temperature on Equilibrium

Changes in concentration, pressure, or volume, for gas phase reactions, shift the position of an equilibrium system, but do not change the value of the equilibrium constant. A change in the reaction temperature not only shifts the equilibrium, it also changes the numerical value of the equilibrium constant.

Consider the following *exothermic* reaction at equilibrium.

$$A + B \rightleftharpoons C + D + \text{heat} \qquad \text{(Eq. 10)}$$

Because the reaction is exothermic, heat is a product of the reaction. Increasing the reaction temperature has the same effect as increasing the concentration of C or D. The equilibrium responds by shifting to the left (favors the reverse reaction). The additional heat is absorbed by C and D and they react to produce A and B. The concentrations of A and B increase while the concentrations of C and D decrease until equilibrium is reestablished. Lowering the reaction temperature shifts the equilibrium to the right (favors the forward reaction). A and B react to produce C and D and to replace the heat that is removed when the reaction temperature is lowered. The concentrations of C and D increase while the concentrations of A and B decrease until equilibrium is reestablished.

Endothermic equilibrium reactions absorb heat as represented by Equation 11.

$$A + B + \text{heat} \rightleftharpoons C + D \qquad \text{(Eq. 11)}$$

Because heat is a reactant in endothermic reactions, increasing the reaction temperature has the same effect as increasing the concentration of A or B. The equilibrium responds by shifting to the right (favors the forward reaction). The additional heat is absorbed by A and B and they react to produce C and D. The concentrations of C and D increase while the concentrations of A and B decrease until equilibrium is reestablished. Lowering the reaction temperature shifts the equilibrium to the left (favors the reverse reaction). C and D react to produce A and B and to replace the heat that is removed when the reaction temperature is lowered. The concentrations of A and B increase while the concentrations of C and D decrease until equilibrium is reestablished.

We can summarize the effects of changing the reaction temperature of a system at equilibrium as follows:

For *exothermic* reactions

increasing the reaction temperature favors the *reverse reaction*
decreasing the reaction temperature favors the *forward reaction*

For *endothermic* reactions

increasing the reaction temperature favors the *forward reaction*
decreasing the reaction temperature favors the *reverse reaction*

In Part C of this experiment, we will reexamine the reaction presented in Eq. 9 for the effects of changing reaction temperature. From the color changes, we can determine if this is an exothermic or endothermic reaction.

PROCEDURE

CAUTION

Students must wear departmentally approved eye protection while performing this experiment. Wash your hands before touching your eyes and after completing the experiment.

Part A – Effect of Concentration Changes on Systems at Equilibrium

1. Add one or two crystals of antimony trichloride ($SbCl_3$) and 2 mL of distilled water to a 50-mL beaker. Stir the mixture with a stirring rod. Should you record your observations in the Lab Report?

Chemical Alert

Concentrated HCl is extremely corrosive. Do not allow it to contact your skin. If it does contact your skin, wash the affected area with copious quantities of water and inform your laboratory instructor.

Do not inhale concentrated HCl vapors. Perform this experiment in a hood or well-ventilated area.

2. Using a beral pipet, add 12 *M* hydrochloric acid (HCl) drop-wise, with stirring, until you observe a chemical change. Should you record your observations in the Lab Report?
3. Did the addition of HCl favor the products or reactants? Did the relative concentrations of $SbCl_3$, H_2O, and SbOCl increase or decrease? Justify your answer based on your observations from the previous step.
4. To the same beaker used in Step 2, add distilled water drop-wise, with stirring, until you observe a chemical change. Should you record your observations in the Lab Report?
5. How does the addition of H_2O affect the equilibrium? How did the relative concentrations of $SbCl_3$, SbOCl, and HCl change after the addition of H_2O? Justify your answer based on your observations from the previous step.
6. Decant the reaction mixture into the designated waste container.

Part B – Effect of Changing pH on a Complex Ion Equilibrium

7. Obtain two clean, dry 25 × 150 mm test tubes, and label them 1 and 2. Add 2 mL of 1.0 *M* $CoCl_2$ solution to test tubes 1 and 2.
8. Add 12 *M* HCl (concentrated) solution drop-wise, with stirring, to test tube 1 until you observe a chemical change. Should you record your observations in the Lab Report?
9. How did the addition of 12 *M* HCl affect the equilibrium (Eq. 9)?
10. How did the relative concentrations of $Co(OH_2)_6^{2+}$ and $CoCl_4^{2-}$ change after the addition of 12 *M* HCl? Justify your answer based on your observations from the previous step.
11. Decant the reaction mixture into the Waste Container.
12. Add 0.1 *M* $AgNO_3$ solution drop-wise, with stirring, to test tube 2 until you observe a chemical change. Should you record your observations in the Lab Report?
13. Is the equilibrium affected by the addition of 0.1 *M* $AgNO_3$ (Eq. 9)?

14. How did the relative concentrations of $Co(OH_2)_6^{2+}$ and $CoCl_4^{2-}$ change after the addition of 0.1 *M* $AgNO_3$? Justify your answer based on your observations from the previous step.

15. Decant the reaction mixture into the designated waste container.

Part C – Effect of Changing Reaction Temperature on an Equilibrium System

16. Obtain two clean, dry 25 × 75 mm test tubes, and label them 1 and 2. Add 2 mL of 1.0 *M* $CoCl_2$ solution to test tube 1. Should you record the color of the solution in the Lab Report? Add 2 mL of 1.0 *M* $CoCl_2$ solution and 2 mL of 12 *M* HCl solution to test tube 2. Why is HCl added to test tube 2? Should you record the color of the solution in the Lab Report?

17. Test tubes 1 and 2 are to be used for color comparison purposes in Step 21.

18. Obtain two clean, dry 50 × 150 mm test tubes. Add 3 mL of 1 *M* aqueous cobalt(II) chloride solution to each of the test tubes. Add 12 *M* HCl drop-wise to each test tube until the solutions turn purple. The purple color indicates an equilibrium mixture of $Co(OH_2)_6^{2+}$ and $CoCl_4^{2-}$ ions.

Chemical Alert

Note if too much HCl is added, the solution from Step 18 will turn blue. If that happens, pour the solutions into the Waste Container and repeat the process.

19. Prepare an ice bath by half filling a 250-mL beaker with ice. Add 100 mL of water to the beaker. Place one of the test tubes from Step 18 into the ice bath for 10 minutes. Remove the test tube from the ice bath. Should you record the color of the solution in the Lab Report?

20. If a microwave oven is available, place the remaining test tube from Step 18 into a 400-mL beaker. Heat the beaker and test tube in a microwave for 15 seconds. Remove the test tube from the boiling bath. Should you record the color of the solution in the Lab Report?

If a microwave oven is not available, add 200 mL of water to a 400-mL beaker. Place the beaker on a hot plate and bring the water to a gentle boil. Place the remaining test tube from Step 18 into the boiling water bath for 5 minutes. Remove the test tube from the boiling bath. Should you record the color of the solution in the Lab Report?

21. Compare the colors of the solutions from Steps 19 and 20 to the test tubes from Step 16. Based upon your observations, is this reaction endothermic or exothermic? Justify your answer with an explanation.

22. Decant the solutions prepared in Steps 16 and 18 into the designated waste container.

Name

Section

Date

Instructor

24 EXPERIMENT 24

Lab Report

Part A – Effect of Concentration Changes on Systems at Equilibrium

Observations for the reaction of $SbCl_3$ and H_2O.

Observations for the addition of HCl to the $SbCl_3$ reaction mixture.

Did the addition of HCl favor the products or reactants? Did the relative concentrations of $SbCl_3$, H_2O, and SbOCl increase or decrease? Justify your answer based on your observations from the previous step.

Observations for the addition of distilled water to the $SbCl_3$ reaction mixture.

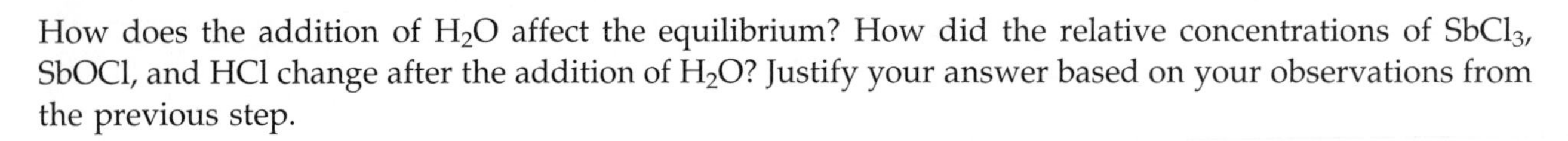

How does the addition of H_2O affect the equilibrium? How did the relative concentrations of $SbCl_3$, SbOCl, and HCl change after the addition of H_2O? Justify your answer based on your observations from the previous step.

Part B – Effect of Changing pH on a Complex Ion Equilibrium

Observations for the addition of HCl to the $Co(OH_2)_6^{2+}$ reaction mixture.

How did the addition of 12 *M* HCl affect the equilibrium?

How did the relative concentrations of $Co(OH_2)_6^{2+}$ and $CoCl_4^{2-}$ change after the addition of 12 *M* HCl? Justify your answer based on your observations from the previous step.

Observations for the addition of 0.1 *M* $AgNO_3$ to the $CoCl_2$ reaction mixture.

Is the equilibrium affected by the addition of 0.1 *M* $AgNO_3$?

How did the relative concentrations of $Co(OH_2)_6^{2+}$ and $CoCl_4^{2-}$ change after the addition of 0.1 *M* $AgNO_3$? Justify your answer based on your observations from the previous step.

Part C – Effect of Changing Reaction Temperature on an Equilibrium System

Observations of $CoCl_2$ solution and $CoCl_2$ + 12 *M* HCl solution

Why is HCl added to test tube 2?

Observations of $CoCl_2$ + 12 *M* HCl solution in ice bath

Observations of $CoCl_2$ + 12 *M* HCl solution in boiling water bath

Based on your observations, is this reaction endothermic or exothermic? Justify your answer with an explanation.

Name

Instructor

Section

Date

24 EXPERIMENT 24

Pre-Laboratory Questions

1. Write the equilibrium constant expression for the following reaction.

$$2\ CO_{2(g)} + \text{heat} \rightleftharpoons 2\ CO_{(g)} + O_{2(g)}$$

K_c = ____________________

2. Predict the effect on the equilibrium system in **Question 1** if the reaction temperature is decreased.

3. Predict the effect on the equilibrium system in **Question 1** if the CO_2 gas concentration is increased.

4. How would the relative amounts of O_2 and CO_2 change after the removal of some CO gas from the equilibrium reaction in **Question 1**.

5. Consider the following system at equilibrium.

$$\underset{\text{brown}}{2\ NO_{2(g)}} \rightleftharpoons \underset{\text{colorless}}{N_2O_{4(g)}}$$

This solution is brown at elevated temperatures and colorless below 0 °C.

A. Predict the color of the reaction mixture at −15 °C.

____

B. Is the forward reaction endothermic or exothermic at −15 °C? Justify your answer with an explanation.

Name

Section

Date

Instructor

24 EXPERIMENT 24

Post-Laboratory Questions

1. Write the equilibrium equations that result when solid NH_4F is dissolved in sufficient water to produce 5.0 mL of 0.5 *M* NH_4F solution.

2. How would the addition of 5 drops of 0.1 *M* HCl affect the equilibrium systems in Question 1? Justify your answer with an explanation.

3. How would the addition of 5 drops of 0.1 *M* NaOH affect the equilibrium systems in Question 1? Justify your answer with an explanation.

4. Some inexpensive humidity detection systems consist of a piece of paper saturated with Na_2CoCl_4 that changes color in dry or humid air. What color is the piece of paper in dry air? What color is the piece of paper in humid air? Justify your answer with an explanation.

Determination of a Reaction Equilibrium Constant Using Absorption Spectroscopy

PURPOSE

Determine the reaction equilibrium constant for the formation of the penta-aquathiocyanatoiron(III) ion, $Fe(SCN)^{2+}$, using absorption spectroscopy.

INTRODUCTION

The concept of dynamic equilibrium is one of the most fundamental ideas in chemistry. Chemical reactions attain a reaction rate that depends upon the nature and concentration of the reactants and the reaction temperature. For a given reaction performed at a constant temperature, the reaction rate depends solely on the concentrations of the species. To understand chemical equilibrium, we must realize that a chemical reaction involves two opposing processes: the reaction in the forward direction in which the reactants react to form the products, and the reaction in the reverse direction in which the products react to form reactants. For example, consider the hypothetical reaction

$$aA \rightleftharpoons bB \qquad \text{(Eq. 1)}$$

where a and b represent the stoichiometric coefficients and A and B represent the reactants and products involved in the reaction. If we assume that the reaction is an elementary reaction, the forward reaction rate (which describes how quickly A forms B) has the mathematical form

$$\text{rate}_{\text{forward}} = k_f[A]^a \qquad \text{(Eq. 2)}$$

The reverse reaction rate (which describes how quickly B reforms A) has the mathematical form

$$\text{rate}_{\text{reverse}} = k_r[B]^b \qquad \text{(Eq. 3)}$$

Notice that the reaction rates depend on the concentrations of each species. Thus, if the concentrations are changed, the rates of formation of the products and reactants also change.

At equilibrium, the forward reaction rate equals the reverse reaction rate. Externally, it appears that nothing is happening in chemical reactions at equilibrium. However, if we could see the atoms, ions, or molecules involved in a reaction at equilibrium, they are far from static. Reactants are forming products and products are forming reactants at the same rate.

It should be noted that all chemical reactions, even those that "go to completion", attain equilibrium. In those cases, the product equilibrium concentrations are very large compared to the reactant equilibrium concentrations.

Because the forward and reverse reaction rates are equal, we can set Eq. 2 equal to Eq. 3 and derive the equilibrium constant expression.

$$\text{rate}_{\text{forward}} = \text{rate}_{\text{reverse}}$$
$$k_f[A]^a = k_r[B]^b$$
$$\frac{k_f}{k_r} = \frac{[B]^b}{[A]^a} \qquad \text{(Eq. 4)}$$
$$K_c = \frac{[B]^b}{[A]^a}$$

Because k_f and k_r (reaction rate constants) are constant at a given temperature, their ratio, k_f/k_r, is also a constant. This constant, K_c, is the called the **equilibrium constant**. Notice that K_c is a ratio of the product concentrations, raised to their stoichiometric powers, divided by the reactant concentrations raised to their stoichiometric powers.

For a more complex reaction, such as the hypothetical reaction given in Eq. 5., the equilibrium constant expression is written according to Eq. 6.

$$aA + bB \rightleftharpoons cC + dD \qquad \text{(Eq. 5)}$$

$$K_c = \frac{[C]^c[D]^d}{[A]^a[B]^b} \qquad \text{(Eq. 6)}$$

The magnitude of the value of K_c is a measure of the extent to which a reaction occurs.

If $K_c > 10$, equilibrium product concentrations >> reactant concentrations

If $K_c < 0.1$, equilibrium reactant concentrations >> product concentrations

If $0.1 < K_c < 10$, neither equilibrium product or reactant concentrations predominate

In this experiment, we will determine the value of K_c for the reaction of hexaaquairon(III) ions, $Fe(H_2O)_6^{3+}{}_{(aq)}$, with thiocyanate ions, $SCN^-{}_{(aq)}$. When iron(III) nitrate, $Fe(NO_3)_3$, is added to water, the highly charged iron(III) ions are hydrated and form yellow colored hexaaquairon(III) ions (Eq. 7).

$$Fe^{3+} + 6\,H_2O \rightarrow [Fe(H_2O)_6]^{3+} \qquad \text{(Eq. 7)}$$

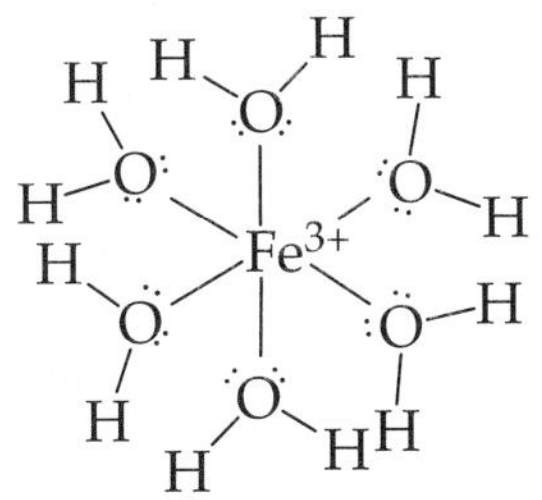

yellow solution

If potassium thiocyanate, KSCN, is added to an iron(III) nitrate solution, a thiocyanate ion, SCN^-, is substituted for one of the water molecules in the hexaaquairon(III) ion complex to form the blood-red colored pentaaquathiocyanatoiron(III) ion (Eq. 8). It can be written in the simplified form, $Fe(SCN)^{2+}$.

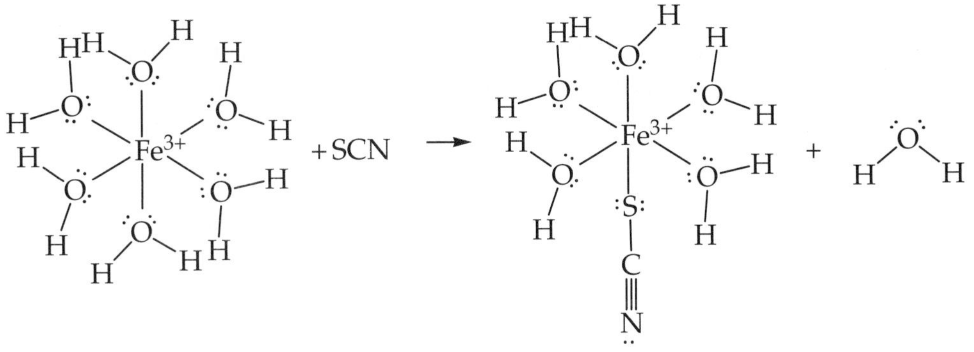

$$[Fe(H_2O)_6]^{3+} + SCN \rightarrow [Fe(H_2O)_5(SCN)]^{2+} + H_2O \qquad \text{(Eq. 8)}$$

yellow   blood red color

Equation 8 can be simplified to Equation 9.

$$Fe^{3+} + SCN^- \rightleftharpoons Fe(SCN)^{2+} \qquad \text{(Eq. 9)}$$

The equilibrium constant expression for the reaction is written as

$$K_c = \frac{[Fe(SCN)^{2+}]}{[Fe^{3+}][SCN^-]} \qquad \text{(Eq. 10)}$$

By determining the Fe^{3+}, SCN^-, and $Fe(SCN)^{2+}$ equilibrium concentrations in solution, we can calculate the value of K_c. As we have seen in Experiments 11 and 14, absorption spectroscopy is a convenient method for determining concentrations of colored solutions.

Colored aqueous solutions contain chemical species that absorb specific wavelengths of light. Transition metals that contain *3d* or *4d* valence electrons produce brightly colored, aqueous solutions. These metals (often-referred to as *heavy metals*) can be identified by the wavelengths of light that they absorb. Furthermore, the amount of light absorbed is directly proportional to the concentration of the metal ion in solution. Transition metals are typically reacted with complexing agents (KSCN in this experiment) to intensify the color of their solutions. By intensifying the color of their solutions, metal ions absorb greater quantities of light that permits their detection at lower solution concentrations.

Absorption spectroscopy measures the amount of light before and after it has passed through an aqueous solution. The difference in the amount of light before it enters the sample and after it exits the sample is the amount of light absorbed by the chemical species in the sample. For light to be

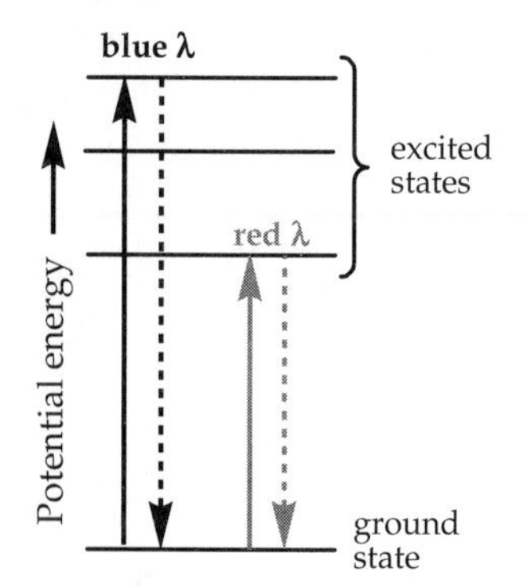

Figure 1
Absorption and emission of energy by electrons. λ is the symbol for the wavelength of light

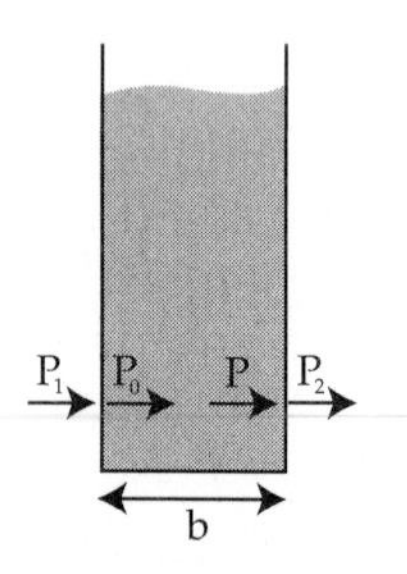

Figure 2
Light passing through a sample

absorbed by a chemical species, the light must have a wavelength, or energy, that exactly matches an energy transition in the absorbing species. Absorption of light by a metal ion promotes an electron from its ground state to an excited state (Figure 1). Shortly after the electron reaches one of the excited states, typically 10^{-9}–10^{-6} seconds later, the electron will return to the ground state by emitting energy. Every chemical species has a unique set of excited states, and consequently, absorbs different wavelengths of light.

Light is passed through sample solutions contained in an optically transparent cell of known path length. These optically transparent cells are called **cuvettes** (Figure 2). As light passes through a cuvette containing a sample solution, it can be reflected, refracted, diffracted, or absorbed. Only the absorption of light is directly proportional to the solution's concentration.

Reflection, diffraction and refraction (essentially scattering of light from the walls of the solution container) can be nullified by the use of a blank solution. A **blank solution** contains all of the species present in the sample solution *except* the absorbing species. The spectrophotometer is designed to subtract the spectrum recorded for the blank solution from the absorbance spectrum of the sample, thus, nullifying the reflection, refraction, and diffraction caused by the walls of the cuvette.

Mathematically, this process can be defined as follows. P_1 is the power of the light before it enters the sample container (Figure 2). P_0 is the power of the light immediately after it passes through the first wall of the sample container, but before it passes through the sample. P is the power of the light after it has passed through the sample. P_2 is the power of the light after it exits the second wall of the cuvette. Finally, b is the path length traveled by the light. The difference in power between P_1 and P_0 (or P and P_2) is due to reflection and/or refraction of the light from the cuvette walls. Absorption spectroscopy is only interested in the ratio of P to P_0 (*light absorbed by the sample*). The reflection/refraction effect can be nullified by measuring the power difference between P_1 and P_0, and P and P_2, for a blank solution and subtracting that from the sample's spectrum. After the subtraction is performed, the ratio of P divided by P_0 can be determined. This ratio is called the **transmittance** of the solution, T.

$$T = \frac{P}{P_0} \qquad \text{(Eq. 11)}$$

While the spectrophotometer actually measures transmittance, we need to ascertain the amount of light absorbed by the solution to determine its concentration. Absorbance, A, and transmittance, T, are related by the following equation.

$$A = \log\frac{1}{T} = -\log T \qquad \text{(Eq. 12)}$$

Note that if more light is transmitted, less light is being absorbed by the sample, and vice versa.

There are two additional factors besides the solution's concentration that affects absorbance. First, every absorbing species only absorbs a fraction of the light that passes through the solution. The **molar absorptivity coefficient** is a measure of the fraction of light absorbed by a given species. Each absorbing species has its own unique molar absorptivity coefficient. Second, as the light's path through the solution is increased (determined by the width of the cuvette), more light is absorbed. The concentration, molar absorptivity coefficient, and the path length of light are directly related to the absorbance of a solution via Beer-Lambert's Law,

$$A = \varepsilon bc \qquad \text{(Eq. 13)}$$

where A is the absorbance of the solution, ε is the molar absorptivity coefficient of the absorbing species, b is the path length of the light, and c is the solution's concentration. Beer-Lambert's Law can be simplified if the same absorbing species and same sample container are used in a series of experiments. In that case, ε and b are constant simplifying Beer-Lambert's Law to

$$A = (\text{constant})c \qquad \text{(Eq. 14)}$$

From Eq. 14, we see that the absorbance of a species is directly proportional to its concentration in solution. This convenient, linear relationship between absorbance and concentration makes absorption spectroscopy one of the most popular analytical techniques for measuring concentrations of dissolved species.

Concentrations of Metals in Solution

Figure 3 represents a plot of the absorbance spectrum for $Fe(SCN)^{2+}$. The wavelength at which maximum absorbance occurs (the highest point on the curve) in an absorption band is designated as the **lambda max**, λ_{max}. $Fe(SCN)^{2+}{}_{(aq)}$ is red colored because it absorbs blue-green light in the 430–490 nm region of the visible spectrum. From Figure 3, we see that the λ_{max} for $Fe(SCN)^{2+}{}_{(aq)}$ is 460 nm.

The concentration of the metal solution is determined by monitoring changes in its absorbance as a function of concentration. A series of **standard solutions**, in which the species concentration is known, are prepared and their absorbance spectra recorded. Typically, 4 to 5 standard solutions are prepared that bracket the concentration of the unknown solution. To bracket the concentration of the unknown solution, at least one standard solution must have a lower concentration than the unknown solution, and at least one standard solution must have a higher concentration than the unknown solution. The absorbance value for each standard solution is

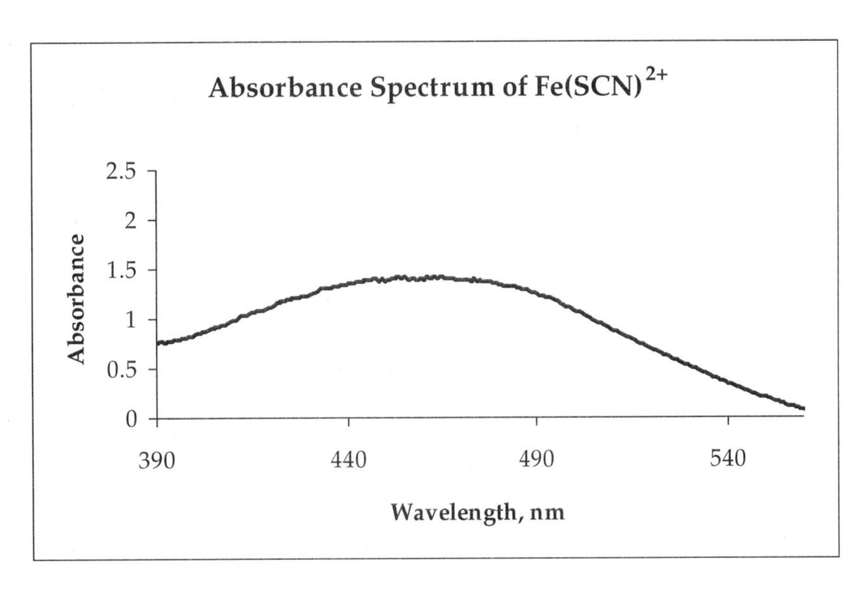

Figure 3
An absorbance spectrum of an aqueous $Fe(SCN)^{2+}$ *solution*

determined at λ_{max} for one of the species absorbance bands. If necessary, more than one λ_{max} can be used to determine each standard solution's absorbance. Typically, λ_{max} of the most intense absorbance band is used to determine the solution's concentration. Figure 4 shows absorbance spectra for five $Fe(SCN)^{2+}$ standard solutions.

Once the absorbance for each standard solution has been determined, a plot of absorbance (y-axis) versus the standard solution concentrations (x-axis) is prepared. In accordance with the Beer-Lambert law, the plot should be linear (or very close to linear). Linear regression analysis is performed, using a spreadsheet program such as Excel, to determine the *linear best-fit* for the absorbance versus concentration data (Figure 5, using absorbance at λ_{max} of 460 nm). The R^2 value shown in Figure 5 indicates how well the regression analysis fits the absorbance-concentration data.

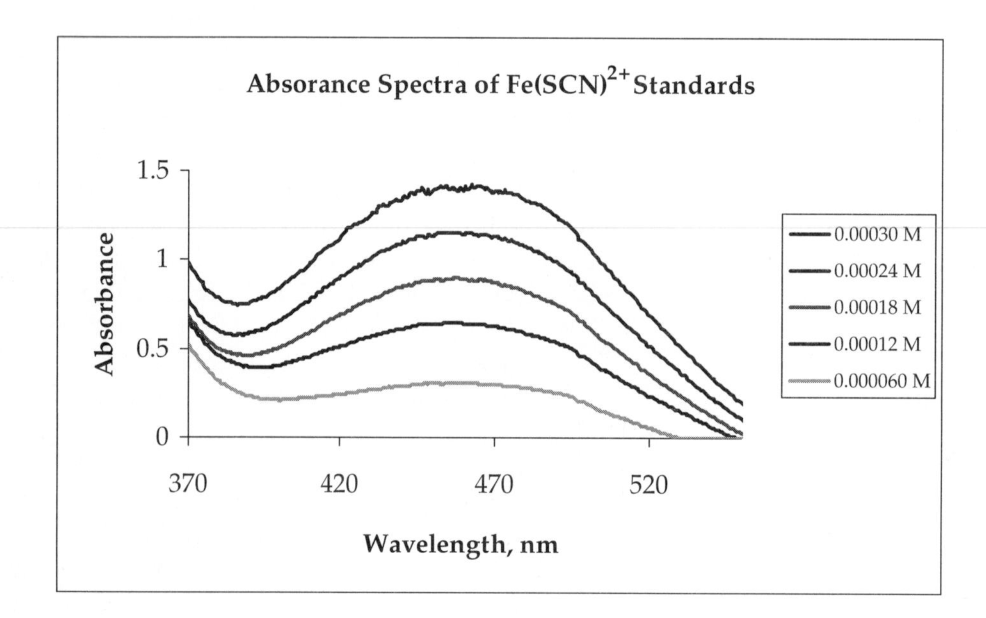

Figure 4
Absorbance Spectra of $Fe(SCN)^{2+}$ Standard Solutions, λ_{max} = 460 nm

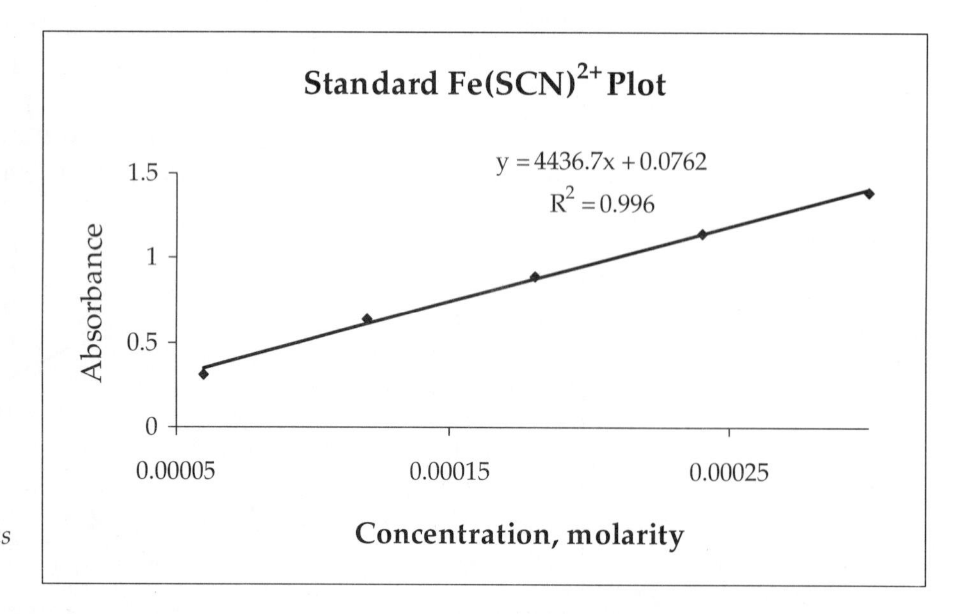

Figure 5
Plot of Absorbance versus Concentration for the $Fe(SCN)^{2+}$ Standard Solutions from Figure 4

The closer the R^2 value is to 1.00, the better the linear regression analysis has fit the data.

Finally, a spectrum of the unknown solution is recorded. From the absorbance value of the unknown solution (determined at the same λ_{max} used to prepare the standard solutions plot), its concentration can be determined either directly from a plot similar to Figure 5, or more precisely by using linear regression analysis. The line determined from the regression analysis will be in the form $y = mx + b$, where *y is the absorbance value* and *x is the solution concentration.* Algebraic substitution of the absorbance value (y) for the unknown solution into the linear regression equation for the line permits the determination of the unknown solution's concentration (x).

In Part A of this experiment, five standard solutions of known $Fe(SCN)^{2+}$ ion concentration will be prepared. In Part B, the absorbance for each standard solution will be determined and a plot of absorbance versus the $Fe(SCN)^{2+}$ ion concentration will be constructed. In accordance with the Beer-Lambert law, the plot should be linear (or very close to linear).

The standard solutions will be prepared using the following aqueous solutions: 0.00150 *M* KSCN, 0.150 *M* $Fe(NO_3)_3$, and 0.050 *M* HNO_3. If these solutions were prepared in distilled water, the Fe^{3+} ions would readily hydrolyze and form a precipitate of $Fe(OH)_3$. This compound would interfere with the reaction of interest. Consequently, nitric acid (HNO_3), a stronger acid than Fe^{3+}, is used as the solvent to retard the hydrolysis of Fe^{3+}.

Note that Fe^{3+} is 100 times more concentrated than SCN^- in each standard solution. When the two solutions are mixed, Le Châtelier's principle indicates that the high Fe^{3+} concentration will shift the equilibrium strongly to the product side of the reaction, forming stoichiometric amounts of $Fe(SCN)^{2+}$. Since the reaction stoichiometry is 1:1 for SCN^- and $Fe(SCN)^{2+}$ (see Eq. 9), the equilibrium concentration of $Fe(SCN)^{2+}$ in each solution will be equal to the initial SCN^- concentration (the limiting reagent).

In Part C of the experiment, three equilibrium solutions will be prepared. The absorbance of each equilibrium solution (at the same λ_{max} that was used in Part A) will be measured. Using the results of the linear regression analysis performed on the standard curve prepared in Part B of the experiment, the equilibrium $Fe(SCN)^{2+}$ concentration will be determined for each solution. From the equilibrium $Fe(SCN)^{2+}$ concentration, the reaction stoichiometry (Eq. 9), and the initial concentrations of Fe^{3+} and SCN^-, the equilibrium concentrations of Fe^{3+} and SCN^- can be calculated. Finally, K_c for the reaction is calculated by inserting the equilibrium concentrations of Fe^{3+}, SCN^-, and $Fe(SCN)^{2+}$ into Eq. 10.

Use of "ICE" Tables for Calculating Equilibrium Concentrations of Reactants and Products

It is recommended that students use "ICE" tables to calculate equilibrium concentrations of reactants and products. An ICE table gives the initial concentrations ("I") of the reactants and products at the moment the reactants are mixed. It also provides the change in concentrations ("C") of the reactants and products as a result of the system establishing equilibrium. Finally, an ICE table provides the new reactant and product concentrations at equilibrium ("E"). The reactant and product equilibrium concentrations are the difference between their initial concentrations and the change in concentrations they undergo for the system to establish equilibrium.

Table 1 *ICE table for the reaction of A^+ and B^- yielding AB*

	A^+	B^-	AB
Initial concentrations	6.0×10^{-4} *M*	1.2×10^{-3} *M*	0.0 *M*
Change in concentrations	-9.4×10^{-5} *M*	-9.4×10^{-5} *M*	$+9.4 \times 10^{-5}$ *M*
Equilibrium concentrations	5.1×10^{-4} *M*	1.1×10^{-3} *M*	9.4×10^{-5} *M*

For example, consider the general equilibrium reaction shown below:

$$A^+ + B^- \rightleftharpoons AB$$

4.0 mL of 0.00150 *M* A^+ are mixed with 6.0 mL of 0.00200 *M* B^-. The initial concentrations of A^+ and B^- are 6.0×10^{-4} *M* and 1.2×10^{-3} *M*, respectively (Table 1). At the instant of mixing A and B, the initial concentration of AB is 0.

After the reaction has attained equilibrium, the absorbance of AB is measured and the concentration of AB is determined to be 9.4×10^{-5} *M*. Therefore, the change in the concentration of AB is $+9.4 \times 10^{-5}$ *M* because it is being produced. Since the stoichiometric ratios of AB to A^+ and AB to B^- are both 1:1, the change in concentrations of A^+ and B^- are -9.4×10^{-5} *M* because they are being consumed. The equilibrium concentrations for A^+, B^-, and AB are the difference between their initial concentrations and the change in concentration they undergo to establish equilibrium (Table 1).

By inserting the equilibrium concentrations of A^+, B^-, and AB into the equilibrium constant expression (Eq. 15), the equilibrium constant, K_c, can be determined.

$$K_c = \frac{[AB]}{[A^+]\,[B^-]} \qquad \text{(Eq. 15)}$$

PROCEDURE

CAUTION

Students must wear departmentally approved eye protection while performing this experiment. Wash your hands before touching your eyes and after completing the experiment.

Part A – Preparation of Five Standard $Fe(SCN)^{2+}$ Solutions

1. Obtain five clean, dry beakers, and label them 1, 2, 3, 4, and 5.
2. Add the indicated volumes of 0.0015 *M* SCN^-, 0.050 *M* HNO_3, and 0.150 *M* Fe^{3+} in the table below to each of the labeled beakers. Thoroughly mix the contents of each beaker.

Beaker	*0.00150 M KSCN*	*0.050 M HNO_3*	*0.150 M $Fe(NO_3)_3$*
1	5.0 mL	15 mL	5.0 mL
2	4.0 mL	16 mL	5.0 mL
3	3.0 mL	17 mL	5.0 mL
4	2.0 mL	18 mL	5.0 mL
5	1.0 mL	19 mL	5.0 mL

3. Is it necessary to calculate the final $Fe(SCN)^{2+}$ concentration for the solution in each beaker? If so, should you record the concentrations in the Lab Report, and to how many significant figures?

Part B - Absorption Measurements for the Standard Solutions and Preparation of the Beer-Lambert Curve

4. Record an absorbance spectrum for each of the five solutions prepared in Step 2. See **Appendix D** – *Instructions for Recording an Absorbance Spectrum using the MeasureNet Spectrophotometer*. Of the three solutions added to each of the five beakers, which solution should be used as the "blank" solution?
5. Steps 6 and 7 are to be completed at the end of the laboratory period. *Proceed to Step 8*.
6. Should you determine the absorbance of each standard solution from the tab delimited files saved in Step 4? Should your λ_{max} be in the 450–460 nm region of the absorbance spectrum of each standard solution? Why or why not? Should you record the absorbance of each solution in the Lab Report?
7. Prepare a Beer-Lambert Plot of the absorbance versus $Fe(SCN)^{2+}$ concentration for each of the five standard solutions. See **Appendix B-2** – *Excel Instructions for Performing Linear Regression Analysis*.
8. Pour the solutions in the five beakers and the cuvettes into the Waste Container. Clean and dry the beakers and cuvettes before proceeding to Step 9.

Part C - Equilibrium Solution Preparation and Absorption Measurements: Finding K_c

9. Label three clean, dry beakers 1, 2, and 3.
10. Add the indicated volumes of 0.00150 *M* SCN^-, 0.050 *M* HNO_3, and 0.00150 *M* Fe^{3+} in the table below to each of the labeled beakers. Thoroughly mix the contents of each beaker.

Beaker	*0.00150 M KSCN*	*0.050 M HNO_3*	*0.00150 M $Fe(NO_3)_3$*
1	2.0 mL	3.0 mL	5.0 mL
2	3.0 mL	2.0 mL	5.0 mL
3	4.0 mL	1.0 mL	5.0 mL

11. Record an absorbance spectrum for each of the three solutions prepared in Step 10. See **Appendix D** – Instructions for Recording an Absorbance Spectrum using the MeasureNet Spectrophotometer. Which of the three solutions should you use as the "blank?"
12. Pour the remaining solutions in the three beakers and the cuvettes into the "Waste container." Clean and dry the beakers and cuvettes.
13. Should you determine the absorbance of each equilibrium mixture from the tab delimited files saved in Step 11? Should your λ_{max} be in the 450–460 nm region of the absorbance spectrum of each equilibrium mixture? Why or why not? Should you record the absorbance of each solution at the λ_{max} you selected?
14. Prepare an "ICE" table for each equilibrium mixture. Include the initial concentrations, changes in concentrations, and the equilibrium concentrations of Fe^{3+}, SCN^- and $Fe(SCN)^{2+}$. Should you include the ICE tables for each equilibrium mixture in the Lab Report?
15. Determine K_c for each of the three equilibrium solutions. Should you record the K_c values in the Lab Report?
16. Determine the average K_c value for the equilibrium mixtures.
17. Include the Beer's Law plot of the absorbance versus $Fe(SCN)^{2+}$ concentration when you submit your Lab Report.

Name | Section | Date

Instructor

25 EXPERIMENT 25

Lab Report

Part A – Preparation of Five Standard $Fe(SCN)^{2+}$ Solutions

Is it necessary to calculate the final $Fe(SCN)^{2+}$ concentration for the solution in each beaker?

Part B – Absorption Measurements for the Standard Solutions and Preparation of the Beer-Lambert Curve

Should you determine the absorbance of each standard solution from the tab delimited files saved in Step 4? Should your λ_{max} be in the 450–460 nm region of the absorbance spectrum of each standard solution? Why or why not?

Part C – Equilibrium Solution Preparation and Absorption Measurements: Finding K_c

Should you determine the absorbance of each equilibrium mixture from the tab delimited files saved in Step 11? Should your λ_{max} be in the 450–460 nm region of the absorbance spectrum of each equilibrium mixture? Why or why not?

Prepare an ''ICE'' table for each equilibrium mixture.

Determine K_c for each of the three equilibrium solutions.

Determine the average K_c value for the equilibrium mixtures.

Name ____________________ Section __________ Date __________

Instructor ____________________

25 EXPERIMENT 25

Pre-Laboratory Questions

1. Complete the table below by calculating the initial SCN^- and equilibrium $Fe(SCN)^{2+}$ concentrations for each of the five standard solutions. In addition to the volume of KSCN indicated in the table below, each solution contains 5.00 mL of $Fe(NO_3)_3$ and sufficient 0.050 *M* HNO_3 to produce a total volume of 25.00 mL of solution.

$$Fe^{3+} + SCN^- \rightleftharpoons Fe(SCN)^{2+}$$

Molarity of KSCN 0.00165 *M* Molarity of $Fe(NO_3)_3$ 0.165 *M*

Solution	*Vol. SCN⁻*	*Initial SCN⁻ M*	*Fe(SCN)²⁺ M*	*Absorbance*
1	5.00 mL			1.590
2	4.00 mL			1.320
3	3.00 mL			0.990
4	2.00 mL			0.690
5	1.00 mL			0.335

2. Using Excel, plot the absorbance versus the $Fe(SCN)^{2+}$ concentration data in Question 1 to prepare a standard curve. Perform a linear regression analysis to determine the equation for the line. The equation for the line must appear on the plot, along with the R^2 value. Submit the plot along with your Pre-Lab Questions.

3. An equilibrium solution is prepared by mixing 2.75 mL of 0.00165 *M* SCN^-, 5.00 mL of 0.00165 *M* Fe^{3+}, and 2.75 mL of 0.050 M HNO_3. The equilibrium solution's absorbance is determined to be 0.915. Prepare an ICE table for the equilibrium mixture. Include the initial concentrations, changes in concentrations, and the equilibrium concentrations of Fe^{3+}, SCN^- and $Fe(SCN)^{2+}$.

4. Determine K_c for the equilibrium mixture.

Name | Section | Date

Instructor

25 EXPERIMENT 25

Post-Laboratory Questions

1. In Step 2, a student added more KSCN solution than instructed when making standard solutions. Assuming no other errors were made in the experiment, would the value of equilibrium constant determined in the experiment higher or lower than it should be? Justify your answer with an explanation.

2. In Step 4, why were 0.00150 *M* KSCN or 0.150 M $Fe(NO_3)_3$ not used as the *"blank solution?"*

.issolved in water, hexaaquairon(III) ions, $Fe(H_2O)_6^{3+}$, form. $Fe(H_2O)_6^{3+}$ ions
.ous solution according to the equation given below.

$$Fe(H_2O)_6^{3+} + H_2O \rightleftharpoons [Fe(H_2O)_5(OH)]^{2+} + H_3O^+$$

.ould the addition of 0.050 *M* nitric acid to the solution affect the equilibrium constant for the
.tion? Justify your answer with an explanation.

Determination of K_a or K_b for an Acid or Base

PURPOSE

Determine the acid ionization constant, K_a, for a weak acid, or the base ionization constant, K_b, for a weak base by titration, and by measuring the pH of the weak acid or weak base solution.

INTRODUCTION

The relative acidity or basicity of a substance or a system is of critical importance in many situations, such as the quality of drinking water, food preservation, soil conditions for agriculture, and physiological functions. One measure of the strength of an acid its ability to donate protons to a base. The **acid ionization constant, K_a**, is a quantitative measure of the strength of an acid. The ionization of a generic acid, HA, can be represented by the following equation

$$HA_{(aq)} + H_2O_{(\ell)} \rightleftharpoons H_3O^+_{(aq)} + A^-_{(aq)} \quad \text{(Eq. 1)}$$

in which the weak acid donates a proton, H^+, to a water molecule, to form the hydronium ion, H_3O^+, and the conjugate base of the weak acid, A^-. The corresponding acid ionization constant expression, K_a, can be written as

$$K_a = \frac{[H_3O^+][A^-]}{[HA]} \quad \text{(Eq. 2)}$$

The K_a value is characteristic of an acid and can be used to identify an unknown acid. The K_a value indicates the relative strength of an acid. The larger the K_a value, the stronger the acid. The smaller the K_a value, the weaker the acid.

The **base ionization constant, K_b,** is a quantitative measure of the strength of a base. The ionization of a generic base, B, can be represented by the following equation

$$B_{(aq)} + H_2O_{(\ell)} \rightleftharpoons BH^+_{(aq)} + OH^-_{(aq)} \quad \text{(Eq. 3)}$$

in which the weak base accepts a proton, H^+, from a water molecule, to form the hydroxide ion, OH^-, and the conjugate acid of the weak base, BH^+. The corresponding base ionization constant expression, K_b, can be written as

$$K_b = \frac{[BH^+][OH^-]}{[B]} \quad \text{(Eq. 4)}$$

The K_b value is characteristic of a base and can be used to identify an unknown base. The K_b value indicates the relative strength of a base. The larger the K_b value, the stronger the base. The smaller the K_b value, the weaker the base.

In this experiment, two methods will be used to determine the K_a or K_b value of a weak acid or weak base: 1) titration, and 2) measuring the pH of the solution. In the first method, the weak acid is titrated with sodium hydroxide, or the weak base is titrated with HCl. A titration curve is produced by plotting the pH of the acid solution versus the volume of NaOH added, or by plotting the pH of the base solution versus the volume of HCl added. The **equivalence point** of the titration is reached when all of the weak acid (HA) has completely reacted with NaOH, or all of the base has completely reacted with HCl. On the titration curve, the equivalence point is read at the center of the region where the pH increases or decreases sharply. The **half-equivalence point** for the titration is reached when exactly one half of the acid or base has been neutralized. At this point, the concentration of the acid in the solution, [HA], is equal to the concentration of its conjugate base, $[A^-]$ (Eq. 5), or the concentration of the base in the solution, [B], is equal to the concentration of its conjugate acid, $[BH^+]$ (Eq. 6).

$$[HA] = [A^-] \quad \text{(Eq. 5)}$$

$$[B] = [BH^+] \quad \text{(Eq. 6)}$$

Therefore, Equation 2 can be simplified to yield Equation 7.

$$K_a = \frac{[H_3O^+]\cancel{[A^-]}}{\cancel{[HA]}} \quad \text{(Eq. 7)}$$

$$K_a = [H_3O^+]$$

Taking the negative logarithm of each side of Eq. 7, we can derive Eq. 8.

$$-\log(K_a) = -\log[H_3O^+]$$
$$pK_a = pH \quad \text{(Eq. 8)}$$

Equation 8 indicates that the pK_a for the acid is equal to the pH of the solution at the half-equivalence point. The K_a of the acid is determined from the pK_a value as follows.

$$K_a = 10^{-pK_a} \qquad \text{(Eq. 9)}$$

Similarly, Equation 4 can be simplified to yield Equation 10.

$$K_b = \frac{[BH^+]\cancel{[OH^-]}}{\cancel{[B]}} \qquad \text{(Eq. 10)}$$
$$K_b = [OH^-]$$

Taking the negative logarithm of each side of Eq. 10, we can derive Eq. 11.

$$-\log(K_b) = -\log[OH^-] \qquad \text{(Eq. 11)}$$
$$pK_b = pOH$$

Equation 11 indicates that the pK_b for the base is equal to the pOH of the solution at the half-equivalence point. The K_b of the base is determined from the pK_b value as follows.

$$K_b = 10^{-pK_b} \qquad \text{(Eq. 12)}$$

The second method for determining the K_a or K_b of a weak acid or weak base requires that we know the pH and the initial weak acid or weak base concentration in the solution. From the pH of the acid solution (HA), we can determine the H^+ and A^- ion concentrations at equilibrium. The H^+ ion concentration is related to the pH of a solution by Equation 13.

$$[H_3O^+] = 10^{-pH} \qquad \text{(Eq. 13)}$$

By substituting [HA], $[H_3O^+]$, and $[A^-]$ at equilibrium into Eq. 2, we can calculate the K_a value for the weak acid.

From the pOH of the base solution (B), we can determine the BH^+ and OH^- ion concentrations at equilibrium. The OH^- ion concentration is related to the pOH of a solution by Equation 14.

$$[OH^-] = 10^{-pOH} \qquad \text{(Eq. 14)}$$

By substituting the [B], $[BH^+]$, and $[OH^-]$ at equilibrium into Eq. 4, we can calculate the K_b value for the weak base.

Acidity or Basicity of Salt Solutions

Salts are ionic compounds that are produced in reactions between acids and bases. Salts that contain an anion that is the conjugate base of a weak acid or a cation that is the conjugate acid of a weak base undergo hydrolysis. **Hydrolysis** is the reaction of an anion or cation with water. A conjugate acid of a weak base or the conjugate base of a weak acid typically undergo hydrolysis. Conjugate acids and bases of strong acids or strong bases generally do not undergo hydrolysis.

Sodium fluoride is the salt produced by the reaction of hydrofluoric acid with sodium hydroxide. The F^- ion is the conjugate base of the weak acid

HF. The Na^+ ion is the conjugate acid of the strong base NaOH. Therefore, F^- undergoes hydrolysis but Na^+ does not. The hydrolysis of F^- is represented (in its simplest form) by Equation 15. The resulting solution is basic.

$$F^-_{(aq)} + H_2O_{(\ell)} \rightleftharpoons HF_{(aq)} + OH^-_{(aq)} \qquad \text{(Eq. 15)}$$

The base ionization expression constant, K_b, is written as

$$K_b = \frac{[HF][OH^-]}{[F^-]}$$

Ammonium chloride is the salt produced by the reaction of ammonia with hydrochloric acid. The NH_4^+ ion is the conjugate acid of the weak base NH_3. The Cl^- ion is the conjugate base of the strong acid HCl. Therefore, NH_4^+ undergoes hydrolysis but Cl^- does not. The hydrolysis of NH_4^+ is represented (in its simplest form) by Equation 16. The resulting solution is acidic.

$$NH^+_{4\,(aq)} + H_2O_{(\ell)} \rightleftharpoons NH_{3(aq)} + H_3O^+_{(aq)} \qquad \text{(Eq. 16)}$$

The acid ionization expression constant, K_a, is written as

$$K_a = \frac{[H_3O^+][NH_3]}{[NH_4^+]}$$

Sample Calculation to Determine the K_a Value of a Weak Acid from Titration with NaOH

A weak acid is titrated with 0.10 *M* NaOH. The titration curve is shown in Figure 1. Determine the K_a of the weak acid.

Equivalence point = 11.62 mL, determined from the titration curve.

$$\text{half-equivalence point} = \frac{11.62\text{ mL}}{2} = 5.81\text{ mL}$$

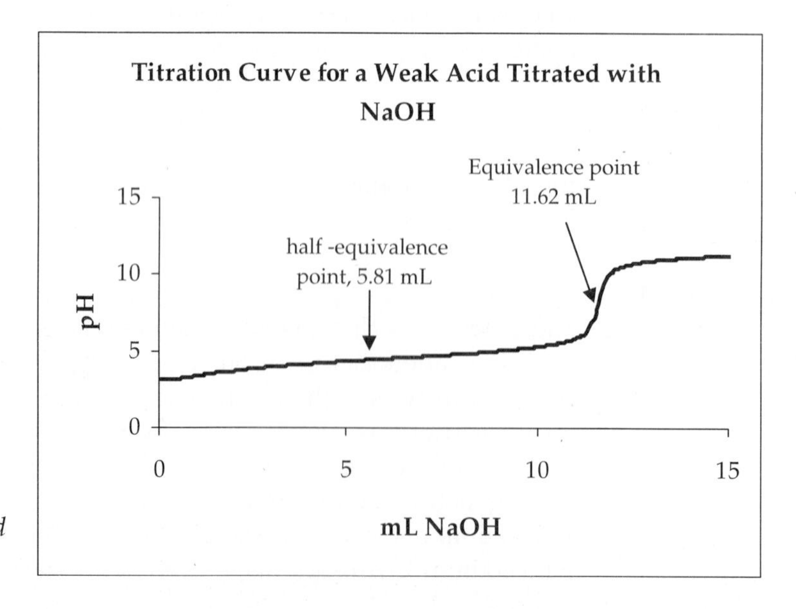

Figure 1
Titration curve for a weak acid titrated with NaOH

The pH value corresponding to 5.81 mL is 4.20, determined from the titration curve.

$$pK_a = pH = 4.20 \text{ at the half-equivalence point}$$

$$K_a = 10^{-pK_a} = 10^{-4.20} = 6.3 \times 10^{-5}$$

Sample Calculation for the Determination of K_a from the Initial Concentration and pH of a Weak Acid Solution

The pH of a 0.10 *M* weak acid solution is 2.52 at 25 °C. Calculate the K_a of the weak acid at 25 °C. Initially, only the weak acid HA is present in the solution. At equilibrium, a fraction of the HA molecules ionize, forming H_3O^+ and A^-. The initial HA concentration decreases while the concentrations of H_3O^+ and A^- increase until equilibrium is attained. An "ICE" table is used to aid in calculating the equilibrium concentrations of HA, H_3O^+, and A^-.

$$pH = 2.52$$

$$[H_3O^+] = 10^{-pH} = 10^{-2.52} = 3.0 \times 10^{-3} M$$

$$HA + H_2O \rightleftharpoons H_3O^+ + A^-$$

	HA	*H_3O^+*	*A–*
Initial concentrations	0.10 *M*	0.0 *M*	0.0 *M*
Change in concentrations	$-3.0 \times 10^{-3} M$	$+3.0 \times 10^{-3} M$	$+3.0 \times 10^{-3} M$
Equilibrium concentrations	$9.7 \times 10^{-2} M$	$+3.0 \times 10^{-3} M$	$+3.0 \times 10^{-3} M$

$$K_a = \frac{[H_3O^+][A^-]}{[HA]}$$

$$K_a = \frac{[3.0 \times 10^{-3}][3.0 \times 10^{-3}]}{[9.7 \times 10^{-2}]} = 9.3 \times 10^{-5}$$

The calculated K_a value of the unknown acid is closest to that of benzoic acid, 6.3×10^{-5} (Table 1).

PROCEDURE

CAUTION 

Students must wear departmentally approved eye protection while performing this experiment. Wash your hands before touching your eyes and after completing the experiment.

Chemical Alert

Both NaOH and HCl are corrosive. If NaOH or HCl contacts your skin, wash the affected area with copious quantities of water and inform your lab instructor.

Table 1 *Ionization constants for some weak acids and bases at 25 °C*

Acid	*Formula*	K_a
Acetic acid	CH_3COOH	1.8×10^{-5}
Benzoic acid	C_6H_5COOH	6.3×10^{-5}
Carbonic acid	H_2CO_3	4.2×10^{-7}
Formic acid	$HCOOH$	1.8×10^{-4}
Hypochlorous acid	$HOCl$	3.5×10^{-8}
Dihydrogen phosphate ion	$H_2PO_4^-$	6.2×10^{-8}
Hydrogen phosphate ion	HPO_4^{2-}	3.6×10^{-13}
Hydrogen carbonate ion	HCO_3^-	4.8×10^{-11}
Nitrous acid	HNO_2	4.8×10^{-11}
Phenol	C_6H_6O	1.0×10^{-10}
Potassium hydrogen phthalate	$KC_8H_5O_4$	5.3×10^{-6}
Ammonium chloride	NH_4Cl	5.6×10^{-10}
Base	*Formula*	K_b
Aqueous ammonia	NH_3	1.8×10^{-5}
Acetate ion	CH_3COO^-	5.6×10^{-10}
Formate ion	$HCOO^-$	5.9×10^{-11}
Hydrogen carbonate ion	HCO_3^-	2.4×10^{-8}
Carbonate ion	CO_3^{2-}	2.1×10^{-4}
Hypochlorite ion	OCl^-	2.9×10^{-7}
Hypoiodite ion	IO^-	4.3×10^{-4}

Part A – Determination of the Ionization Constant (K_a or K_b) of an Unknown Solution

1. Obtain approximately 40 mL of an unknown acid, base, or salt solution from your lab instructor. Should you record the Unknown Number in the Lab Report?
2. Should you determine whether the solution is acidic or basic? How would you make this determination? Would this determination involve the use of MeasureNet? If this determination involves MeasureNet, how do you calibrate the appropriate MeasureNet probe to

make the determination? Should you record the results of the determination in the Lab Report?

3. Titrate 10.00 mL of the unknown solution with either 0.100 *M* HCl or 0.100 *M* NaOH.
4. See Appendix F – **Instructions for Recording a Titration Curve Using the MeasureNet pH Probe and Drop Counter.** Complete steps 1–22 in Appendix F before proceeding to Step 5 below with one notable exception. Appendix F instructs you to fill the buret with NaOH. *In this experiment, you will fill the buret with either NaOH or HCl, depending on your determination in Step 2 above.*
5. Pour the reaction mixture into a laboratory sink. Be sure to flush the sink with copious quantities of water.
6. Repeat Steps 2–5 to perform a second titration of the unknown solution.
7. *Steps 7–13 are to be completed after the laboratory period is concluded (outside of lab). Proceed to Step 16.*
8. From the tab delimited files you saved, prepare titration curves (plots of the pH versus volume of solution added) using Excel (or a comparable spreadsheet program). Instructions for preparing titration curves using Excel are provided in **Appendix B-4.**
9. What are the equivalence and half equivalence points (pH and mL added) for each titration? Should these values be recorded in the Lab Report, and to how many significant figures?
10. What is the pH of the unknown solution that will be used to determine the ionization constant (K_a or K_b)?
11. What are the ionization constants determined from each titration?
12. What is the average ionization constant for the unknown?
13. Using Table 1, identify the unknown solution.
14. What is the molarity of the unknown solution used in each titration?
15. What is the average molarity of the unknown solution?

Part B – Determination of the Ionization Constant (K_a or K_b) of an Unknown Solution from the Initial Concentration and pH or pOH of the Solution

16. Add 20 mL of the unknown solution to a clean, dry 50-mL beaker. Determine the pH of the solution. If the unknown solution is a base, how will you determine the pOH of the solution (see Note)?

NOTE: The pH and the pOH of a solution sum to 14.

$$pH + pOH = 14$$

For the complete derivation of this formula, see your textbook for a discussion of pH and pOH.

---

17. Be sure to rinse the electrode with distilled water after removing it from the pH 7 buffer solution. *Gently* dry the tip of the probe with a Kimwipe®.
18. Insert the electrode into the beaker containing the unknown solution from Step 16. Gently stir the solution with a stirring rod until the pH reading stabilizes. Should you record the pH of the solution in the Lab Report?

19. Decant the weak acid or weak base solution into a Waste Container.
20. Repeat Steps 16 to 19 to perform a second pH measurement (you can use the same 20 mL from Step 16).
21. Rinse the pH probe with distilled water. Return the pH probe to the beaker containing the pH 7 buffer solution.
22. Given the initial concentration (molarity determined in Step 15) and the pH of the unknown solution, what is the ionization constant (K_a or K_b) determined from each trial?
23. What is the average ionization constant for the unknown?
24. Using the information provided in Table 1 and the results obtained in Parts A and B of this experiment, identify the unknown solution.

Name

Section

Date

Instructor

26 EXPERIMENT 26

Lab Report

Part A – Determination of the Ionization Constant (K_a or K_b) of an Unknown Solution

Should you record the unknown number of the unknown solution?

Is the solution acidic or basic? How would you make this determination? Would this determination involve the use of MeasureNet? If this determination involves MeasureNet, how do you calibrate the appropriate MeasureNet probe to make the determination?

What are the equivalence and half equivalence points (pH and mL added) for each titration?

What is the pH of the unknown solution that will be used to determine the ionization constant (K_a or K_b)?

What are the ionization constants determined from each titration?

What is the average ionization constant for the unknown?

Using Table 1, identify the unknown solution.

What is the molarity of the unknown solution used in each titration?

What is the average molarity of the unknown solution?

Part B – Determination of the Ionization Constant (K_a or K_b) of an Unknown Solution from the Initial Concentration and pH or pOH of the Solution

Determine the pH of the solution. If the unknown solution is a base, how will you determine the pOH of the solution?

Given the initial concentration (molarity determined in Step 15) and the pH of the unknown solution, what is the ionization constant (K_a or K_b) for each trial?

What is the average ionization constant for the unknown?

Using the information provided in Table 1 and the results obtained in Parts A and B of this experiment, identify the unknown solution.

Name

Section

Date

Instructor

26

EXPERIMENT 26

Pre-Laboratory Questions

1. 10.00 mL of an unknown base solution is titrated with 0.100 *M* HCl solution. The pH versus the volume of NaOH added is shown below.

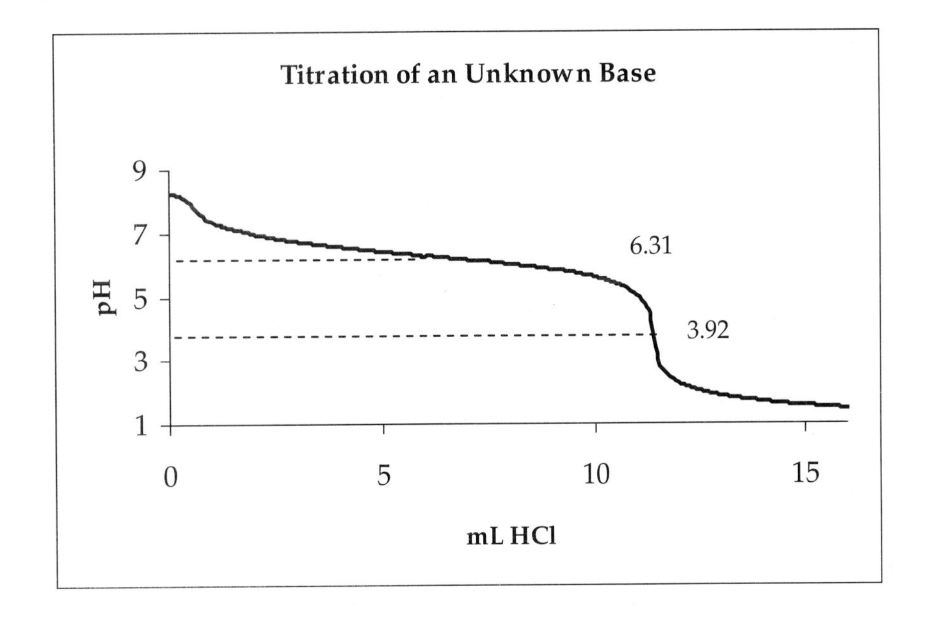

A. What is the pOH of the solution at the half equivalence point?

B. What is the ionization constant, K_b, for the unknown base?

C. Using Table 1, identify the unknown base.

2. The pH of a 0.100 *M* weak acid solution is 4.07. Calculate the K_a value for the weak acid.

Name

Section

Date

Instructor

26

EXPERIMENT 26

Post-Laboratory Questions

1. In Steps 3 through 5 of this experiment, a student accidentally titrated an unknown acid with 0.050 *M* NaOH solution instead of 0.100 *M* NaOH solution.

   A. How would this error affect the general appearance of the titration curve in terms of the volume of based added at the half-equivalence point and the equivalence point?

   B. Draw a sketch illustrating how the titration curve in Question 1.A would appear different from the one you plotted in Step 8.

2. In this experiment, a student used a pH probe that was not properly calibrated. The probe reads the pH one unit higher than it should be. Would the calculated K_a value of the unknown acid be higher or lower than the correct value? Justify your answer with an explanation.

EXPERIMENT 27

pH and Buffer Solutions

PURPOSE

Determine the pH of various common household substances and several buffer solutions. Use the Henderson-Hasselbalch equation to prepare acidic and basic buffer solutions. Calculate the changes in pH after the addition of a strong acid or a strong base to a buffer solution.

INTRODUCTION

From a chemical point of view, acids and bases differ in their ability to donate or accept hydrogen ions. According to the **Brønsted-Lowry definition**, an **acid** is a species that donates hydrogen ions (H^+, a proton), and a **base** is a species that accepts H^+ ions. Consider the acid-base reaction that occurs when hydrogen chloride gas (HCl) dissolves in water.

$$HCL_{(g)} + H_2O_{(\ell)} \rightarrow H_3O^+_{(aq)} + CI^-_{(aq)} \qquad \text{(Eq. 1)}$$

HCl acts an acid because it donates a proton (H^+) to water. Water acts as a base because it accepts a proton from HCl.

Brønsted-Lowry acid-base reactions form conjugate acid-base pairs. A **conjugate base** of an acid is the species that remains after the acid has donated a proton (an acid minus an H^+). A **conjugate acid** of a base is a species that forms when a base accepts a proton (a base plus an H^+). In Eq. 1 when HCl donates its H^+ to water, its conjugate base Cl^- is formed. Similarly, when water accepts H^+ from HCl, its conjugate acid H_3O^+ (known as the hydronium ion) is formed (see Note).

HCl and Cl^- an acid-conjugate base pair
H_2O and H_3O^+ a base-conjugate acid pair

NOTE: Free H^+ ions do not exist in aqueous solution. H^+ ions readily react with water molecules to form hydrated H^+ ions, represented as H_3O^+. H_3O^+ and H^+ are used interchangeably when referring to hydrogen ions in aqueous solution

It is the competition for H^+ ions between conjugate acid-base pairs that ultimately defines the relative strengths of acids and bases in aqueous solution.

Strong acids, such as HCl, completely ionize in aqueous solutions (Eq. 1). In other words, as HCl dissolves in water, essentially all of the HCl molecules separate into H^+ and Cl^- ions, leaving relatively few HCl molecules remaining in solution. The Cl^- ions tend not to recombine with H^+ to form HCl molecules.

Strong bases also completely dissociate when dissolved in water. For example, when KOH dissolves in water, essentially all of the KOH formula units separate into K^+ and OH^- ions, leaving relatively few KOH ion pairs remaining in solution.

$$KOH_{(s)} + H_2O_{(\ell)} \rightarrow K^+{}_{(aq)} + OH^-{}_{(aq)} \qquad \text{(Eq. 2)}$$

Weak acids like acetic acid only *slightly* ionize in aqueous solution. When acetic acid (CH_3COOH, also called ethanoic acid) dissolves in water, relatively small numbers of H_3O^+ ions and acetate (CH_3COO^-, also called ethanoate) ions are formed. The CH_3COO^- ions readily react with H_3O^+ ions to produce CH_3COOH molecules. Thus, the reaction is reversible, and chemical equilibrium is established.

$$CH_3COOH_{(aq)} + H_2O_{(\ell)} \rightleftharpoons H_3O^+{}_{(aq)} + CH_3COO^-{}_{(aq)} \qquad \text{(Eq. 3)}$$

The solution primarily contains CH_3COOH molecules and relatively few H_3O^+ and CH_3COO^- ions.

Like weak acids, **weak bases** only slightly ionize when dissolved in water. For example, when NH_3 dissolves in water, relatively few $NH_4{}^+$ and OH^- ions form. Most of the NH_3 molecules remain in solution.

$$NH_{3(aq)} + H_2O_{(\ell)} \rightleftharpoons NH^+_{4(aq)} + OH^-_{(aq)} \qquad \text{(Eq. 4)}$$

pH Measurements

Water autoionizes (dissociates) to produce hydronium ions and hydroxide ions according to Eq. 5.

$$H_2O_{(\ell)} + H_2O_{(\ell)} \rightleftharpoons H_3O^+_{(aq)} + OH^-_{(aq)} \qquad \text{(Eq. 5)}$$

One water molecule acts as an acid and donates an H^+, while the second water molecule acts as a base and accepts an H^+. Relatively few H_3O^+ and OH^- ions are produced. The $[H_3O^+]$ and $[OH^-]$ in pure water at 25 °C have been measured as $1.00 \times 10^{-7} M$.

The ionization constant (K) for the autoionization of water can be expressed as

$$K = \frac{[H_3O^+][OH^-]}{[H_2O]^2} \qquad \text{(Eq. 6)}$$

Equilibrium constants are defined based upon a concept called activity. For ions dissolved in solution, the activity is approximately equal to the ion's molar concentration. For pure liquids and solids, like water, the activity is 1. Consequently, Eq. 6 can be simplified to the following expression.

$$K_w = [H_3O^+][OH^-] \qquad \text{(Eq. 7)}$$

Substituting $[H_3O^+]$ and $[OH^-]$ for water at 25 °C into the K_w expression yields

$$K_w = [H_3O^+][OH^-] = (1.00 \times 10^{-7}\ M)(1.00 \times 10^{-7}\ M)$$
$$K_w = 1.00 \times 10^{-14}$$

(K_w will have different numerical values at other temperatures). Eq. 7 gives the relationship between the $[H_3O^+]$ and $[OH^-]$ in aqueous solution.

In a neutral solution, the hydronium ion concentration is equal to the hydroxide ion concentration, $[H_3O^+] = [OH^-]$. In an acidic solution, the hydronium ion concentration is greater than the hydroxide ion concentration, $[H_3O^+] > [OH^-]$. In a basic solution, $[OH^-] > [H_3O^+]$. If the $[H_3O^+]$ for a solution is known, Eq. 7 can be used to calculate the $[OH^-]$ in solution, and vice versa. For example, if a solution has $[H_3O^+] = 1.00 \times 10^{-4} M$, then its $[OH^-]$ must equal $1.00 \times 10^{-10} M$, and the solution is acidic.

In aqueous solutions, the concentrations of $[H_3O^+]$ and $[OH^-]$ can be quite small. The pH scale is used as a convenient (short hand) method of expressing the acidity or basicity of a solution. **pH** is defined as the negative of the logarithm of the hydrogen ion ($[H_3O^+]$ or $[H^+]$) concentration.

$$pH = -\log[H_3O^+] \qquad \text{(Eq. 8)}$$

The common logarithm of a number is the power to which 10 must be raised to equal that number. For example, the logarithm of 1.00×10^{-5} is −5. The pH of a solution is normally a number between 0 and 14. A solution with a $pH < 7$ is acidic, a $pH = 7$ is neutral, and a $pH > 7$ is basic. For example, a solution having $[H_3O^+] = 1.00 \times 10^{-4} M$ would have a pH of 4.

Similarly, it is possible to define a scale that expresses the $[OH^-]$ in a solution. **pOH** is defined as the negative of the logarithm of the hydroxide ion concentration.

$$pOH = -\log[OH^+] \qquad \text{(Eq. 9)}$$

By taking the log of both sides of Eq. 7 and multiplying each side by −1, we can derive an important relationship between pH and pOH.

$$[H_3O^+][OH^-] = 1.00 \times 10^{-14}$$
$$-\log\{[H_3O^+][OH^-]\} = -\log(1.00 \times 10^{-14})$$
$$-\{\log[H_3O^+] + \log[OH^-]\} = -\log(1.00 \times 10^{-14})$$
$$-\log[H_3O^+] - \log[OH^-] = -\log(1.00 \times 10^{-14})$$
$$pH + pOH = 14 \qquad \text{(Eq. 10)}$$

Buffer Solutions

Many biological and chemical reactions must occur within a certain pH range. For example, human blood must maintain a pH of 7.35 to 7.45 for normal biochemical reactions to occur. Blood pH is maintained by a buffer solution. **Buffer solutions** resist changes in pH when acids or bases are added to the solution. There are two types of buffer solutions. An **acidic buffer** solution is a mixture of a weak acid and the salt of its conjugate base. A **basic buffer** solution is a mixture of a weak base and the salt of its conjugate acid.

An example of an acidic buffer is acetic acid solution mixed with sodium acetate ($NaCH_3COO$, also called sodium ethanoate). Acetic acid is a weak acid, only a small amount of the acetic acid molecules ionize to form acetate ions.

$$CH_3COO_{(aq)} + H_2O_{(\ell)} \rightleftharpoons H_3O^+_{(aq)} + CH_3COO^-_{(aq)} \quad \text{(Eq. 11)}$$

Sodium acetate is a water soluble salt containing the conjugate base of acetic acid, CH_3COO^-. Adding sodium acetate to the acetic acid solution greatly increases the acetate ion concentration. Thus, the buffer solution contains both acidic (CH_3COOH) and basic (CH_3COO^-) components. It has the capacity to neutralize both acids and bases added to the solution.

For example, if a small amount of HCl is added to the CH_3COOH/CH_3COO^- buffer solution, the H_3O^+ ions from HCl react with CH_3COO^- ions (a base) to form CH_3COOH molecules. The equilibrium shifts to the reactant side, in accordance with Le Châtelier's principle, to reestablish equilibrium with only a slight reduction in the pH of the solution (typically 0.1-0.2 pH units). If a small amount of NaOH is added to the CH_3COOH/CH_3COO^- buffer solution, the OH^- ions from NaOH react with the CH_3COOH molecules to form CH_3COO^- ions. The equilibrium shifts to the product side to reestablish equilibrium with only a slight increase in pH of the solution.

A basic buffer can be prepared by mixing a weak base with its conjugate acid. Consider the buffer formed when aqueous ammonia (NH_3) and ammonium chloride (NH_4Cl) are mixed.

$$NH_{3(aq)} + H_2O_{(l)} \rightleftharpoons NH^+_{4(aq)} + OH^-_{(aq)} \quad \text{(Eq. 12)}$$

Ammonia, NH_3, is a weak base. Only a small amount of ammonia molecules react with water to form ammonium and hydroxide ions. Ammonium chloride, a water soluble salt, is added to increase the concentration of NH_4^+ ion, the conjugate acid of NH_3. Thus, the buffer solution contains both an acidic component (NH_4^+) and a basic component (NH_3). If a small amount of HCl is added to the NH_3/NH_4^+ buffer solution, the H_3O^+ ions from HCl react with NH_3 molecules (a base) to form NH_4^+ ions. The equilibrium shifts to the product side to reestablish equilibrium. If a small amount of NaOH is added to the NH_3/NH_4^+ buffer solution, the OH^- ions from NaOH react with NH_4^+ ions (an acid) to form NH_3 molecules. The equilibrium shifts to the reactant side to reestablish equilibrium.

Henderson-Hasselbalch Equation

When a weak acid, HA, is added to water, its ionization can be represented by the reaction given below.

$$HA + H_2O \rightleftharpoons H_3O^+ + A^- \quad \text{(Eq. 13)}$$

The **ionization constant**, K_a, for the acid can be expressed as

$$K_a = \frac{[H_3O^+][A^-]}{[HA]} \quad \text{(Eq. 14)}$$

Solving Eq. 14 for $[H_3O^+]$, taking the negative of the logarithm of both sides of the equation, and expressing $-\log K_a$ as pK_a, gives the Henderson-Hasselbalch equation for an acidic buffer solution.

$$[H_3O^+] = K_a\frac{[HA]}{[A^-]}$$

$$-\log[H_3O^+] = -\log K_a - \log\frac{[HA]}{[A^-]}$$

$$pH = pK_a + \log\frac{[A^-]}{[HA]} \quad \text{(Eq. 15)}$$

This form of the Henderson-Hasselbalch equation is used to calculate the pH of an acidic buffer solution. $[A^-]$ is the initial concentration of the salt (conjugate base) and [HA] is the initial concentration of the weak acid.

From Eq. 15, the pH of an acidic buffer solution depends on the p*K*a value of the weak acid, and the ratio of the conjugate base concentration to the acid concentration. When preparing a buffer solution with a specific pH, it is important to choose a weak acid with a pK_a value within W 1 pH unit of the desired pH of the solution. By varying the ratio of the concentration of the conjugate base to that of the weak acid ($[A^-]/[HA]$), a buffer solution with the desired pH can be attained.

A similar form of the Henderson-Hasselbalch equation can be derived to calculate the pOH of a basic buffer solution.

$$pOH = pK_b + \log\frac{[BH^+]}{[B]} \quad \text{(Eq. 16)}$$

$[BH^+]$ is the initial concentration of the conjugate acid, [B] is the initial concentration of the weak base. The pOH of a basic buffer solution depends on the pK_b value of the weak base and the ratio of $[BH^+]/[B]$.

Buffer solutions lose their ability to resist changes in pH once one component of the conjugate acid-base pair is consumed. If sufficient acid or base is added to a buffer solution to consume one of the buffer components, the **buffering capacity** of the solution is exceeded. For example, a buffer composed of 0.1 *M* acetic acid and 0.1 *M* sodium acetate will have the same pH as a buffer composed of 1.0 *M* acetic acid and 1.0 *M* sodium acetate. However, ten times more HCl must be added to the 1.0 *M* acetic acid/sodium acetate solution to consume the acetate ions than would be needed to consume the acetate ions in the 0.1 *M* acetic acid/sodium acetate solution. Thus, 1.0 *M* acetic acid/sodium acetate solution has a larger buffering capacity than a 0.1 *M* acetic acid/sodium acetate solution.

In this experiment the pH of various household products will be measured and used to determine whether they are acidic, basic, or neutral. The Henderson-Hasselbalch equation will be utilized to prepare buffer solutions with a specific pH, and to calculate the changes in pH after the addition of a strong acid or a strong base to a buffer solution.

Preparing a Buffer Solution with a Specific pH

Prepare 150.0 mL of an acidic buffer solution with a pH of 3.50 and a weak acid concentration of 0.10 *M*. To prepare the buffer, choose the appropriate weak acid and conjugate base pair (salt of the weak acid) from the list of chemicals provided below.

3.0 *M* acetic acid (CH_3COOH) solution ($pK_a = 4.74$)
3.0 *M* formic acid (HCOOH) solution ($pK_a = 3.74$)
solid sodium acetate ($NaCH_3COO$)
solid sodium formate (NaHCOO)

When selecting a conjugate acid-base pair to prepare a buffer solution, the weak acid should have a pK_a value very close to the desired pH of the buffer solution. Of the weak acids listed above, the pK_a value of formic acid is closest to pH 3.50. Sodium formate is a salt containing $HCOO^-$ ions, the conjugate base of formic acid.

Next, substitute the pH and pK_a values into the Henderson-Hasselbalch equation to obtain the [conjugate base]/[weak acid] ratio. Knowing the concentration of the weak acid is 0.10 *M* in the buffer solution, we calculate the [conjugate base] in the buffer solution.

$$\frac{[HCOO^-]}{[HCOOH]} = \frac{[HCOO^-]}{0.10\ M} = 0.58$$

$$[HCOO^-] = 0.058\ M$$

Now we must calculate the volume of 3.0 *M* formic acid and the mass of NaHCOO needed to prepare 150 mL of a buffer solution with a pH of 3.50, a $[HCOO^-]$ of 0.058 *M*, and a [HCOOH] of 0.10 *M*. Add 75 mL of distilled water to a 150-mL volumetric flask. Next, add 5.0 mL of 3.0 *M* formic acid and 0.59 grams of sodium formate to the flask. Finally, add sufficient distilled water to produce 150 mL of the buffer solution with a pH of 3.50.

PROCEDURE

CAUTION

Students must wear departmentally approved eye protection while performing this experiment. Wash your hands before touching your eyes and after completing the experiment.

If acid or base contacts your skin, wash the affected area with copious quantities of water. Be especially cautious with Liquid Plumber®, it is extremely caustic and corrosive.

Part A – Set up the MeasureNet Workstation to Record pH.

1. Press the **On/Off** button to turn on the power to the MeasureNet Workstation.
2. Press **Main Menu**, then press **F3 pH vs. mVolts**, then press **F1 pH vs. Time.**
3. Press **Calibrate**. The MeasureNet pH probe will be stored in a beaker containing pH 7.00 buffer solution. Using a thermometer, determine the temperature of the pH 7.00 buffer solution and enter it at the workstation. Press **Enter**.
4. Enter 7.00 as the pH of the buffer solution at the workstation, press **Enter**.

5. Gently stir the buffer solution with a stirring rod. When the displayed pH value stabilizes, press **Enter**. The pH should be close to 7.00, but it does not have to read exactly 7.00.
6. Remove the MeasureNet pH electrode from the pH 7.00 buffer solution, rinse the tip of the probe with distilled water, and dry it with a Kimwipe®.
7. Press F1 if a one point standardization is to be used. If a two point standardization is to be used, enter the pH (either pH 4.00 or pH 10.00) of the second buffer solution at the workstation, press **Enter**. Insert the MeasureNet pH probe into the buffer solution. Gently stir the buffer solution with a stirring rod. When the displayed pH value stabilizes, press **Enter**.
8. Press **Display** to accept all values.

Part B – pH Measurements

9. Determine the pH of each of the following solutions: lemon juice, Liquid Plumber®, Windex®, Coca Cola®, vinegar, tap water, and distilled water. Use 20 mL in a 50 mL beaker to determine the pH of each solutions. Be sure to rinse the pH electrode with distilled each time it is removed from one solution, and before it is added to a different solution. Be sure to stir each solution while measuring its pH.
10. Should you record the pH of each solution in the Lab Report? Indicate whether each solution is acidic, neutral, or basic.
11. Discard each solution into the sink.
12. Be sure to rinse the pH probe with distilled water before returning the probe to pH 7 buffer solution.

Part C – pH Changes of a Distilled Water Sample before and after the Addition of a Strong Acid or Base

13. Pour 45.0 mL of distilled water into each of two clean 150-mL beakers. Should you record the pH of the water in each beaker? Add 5.0 mL of 0.10 *M* hydrochloric acid (HCl) to one of the beakers and 5.0 mL of 0.10 *M* sodium hydroxide (NaOH) to the other. Should you record the pH of the water containing HCl and the water containing NaOH? Be sure to immerse the pH probe in the pH 7.00 standard buffer solution after the measurements are concluded.
14. What are the differences in pH before and after addition of the HCl and NaOH to the distilled water. Did the pH change significantly (> 1 pH unit) when HCl or NaOH was added to the distilled water? Why or Why not?

Part D – Preparation of an Acidic Buffer Solution

15. Prepare 125 mL of an acidic buffer solution with a pH value specified by your laboratory instructor. The concentration of the weak acid in the buffer solution to be prepared is 0.10 *M*. Choose the appropriate weak acid and conjugate base pair from the list of chemicals provided below to prepare the buffer solution. Should you show all calculations used to prepare the buffer solution in the Lab Report? Should you record all measured pH values for the buffer solution in the Lab Report?

3.0 *M* acetic acid solution ($pK_a = 4.74$)
3.0 *M* formic acid solution ($pK_a = 3.74$)
solid sodium acetate
solid sodium formate

16. Pour 45.0 mL of the buffer solution prepared in Step 15 into a 100-mL beaker. Record a 15 second pH versus time scan to verify the pH of the solution. Should you record the pH in the Lab Report?
17. Be sure to save the pH versus time scan. Press **File Options**, then press **F3**. Enter a 3-digit number to record a file name for the scan. Press **Enter**. Should you record the file name in the Lab Report?
18. Press **Display** to prepare the workstation to record another pH versus time scan.
19. Add 5.0 mL of 0.10 *M* HCl to the buffer solution and thoroughly mix in the 100-mL beaker.
20. Record a 15 second pH vs. time scan. Press **File Options**, then press **F3**. Enter a 3-digit number to record a file name for the scan. Press **Enter**. Should you record the file name in the Lab Report?
21. Press **Display** to prepare the workstation to record another pH versus time scan.
22. Decant the buffer solution into the sink.
23. Pour 45.0 mL of the buffer solution prepared in Step 15 into a 100-mL beaker. Add 5.0 mL of 0.10 *M* NaOH to the buffer solution and thoroughly mix.
24. Record a 15 second pH vs. time scan. Press **File Options**, then press **F3**. Enter a 3-digit number to record a file name for the scan. Press **Enter**. Should you record the file name in the Lab Report?
25. Press **Display** to prepare the workstation to record another pH versus time scan. Be sure to immerse the pH probe in the pH 7.00 standard buffer solution after the measurements are concluded.
26. *Step 26 is to be performed after the experiment is concluded.* Plot pH versus time curves for the files saved in Steps 17, 20, and 24 using the Excel instructions provided in **Appendix B–4**.

Part E – Preparation of a Basic Buffer Solution

27. Prepare 125 mL of a basic buffer solution with a pH value specified by your laboratory instructor. The concentration of the weak base in the buffer solution to be prepared is 0.10 *M*. Choose the appropriate weak base and conjugate acid pair from the list of chemicals provided below to prepare the buffer solution.

3.0 *M* aqueous ammonia ($pK_b = 4.74$)
3.0 *M* sodium carbonate solution ($pK_b = 3.67$)
solid sodium hydrogen carbonate
solid ammonium chloride

28. Pour 45.0 mL of the buffer solution into a 100-mL beaker. Record a 15 second pH versus time scan to verify the pH of the solution.
29. Once the scan stops, press **File Options**, then press **F3**. Enter a 3-digit number to record a file name for the scan. Press **Enter**. Should you record the file name in the Lab Report?
30. Press **Display** to prepare the workstation to record another pH versus time scan.

31. Add 5.0 mL of 0.10 *M* HCl to the buffer solution prepared in Step 27 in the 100-mL beaker. Thoroughly mix the solution.

32. Record a 15 second pH vs. time scan. Press **File Options**, then press **F3**. Enter a 3-digit number to record a file name for the scan. Press **Enter**. Should you record the file name in the Lab Report?

33. Press **Display** to prepare the workstation to record another pH versus time scan.

34. Decant the buffer solution into the sink.

35. Pour 45.0 mL of the original buffer solution you prepared in Step 27 into a 100-mL beaker. Add 5.0 mL of 0.10 *M* NaOH to the buffer solution and thoroughly mix.

36. Record a 15 second pH vs. time scan. Press **File Options**, then press **F3**. Enter a 3-digit number to record a file name for the scan. Press **Enter**. Should you record the file name in the Lab Report?

37. Press **Display** to prepare the workstation to record another pH versus time scan. Be sure to immerse the pH probe in the pH 7.00 standard buffer solution after the measurements are concluded.

38. Excel Instructions for plotting pH versus times curves are provided in **Appendix B-4**. All pH versus time plots must be submitted to your laboratory instructor along with the Lab Report.

39. Did the acidic and basic buffer solutions maintain a relatively constant pH after the addition of HCl and NaOH? Explain.

Name | Section | Date

Instructor

27 EXPERIMENT 27

Lab Report

Part B – pH Measurements

pH of lemon juice, Liquid Plumber®, Windex®, Coca Cola®, vinegar, tap water, and distilled water. Indicate whether each solution is acidic, neutral, or basic.

Part C – pH Changes of a Distilled Water Sample Before and After Addition of a Strong Acid or Base

What is the pH of the water?

What is the pH of the water containing HCl?

What is the pH of the water containing NaOH?

What are the differences in pH before and after addition of the HCl and NaOH to the distilled water. Did the pH change significantly (> 1 pH unit) when HCl or NaOH was added to the distilled water? Why or Why not?

Part D – Preparation of an Acidic Buffer Solution

Preparation of an acidic buffer solution. What is the pH of the buffer designated by the lab instructor?

Part E – Preparation of a Basic Buffer Solution

Preparation of an basic buffer solution. What is the pH of the buffer designated by the lab instructor?

Did the acidic and basic buffer solutions maintain a relatively constant pH after the addition of HCl and NaOH? Explain.

Name

Section

Date

Instructor

27

EXPERIMENT 27

Pre-Laboratory Questions

1. A buffer solution is prepared by mixing 50.0 mL of 0.300 *M* $NH_{3(aq)}$ with 50.0 mL of 0.300 *M* $NH_4Cl_{(aq)}$. The pK_b of NH_3 is 4.74.

$$NH_3 + H_2O \rightleftharpoons NH_4^+ + OH^-$$

A. Calculate the $[NH_3]$ and $[NH_4Cl]$ in the buffer solution. Calculate the pH of the buffer solution.

B. 7.50 mL of 0.125 *M* HCl is added to the 100.0 mL of the buffer solution. Calculate the new $[NH_3]$ and $[NH_4Cl]$ for the buffer solution. Calculate the new pH of the solution.

C. 7.50 mL of 0.125 *M* NaOH is added to the 100.0 mL of the buffer solution. Calculate the new [NH_3] and [NH_4Cl] for the buffer solution. Calculate the new pH of the solution.

Name

Section

Date

Instructor

27 EXPERIMENT 27

Post-Laboratory Questions

1. A student added 5.00 mL of 0.10 *M* H_2SO_4 instead of 0.10 *M* HCl in Step 19 of the experiment. Would the pH of the resulting solution be higher or lower than the value measured in the experiment? Why or why not?

2. Using the Henderson-Hasselbalch equation, calculate the pH of the solution described Question 1?

3. In Step 35, a student added 10.00 mL, instead of 5.00 mL of 0.1 *M* NaOH to the basic buffer solution. Would the pH of the resulting solution be higher or lower than the value measured in the experiment? Why or why not?

4. Using the Henderson-Hasselbalch equation, calculate the pH of the solution described Question 3?

Identifying an Unknown Weak Acid or Weak Base: A Self-Directed Experiment

PURPOSE

Students will identify a weak acid or weak base by determining its K_a or K_b value via three different experimental methods.

INTRODUCTION

This is a *self-directed* experiment wherein teams of students will write the procedure they will use, and rely on lab skills and techniques acquired this semester, to perform this experiment. Each team will submit a **Procedure Proposal** to their lab instructor two weeks before the experiment is performed. *The Procedure Proposal must address several important questions, a few examples are listed below.*

1. What is the central question to be answered in this experiment?
2. What experimental technique(s) will be utilized to answer the central question?
3. What calculations should be provided in the Procedure Proposal?
4. What safety precautions must be addressed in the Procedure Proposal?
5. What additional questions should be addressed by each team?

Each team may use their lab manual, textbook, reference books, and the internet for assistance in writing the **Procedure Proposal** (the format for writing the Procedure Proposal is provided in Appendix C-1). The graded and annotated Procedure Proposal will be returned to each team one week before the lab period during which the experiment is to be performed.

When the experiment is completed, each team will submit one **Formal Lab Report** (the format for writing the Formal Lab Report is provided in Appendix C-2) addressing the following questions.

1. What conclusions can be drawn from the experimental data collected by each team?
2. Does the experimental data collected answer the central question?
3. Can the team's conclusions be supported by information obtained from reference sources, textbooks, or the internet?
4. What are some possible sources of error in the experimental data?
5. What modifications could be made to the experimental design and procedures to improve the accuracy of the data?
6. Which experimental method yields the more accurate K_a or K_b value? Justify your answer with an explanation.

PROCEDURE

Each team will be given 80 mL of a 0.100 *m* unknown, monoprotic acid or base solution. Each team is to determine the K_a or K_b value of the unknown acid or base using three different experimental techniques: 1) titration; 2) determination of the percent ionization from freezing point depression; and 3) from the determination of the pH of the solution.

Each team will calculate an average K_a or K_b value from the three experimental methods used in this experiment. The average K_a or K_b value will then be compared to the K_a or K_b values for common acids found in any general chemistry textbook, reference books such as *Lange's Handbook* or the *CRC Handbook of Chemistry and Physics*, or via the internet, and used to identify the unknown monoprotic acid.

The effective molality, $m_{effective}$, represents the total number of particles in solution.

$$m_{effective} = i(m_{stated\ concentration}) \qquad \text{(Eq. 1)}$$

The $m_{effective}$ for the solution is equal to the sum of the molalities of HA, H^+, and A^-, or B, BH^+, and OH^-.

LIST OF CHEMICALS

A. 0.100 *M* Unknown acid and base solutions

B. 0.100 *M* Sodium hydroxide solution

C. 0.100 *M* Hydrochloric acid solution

D. Sodium chloride or calcium chloride (solids)

E. Ice

LIST OF SPECIAL EQUIPMENT

A. MeasureNet temperature probe

B. MeasureNet pH probe and drop counter.

C. Normal lab glassware and equipment

Hydrometallurgy to Analyze a Chromite Sample: A Self-Directed Experiment

PURPOSE

To determine the chemical content of a metal ore sample thought to be the mineral chromite.

INTRODUCTION

The mineral chromite is the only ore of chromium. Chromium has many industrial uses. It is used to produce stainless steel, and other nonferrous alloys, to enhance their hardenability and to make these allows more resistant to corrosion and oxidation. Chromium is used in the plating of metals, pigments, leather processing, catalysts, surface treatments, and refractories. **Chromite** is iron magnesium chromium oxide. In some samples, magnesium can substitute for iron, and aluminum and ferric ions can substitute for chromium.

Hydrometallurgy is a form of extractive metallurgy that utilizes aqueous solution chemistry for the recovery of metals from salts, minerals, or ores. The first step in the process is extraction of the metal. **Extraction** is the process of removing metal from ore by dissolving the metal in a suitable solvent, then recovering the metal from the solution, and discarding waste materials. The operations usually involved are leaching (dissolving in water or acid), commonly with additional agents; separating the waste and purifying the leach solution; and precipitating the metal or one of its pure compounds from the leach solution by chemical or electrolytic means.

This is a *self-directed* experiment wherein teams of students will write the procedure they will use, and rely on lab skills and techniques acquired this semester, to perform this experiment. Each team will submit a **Procedure Proposal** to their lab instructor two weeks before the experiment is performed. *The Procedure Proposal must address several important questions, a few examples are listed below.*

1. What is the central question to be answered in this experiment?
2. What experimental technique(s) will be utilized to answer the central question?
3. What calculations should be provided in the Procedure Proposal?
4. What safety precautions must be addressed in the Procedure Proposal?
5. What additional questions should be addressed by each team?

Each team may use their lab manual, textbook, reference books, and the internet for assistance in writing the **Procedure Proposal** (the format for writing the Procedure Proposal is provided in Appendix C-1). The graded and annotated Procedure Proposal will be returned to each team one week before the lab period during which the experiment is to be performed.

When the experiment is completed, each team will submit one **Formal Lab Report** (the format for writing the Formal Lab Report is provided in Appendix C-2) addressing the following questions.

1. What conclusions can be drawn from the experimental data collected by each team?
2. Does the experimental data collected answer the central question?
3. Can the team's conclusions be supported by information obtained from reference sources, textbooks, or the internet?
4. What are some possible sources of error in the experimental data?
5. What modifications could be made to the experimental design and procedures to improve the accuracy of the data?

PROCEDURE

Each team will obtain one gram of an unknown sample from their instructor. The team must determine if the sample is actually chromite. Chromite contains magnesium, iron, and chromium. Each team must determine whether their sample contains each of these metals. Teams may use chemical reactions or instrumental analysis to analyze their sample.

Given magnesium neither absorbs nor emits in the 200–900 nm region of the spectrum, teams will have to utilize the extraction process to isolate and precipitate magnesium. To isolate magnesium, you will need to take advantage of the fact that some metals are amphoteric. **Amphoterism** is the ability of a substance to react with either acids or bases. For example, beryllium is amphoteric. When aqueous beryllium nitrate is reacted with a limited amount of potassium hydroxide, insoluble beryllium hydroxide forms.

$$2\ OH^-_{(aq)} + Be^{2+}_{(aq)} \rightarrow Be(OH)_{2(s)}$$

Beryllium hydroxide behaves as a base when it reacts with acids to form salts and water. In a solution containing excess OH^- ions, $Be(OH)_2$ acts like an acid and dissolves forming the complex ion $[Be(OH)_4]^{2-}$.

$$2\ OH^-_{(aq)} + Be(OH)_{2(aq)} \rightarrow [Be(OH)_2]^{2-}_{(aq)}$$

Iron and magnesium form insoluble precipitates in basic solution, and are not amphoteric. Zinc and chromium are amphoteric and form complex ions in the presence of excess base ($[Zn(OH)_4]^{2-}$ and $[Cr(OH)_4]$).

LIST OF CHEMICALS

A. Iron(III) nitrate, solid

B. Chromium(III) nitrate, solid

C. 1.00 *M* Hydrochloric acid solution

D. 4 *M* Sodium hydroxide solution

E. Zinc metal

LIST OF SPECIAL EQUIPMENT

A. MeasureNet temperature probe

B. MeasureNet pH probe and drop counter.

C. MeasureNet spectrophotometer and cuvettes

D. Normal lab glassware and equipment

Determination of Iron Content in a Food Sample: A Self-Directed Experiment

PURPOSE

Determine the iron content per gram of a food sample.

INTRODUCTION

Iron is essential to human physiology because it is a component of some important proteins and enzymes in the human body. Iron is found in hemoglobin, the protein in red blood cells that carries oxygen to tissues. It is also found in myoglobin, a protein that helps supply oxygen to muscle. A deficiency of iron limits oxygen delivery to cells, resulting in fatigue, poor performance, and decreased immunity.

The Institute of Medicine of the National Academy of Sciences' recommended daily intake of Fe is 8 mg/day for adult males and 18 mg/day for adult females. Iron can be found in red meat, fish, and poultry. It is also added to many iron-enriched food products such as cereals and oatmeal. In this experiment, teams of students will devise a method to determine the iron content (mg of Fe per gram of sample) in a food sample. Preliminary analysis shows that the iron content in the food sample is between 0.1 and 2.0 mg Fe/g food.

This is a self-directed experiment wherein teams of students will write their procedure, relying on lab skills and techniques acquired this semester, to perform the experiment. Each team will submit a **Procedure Proposal** to their lab instructor two weeks before the experiment is to be performed. The procedure proposal must address several important questions; a few examples are given below.

1. What is the central question to be answered in this experiment?
2. What experimental technique(s) will be utilized to answer the central question?

3. How will the team determine the iron content in the cereal sample?
4. What safety precautions must be addressed in the Procedure Proposal?
5. What additional questions should be addressed by each team?

Each team may use their lab manual, textbook, reference books, and the internet for assistance in writing the Procedure Proposal (the format for writing the Procedure Proposal appears in Appendix C-1). The graded and annotated Procedure Proposal will be returned to each team one week before the lab period during which the experiment is to be performed.

When the experiment is completed, each team will submit one **Formal Lab Report** (the format for writing the Formal Lab Report appears in Appendix C-2) addressing the following questions.

1. What conclusions can be drawn from the experimental data collected by the team?
2. Does the experimental data collected answer the central question?
3. What are some possible sources of errors in the experimental data?
4. Compare the experimental result with manufacturer's claim for the iron content. Explain any deviation of the team's result from the manufacturer's claim.
5. What modifications could be made to the experimental design and procedures to improve the accuracy of the data?

PROCEDURE

Each team will be provided 1.5 g of a food sample and 15 mL standardized 0.050 *M* Fe^{2+} solution. The iron in the food sample must be extracted into aqueous solution as Fe^{2+} ions before being analyzed. This process is called **digestion**. To digest the sample, add 20 mL of 6 *M* HCl to the 1.5 g food sample in a beaker, and *gently* heat it for approximately 15 minutes. The beaker should be covered with a watch glass during the digestion process. Let the resulting solution cool to room temperature. Filter the solution via gravity filtration.

Fe^{2+} ions are essentially colorless when dissolved in water. Orange-red complexes of iron form when Fe^{2+} ions react with *o*-phenanthroline. Every 5 mL of digested sample solution (or standardized 0.050 M Fe^{2+} solution) must be reacted with 20 mL of complexing solution containing *o*-phenanthroline. The reaction takes 15–20 minutes to complete. Teams are to report the iron content in the food sample in mg Fe^{2+}/g food.

LIST OF CHEMICALS

A. 1.5 g food sample

B. 20 mL 6 M HCl solution

C. 15 mL standardized 0.050 M Fe^{2+} solution

D. 120 mL complexing solution (20 mL of complexing solution contains 5 mL 1.0 M ammonium acetate, 1 mL 10% hydroxylamine hydrochloride, 10 mL 0.30% o-phenanthroline, and 4 mL deionized water.)

LIST OF SPECIAL EQUIPMENT

A. MeasureNet spectrophotometer, cuvettes, and nichrome wires

B. 50-mL volumetric flasks

C. Qualitative filter paper

Quality Control for GlassOff: A Self-Directed Experiment

PURPOSE

To determine the molarity and the percent by mass of ammonia in GlassOff.

INTRODUCTION

The Prince and Humble Company is seeking to hire a team of four quality control scientists to analyze their new window cleaning product, GlassOff. The ammonia content of GlassOff must be within a narrow concentration range, 0.600–0.800 *M*. Applicant teams will compete for the quality control positions by determining the molarity and the percent by mass of ammonia in GlassOff. The team that submits the more accurate concentration data will be awarded the quality control positions with the Prince and Humble Company.

This is a *self-directed* experiment wherein teams of students will write the procedure they will use, and rely on lab skills and techniques acquired this semester, to perform this experiment. Each team will submit a **Procedure Proposal** to their lab instructor two weeks before the experiment is performed. *The Procedure Proposal must address several important questions; a few examples are listed below.*

1. What is the central question to be answered in this experiment?
2. What experimental technique(s) will be utilized to answer the central question?
3. What calculations should be provided in the Procedure Proposal?
4. What safety precautions must be addressed in the Procedure Proposal?
5. What additional questions should be addressed by each team?

Each team may use their lab manual, textbook, reference books, and the internet for assistance in writing the **Procedure Proposal** (the format for writing the Procedure Proposal is provided in Appendix C-1). The graded and annotated Procedure Proposal will be returned to each team one week before the lab period during which the experiment is to be performed.

When the experiment is completed, each team will submit one **Formal Lab Report** (the format for writing the Formal Lab Report is provided in Appendix C-2) addressing the following questions.

1. What conclusions can be drawn from the experimental data collected by each team?
2. Does the experimental data collected answer the central question?
3. Can the team's conclusions be supported by information obtained from reference sources, textbooks, or the internet?
4. What are some possible sources of error in the experimental data?
5. What modifications could be made to the experimental design and procedures to improve the accuracy of the data?

PROCEDURE

Each team will be provided 60 mL of GlassOff produced by the Prince and Humble Company. Each team is to report the molarity and the percent by mass of ammonia in GlassOff. If the concentration of ammonia is not in the 0.600–0.800 *M* range, each team must adjust the molarity such that it does fall within this range. Teams must verify by experimentation that the concentration of ammonia is in the aforementioned range. Teams will be provided 3 *M* hydrochloric acid and sodium hydroxide pellets to perform this experiment. To perform titrations and adjust the molarity of GlassOff, each team must prepare 0.100 *M* hydrochloric acid or sodium hydroxide solution. Teams should prepare 300 mL of the solution they will use to determine the concentration of ammonia in GlassOff.

LIST OF CHEMICALS

A. 3 *M* Hydrochloric acid solution

B. Sodium hydroxide pellets

C. Sodium carbonate, solid

D. Potassium hydrogen phthalate, solid

E. GlassOff

F. 1.0 *M* Aqueous ammonia

G. pH 4 and pH 7 Buffer solutions

LIST OF SPECIAL EQUIPMENT

A. MeasureNet temperature probe

B. MeasureNet pH probe and drop counter.

C. MeasureNet spectrophotometer

D. Normal lab glassware and equipment

What is in this Container: A Capstone Experiment

PURPOSE

Determine the identity of an unknown compound using all techniques, skills, concepts, and principles acquired in a two semester general chemistry laboratory course.

INTRODUCTION

An Environmental Protection Agency inspector was performing a routine laboratory stockroom inspection. The inspector found a container of unknown material on the shelf. The inspector immediately wants to know what is in this container. The chemical inventory list for this stockroom listed the following 12 chemicals.

1. aluminum$_{(s)}$
2. sodium hydrogen carbonate$_{(s)}$
3. barium chloride$_{(s)}$
4. ammonia$_{(aq)}$
5. formic acid$_{(aq)}$
6. sodium hydroxide$_{(s)}$
7. hydrochloric acid$_{(aq)}$
8. iron(III) chloride$_{(aq)}$
9. ethanol$_{(\ell)}$
10. hexane$_{(\ell)}$
11. urea$_{(s)}$
12. potassium sulfate$_{(s)}$

The inspector has given you three hours to identify the unknown chemical in the container before he issues a fine for violation of the Chemical Specific Right-to-Know Act.

This is a *Capstone* experiment wherein teams of students will write their *Procedure Proposal* relying on techniques, skills, concepts, and principles acquired in a two semester general chemistry laboratory course. Each team will submit a *Procedure Proposal* to their lab instructor two weeks before the experiment is to be performed. The *Procedure Proposal* must provide a series of chemical tests or instrumental analyses that will permit the team to identify any one of the 12 chemicals. Each team must provide a method to identity or confirm the presence of each atom, ion, or molecule that comprises the unknown compound.

Each team may use their lab manual, textbook, reference books, and the internet for assistance in writing the **Procedure Proposal** (the format for writing the Procedure Proposal appears in Appendix C-1). The graded and annotated Procedure Proposal will be returned to each team one week before the lab period during which the experiment is to be performed. When the experiment is completed, each team will submit one **Formal Lab Report** (the format for writing the Formal Lab Report appears in Appendix C-2) that provides the series of chemical tests or instrumental analyses used by the team to identify the unknown compound in the container.

PROCEDURE

One useful method to devise a plan to identify multiple compounds is the use of a flow chart. The flow chart should begin with general tests that can be used to quickly categorize an unknown chemical according to compound class. Teams are limited to the instrumentation, equipment, and chemicals listed below to determine what is in this container.

LIST OF CHEMICALS

A. 0.1 *M* Hydrochloric acid solution

B. pH 4 and pH 7 buffer solutions

C. 0.1 *M* Sodium hydroxide solution

D. 0.1 *M* Silver nitrate solution

E. 0.1 *M* Barium chloride solution

F. 0.1 *M* Iron(III) nitrate solution

G. 0.1 *M* Sodium chloride solution

H. 0.1 *M* Potassium chloride solution

I. Distilled water

LIST OF SPECIAL EQUIPMENT

A. MeasureNet spectrophotometer, cuvettes, and nichrome wires

B. MeasureNet pH probe and drop counter.

C. MeasureNet temperature probe

D. Conductivity detector

E. Styrofoam cup calorimeters and lid

F. Volumetric flasks

G. Kimwipes®

H. Crucible and lid

I. Ordinary lab glassware

Instructions for Initializing the MeasureNet Workstation to Record a Thermogram

1. Press the **On/Off** button to turn on the MeasureNet workstation (Figure 1).
2. Press **Main Menu**, then press **F2 Temperature**, next press **F1 Temperature vs. Time**.
3. Half fill a 150-mL beaker with ice and water. Press **Calibrate** to calibrate the temperature probe (Figure 1). When prompted, enter 0.0 °C as the actual temperature of the constant temperature bath (ice water). Press **Enter**. Insert the temperature probe in the ice water. Swirl the temperature probe until the temperature approaches and stabilizes near 0 °C (*it may not read exactly 0 °C*). Press **Enter**.
4. Remove the temperature probe from the ice water bath.

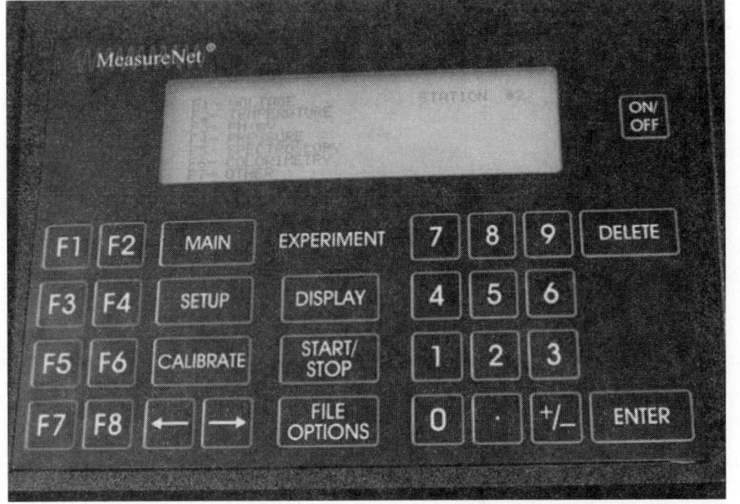

Bobby Stanton/Wadsworth/Cengage Learning

Figure 1
MeasureNet workstation control panel

5. Press **SetUp** to establish scan parameters for the experiment. Press **F1** to set the limits for the scan. Use the ← → keys to move from **min** to **max** or from **X** to **Y**. An * marks the parameter selected to change at any given time. The **Y** axis is for temperature. Set the **min** temperature 2–3 °C below the initial temperature of the substance that the probe is inserted into. Press **Enter**.
6. Set the **max** temperature 2–3 °C above the expected final temperature. Press **Enter**.
7. For example, to calibrate a styrofoam cup calorimeter, the **min** temperature should be set to 15–18 °C, and the **max** temperature should be set to 70–75 °C.
8. The **X** axis is for time. Set scan parameters from 0 seconds (***min***) to 250 seconds (***max***). This should be sufficient time for most experiments that require thermograms. (If the experiment is completed before 250 seconds elapses, press **Stop** *at any time* to end the experiment).
9. When all parameters are entered, press **Display** to accept all values. The MeasureNet workstation is now ready to record a thermogram.

Instructions for Recording a Thermogram to Determine the Styrofoam Cup Calorimeter Constant

1. Obtain two styrofoam cups and a lid from your instructor to serve as a calorimeter. Nest one cup inside the other to construct the calorimeter.
2. 45–50 grams of tap water should be added to the calorimeter to serve as the cool water. Should the exact mass of the cool water be recorded in the lab report? If so, to what number of significant figures should the mass be recorded?
3. If a magnetic stirrer and stir bar are available, they will be used to stir the solution. If a magnetic stirrer is not available, proceed to Step 6.
4. Setup the magnetic stirrer/calorimeter assembly depicted in Figure 1 below.

Bobby Stanton/Wadsworth/Cengage Learning

Figure 1
Magnetic stirrer/calorimeter assembly

5. The temperature probe is inserted into a 2-hole stopper and secured to a ring stand with a utility clamp. The temperature probe is inserted through the hole in the calorimeter lid. Position the probe so that its tip is ~ 1 cm from the bottom of the calorimeter. *Be careful not to poke a hole in the bottom of the styrofoam cup.* A stir bar is placed in the bottom of the calorimeter. Proceed to Step 9.
6. If a magnetic stirrer is not available, assemble the stopper/wire stirrer/probe assembly depicted in Figure 2 below. Spread the slit hole in a 2-hole rubber stopper and insert the temperature probe. Insert the wire stirrer (one end has a loop) through the other hole in the stopper. Make certain that the temperature probe passes through the loop of the wire.
7. Secure the rubber stopper/wire/temperature probe assembly to a ring stand using a utility clamp.
8. Assemble the calorimeter shown in Figure 3. Insert the probe assembly through the hole in the cardboard lid. Position the probe tip so that it is ~ 1 cm from the bottom of the styrofoam cup.
9. Obtain a hot plate and place it at least two feet away from the MeasureNet workstation (Figure 3).
10. Add ~ 60 mL of water to a clean, dry 150-mL or 250-mL beaker. Place the beaker of water in a microwave and heat for two minutes. (If a microwave is not available, heat the beaker of water on a hotplate.)

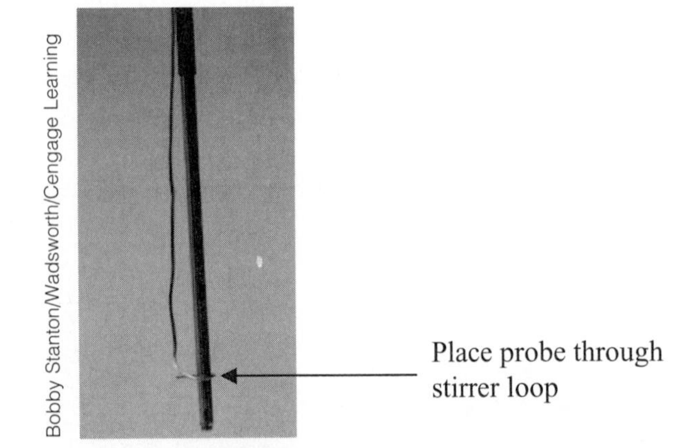

Figure 2
Stirrer/temperature probe assembly

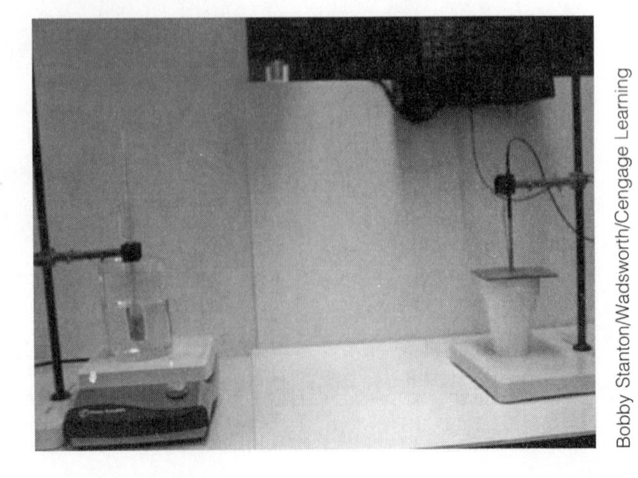

Figure 3
A styrofoam cup calorimeter with a MeasureNet temperature probe and wire stirrer

Remove the beaker from the microwave. Using a thermometer, make sure the water temperature is 45–60 °C above room temperature. If not, heat the water for an additional minute in a microwave. This will serve as the hot water.

11. How do you determine the temperature of the cool water? Should this temperature be recorded in the Lab Report, and to what number of significant figures?
12. Using a cloth towel pour exactly 50.0 mL of hot water into a graduated cylinder. Assume the density of the hot water is 1.00 g/mL. How should the mass of the hot water be determined? Should the mass be recorded in the Lab Report, and to what number of significant figures? Should the temperature of the hot water be measured with a thermometer? Should this temperature be recorded in the lab report, and to what number of significant figures?
13. Press **Start** on the MeasureNet workstation to begin recording the thermogram. Turn on the power to the magnetic stirrer to a low to medium speed. Make sure the stir bar is spinning without contacting the temperature probe or the walls of the calorimeter.
14. After 5–10 seconds have elapsed, raise the calorimeter lid, and *quickly, but carefully*, pour the warm water (using a towel to hold the graduated cylinder) into the calorimeter. Immediately replace the lid on the calorimeter. If a wire stirrer is used, you must constantly move the stirrer up and down to stir the contents of the calorimeter.
15. When the temperature of the water has risen and stabilized at the equilibrium (final) temperature, press **Stop** to end the scan.
16. Press **File Options**. Press **F3** to save the scan as a tab delimited file. You will be prompted to enter a 3 digit code (any 3 digit number you choose). The name of the file will be saved as a 4–5 digit number. The first 1–2 digits represent the workstation number, the last 3 digits are the 3 digit access coded you entered. For example, suppose you were working on Station 6 and you select 543 as your code. Your file name will be saved as 6543 on the computer. If you are working at Station 12 and you choose 123 as you're save code, the file name will be 12123. Press **Enter** to accept your 3 digit number. *You must use a different 3 digit code for every file you save or you will overwrite the previously saved file.*
17. Record the file name in your lab report. Note what type of information is contained in the file in your lab report (i.e., thermogram for determining calorimeter constant).
18. Press **Display** to clear the previous scan. The station is ready to record a new thermogram.
19. Using a magnetic rod, remove the stir bar from the calorimeter, and decant the water in the calorimeter into the sink. Thoroughly dry the calorimeter. Perform a second trial to determine the calorimeter constant by repeating Steps 2–19.
20. When you are finished with the experiment, transfer the files to a flash drive, or email the files to yourself via the internet.
21. When two experimental trials for the calorimeter calibration have been completed, return to the procedure in the corresponding laboratory experiment.

Instructions for Recording a Thermogram to Determine the Enthalpy Change for a Neutralization Reaction

1. If a magnetic stirrer and stir bar are available, they will be used to stir the solution. If not, assemble the copper wire-stirrer assembly depicted in Figures 1 and 2.
2. Nest two styrofoam cups inside each other to prepare a styrofoam cup calorimeter.
3. If a copper wire-stirrer is used to stir the reaction mixture, insert the stirrer assembly inside the calorimeter and lid as shown in Figure 2, then proceed to Step 6. If a magnetic stirrer is used, proceed to Step 4.
4. If a magnetic stirrer and stir bar are used to stir the reaction mixture, place the calorimeter on the magnetic stirrer. Place a 1 cm stir bar inside the calorimeter. Turn the magnetic stirrer power switch to a low setting. (*When spinning, the stirrer bar must not touch the temperature probe or the walls of the calorimeter*).

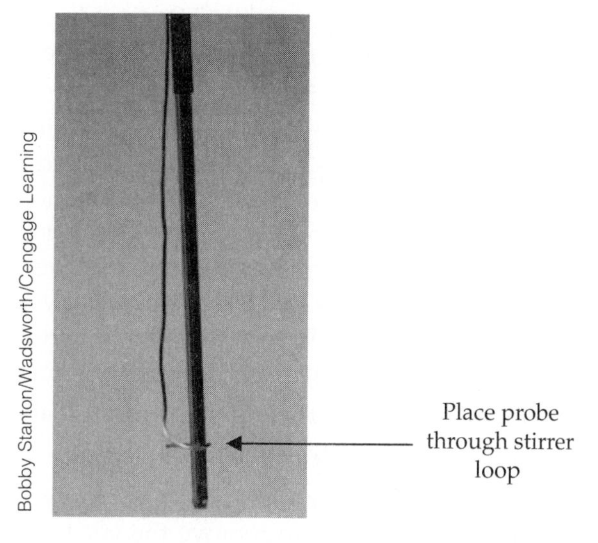

Figure 1
Stirrer/temperature probe assembly

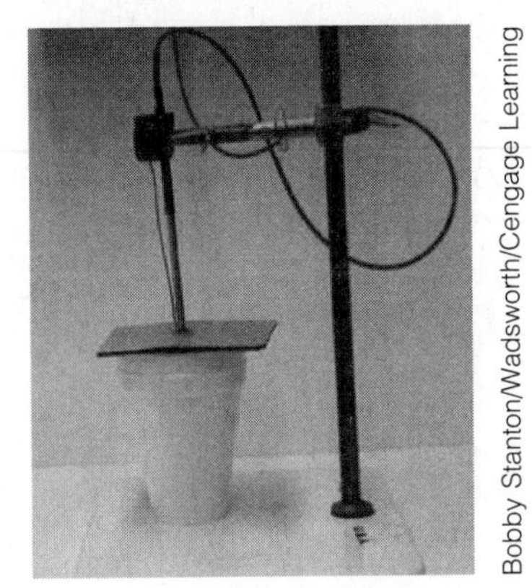

Figure 2
A styrofoam cup calorimeter with a MeasureNet temperature probe and wire stirrer

5. Insert the MeasureNet temperature probe through the cardboard lid and place the lid-temperature probe assembly inside the calorimeter as depicted in Figure 2. (*Be sure the probe is placed to one side of the calorimeter so that it does not contact the spinning stir bar.*)
6. Obtain exactly 25.0 mL of acid solution. Raise the calorimeter lid and *carefully* pour the acid solution into the calorimeter. Must you transfer every drop of the acid to the calorimeter? Be sure the temperature probe is not touching the bottom or sides of the calorimeter.
7. Should you determine the initial temperature of the acid? If so, how would you determine the acid's temperature? Should you record the temperature in the Lab Report?
8. Obtain exactly 25.0 mL of the base. Should you determine the initial temperature of the base? If so, how would you determine the base's temperature? Should you record the temperature in the Lab Report?
9. Press **Start** on the MeasureNet workstation to begin the temperature versus time scan. After 5–10 seconds have elapsed, raise the calorimeter lid, and *quickly, but carefully, pour all* of the base solution into the calorimeter. Must you transfer every drop of the base to the calorimeter?
10. When the temperature of the reaction mixture has risen and stabilized at the equilibrium (final) temperature, press **Stop** to end the scan.
11. Press **File Options**. Press **F3** to *save* the scan as a tab delimited file. You will be prompted to enter a 3 digit code (any 3 digit number you choose). The name of the file will be saved as a 4–5 digit number. The first 1–2 numbers represent the workstation number, the last 3 digits are the 3 digit access code you entered. For example, suppose you were working on Station 6 and you select 543 as your code. Your file name will be saved as 6543 on the computer. If you are working at Station 12 and you choose 123 as your save code, the file name will be 12123. **Press Enter** to accept your 3 digit number.
12. Should you record the file name in the Lab Report?
13. Press **Display** to clear the previous scan.
14. Remove the temperature probe assembly from the reaction mixture. Use a magnetic rod to remove the stir bar from the reaction mixture. Thoroughly rinse the stir bar or copper wire stirrer with distilled water and thoroughly dry it with a towel.

15. Decant the reaction mixture in the calorimeter into the sink, and *thoroughly* dry the calorimeter.

16. Perform a second trial by repeating Steps 6–15.

17. When you are finished with the experiment, transfer the files from the computer to a flash drive, or email the files to yourself via the internet.

18. *When two experimental trials for each acid-base pair have been completed, return to the procedure in the corresponding laboratory experiment.*

Instructions for Recording a Thermogram to Determine the Enthalpy Change for a Physical or Chemical Process

1. If a magnetic stirrer and stir bar are available, they will be used to stir the solution. If not, assemble the copper wire-stirrer assembly depicted in Figures 1 and 2.
2. Nest two styrofoam cups inside each other to prepare a styrofoam cup calorimeter.
3. If a copper wire-stirrer is used to stir the solution, insert the stirrer assembly inside the calorimeter and lid as shown in Figure 2, then proceed to Step 6. If a magnetic stirrer is used, proceed to Step 4.
4. If a magnetic stirrer and stir bar are used to stir the solution, place the calorimeter on the magnetic stirrer. Place a 1 cm stir bar inside the calorimeter. Turn the magnetic stirrer power switch to a low setting. (*When spinning, the stirrer bar must not touch the temperature probe or the walls of the calorimeter*).

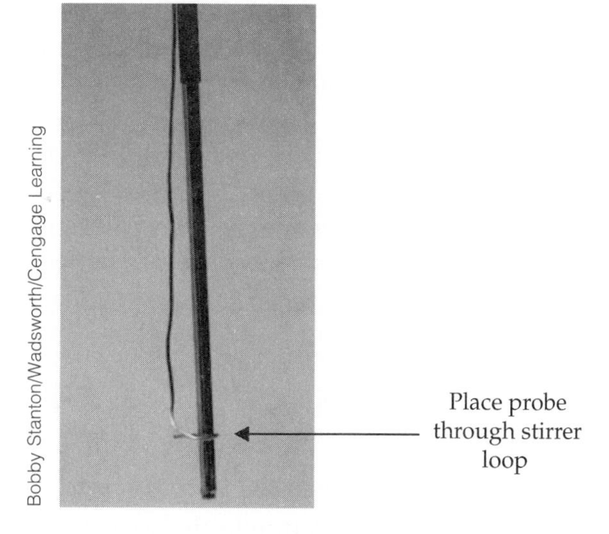

Figure 1
Stirrer/temperature probe assembly

Figure 2
A styrofoam cup calorimeter with a MeasureNet temperature probe and wire stirrer

5. Insert the MeasureNet temperature probe through the cardboard lid and place the lid-temperature probe assembly inside the calorimeter as depicted in Figure 2. (*Be sure the probe is placed to one side of the calorimeter so that it does not contact the spinning stir bar.*)
6. Obtain the specified amount of substance A. Raise the calorimeter lid and pour substance A into the calorimeter (*allof the substance must be transferred to the calorimeter*). Be sure the temperature probe is not touching the bottom or sides of the calorimeter.
7. Should you determine the initial temperature of substance A? If so, how would you determine substance A's temperature? Should you record the temperature in the Lab Report?
8. Obtain the specified mass or volume of substance B. Should you determine the temperature of substance B. If so, how would you determine B's temperature? Should you record the temperature in the Lab Report?
9. Press **Start** on the MeasureNet workstation to begin the thermogram. After 5–10 seconds have elapsed, raise the calorimeter lid and *quickly, but carefully, pour all* of substance B into the calorimeter.
10. When the temperature of the reaction mixture has increased or decreased and stabilized at the equilibrium (final) temperature, press **Stop** to end the scan.
11. Press **File Options**. Press **F3** to *save* the scan as a tab delimited file. You will be prompted to enter a 3 digit code (any 3 digit number you choose). The name of the file will be saved as a 4–5 digit number. The first 1–2 numbers represent the workstation number, the last 3 digits are the 3 digit access code you entered. For example, suppose you were working on Station 6 and you select 543 as your code. Your file name will be saved as 6543 on the computer. If you are working at Station 12 and you choose 123 as your save code, the file name will be 12123. **Press Enter** to accept your 3 digit number.
12. Should you record the file name in the Lab Report?
13. Press **Display** to clear the previous scan.
14. Remove the temperature probe assembly from the mixture. If a magnetic stirrer is used, remove the stir bar from the reaction mixture with a magnetic rod. Thoroughly rinse the stir bar or copper wire stirrer with distilled water and thoroughly dry it with a towel.

15. Decant the reaction mixture in the calorimeter into the sink, and *thoroughly* dry the calorimeter.

16. Perform a second trial by repeating Steps 5–15.

17. When you are finished with the experiment, transfer the files from the computer to a flash drive, or email the files to yourself via the internet.

18. *When two experimental trials have been completed, return to the procedure in the corresponding laboratory experiment.*

Instructions for Plotting Excel XY Thermograms for Calorimetry

1. Open an Excel spreadsheet (i.e., worksheet).
2. Go to File Open, open a MeasureNet tab delimited file containing temperature versus time data. Click Finish.
3. Copy the first two columns (containing time and temperature data) in the tab delimited (*text*) file, and paste it into columns A (time) and B (temperature) in the Excel worksheet. Close the tab delimited file.
4. Click on the "Chart Wizard" Icon.
5. Click on XY scatter plot. Click on the smooth line type.
6. Click Next. Then highlight all cells in columns A and B that contain time and temperature data. An XY plot of the data will appear.
7. Click Next. Click Titles. Chart Title is the name of the plot. Enter a name for your plot (i.e., Calibration of Calorimeter). The x-value box is for labeling the X axis (Time, s), and the y-value box is for labeling the Y axis (Temperature, °C) on your plot.
8. If you wish to remove the gridlines, Click on Gridlines and click on the axes that are checked to turn off the gridlines. This is a <u>suggested</u> cosmetic function, it is not necessary.
9. Click Legend. Click on Show Legend to remove the legend, it is not required for a single curve.
10. Click Next. You can save the plot as a separate sheet which can then be printed. Alternatively, you can save it as an object in the current worksheet. The plot will appear in the Excel worksheet beside columns A and B. If you save the file in this manner, you will have to select and highlight a copy area box (i.e., highlight the plot area) under Page Set-Up to print the plot. The plot should look similar to the one depicted below when you are finished.
11. Alternatively, you can select the plot, copy, and paste it into a "Word" document, and print the plot from within the word document.
12. YOU MUST TYPE OR WRITE the ΔT value on your plot.

13. There are numerous other functions and options available to you to enhance the appearance of your plot. Trial and error is the best way for you to become proficient with Excel.
14. Scaling a Plot – You may scale the plot to exhibit only those portions of a scan that are significant to your experiment. For example, the time scan occurred over the 0–100 second time interval, but you only wish to display the 30–60 seconds portion of the plot.

    Click on any number (0–100) on the time axis, and the Scale Box appears. Press scale, then Enter a new *minimum* number (30 s) and a new *maximum* number (60 s), then click okay. Excel will automatically produce a re-scaled plot.
15. Repeat Steps 1–13 to prepare the remaining thermograms.

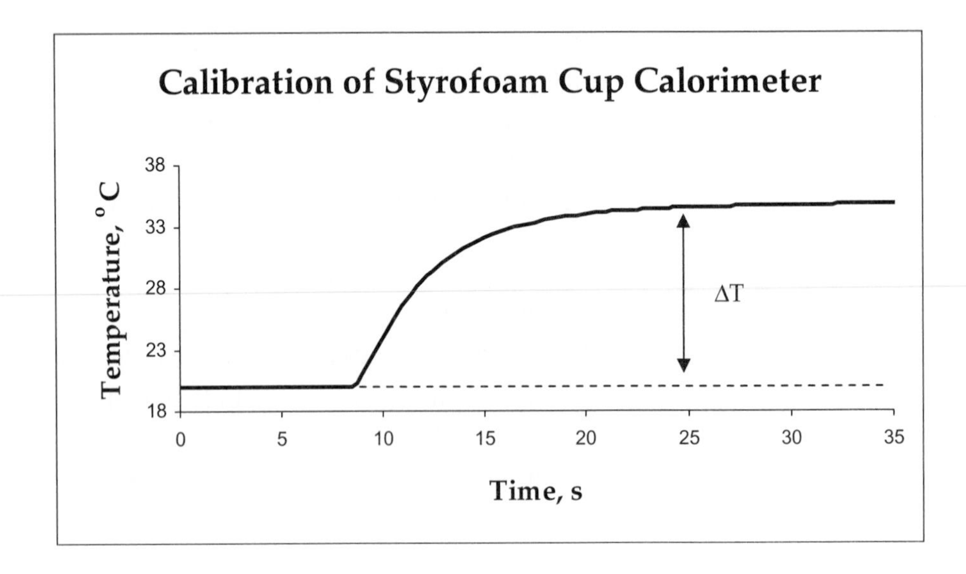

Excel Instructions for Performing Linear Regression Analysis

1. In column A, enter the concentration values for your standards from most to least concentrated. In column B, enter the absorbance value for each standard solution at the λ_{max}.
2. Use Excel's Chart Wizard to prepare an XY Scatter plot of absorbance versus concentration. Click XY Scatter Plot (Chart subtype, *Scatter, compares pairs of values*).
3. Click Next.
4. Enter the cell ranges to plot by highlighting all cells in columns A and B that contain data (Excel will plot your data automatically).
5. Click Next, then click Titles. Enter the title of the plot in the Chart Title slot. Label the concentration axes in the Value (x) Slot. Label the absorbance axes in the Value (y) Slot.
6. Click Legend, Remove the Check from the Show Legend box, a legend is not required for a single line.
7. Click Next.
8. Click "New Sheet."
9. Click Finish.
10. Using the mouse, select the entire Plot, then go to Chart on the Main Menu. Under Chart, select Add a Trendline.
11. Click Type, then select Linear.
12. Click Options, then click Display Equation and r-squared. This process will perform a linear regression, plot a best fit line for the absorbance-concentration data, and print the equation for the line on the plot in the $y = mx + b$ format. It will also print the r-squared value.
13. Click Okay.
14. Print your Plot

Use the equation for the line, $y = mx + b$, to determine the concentration of your unknown.

15. When finished your plot should look similar to the one depicted below.

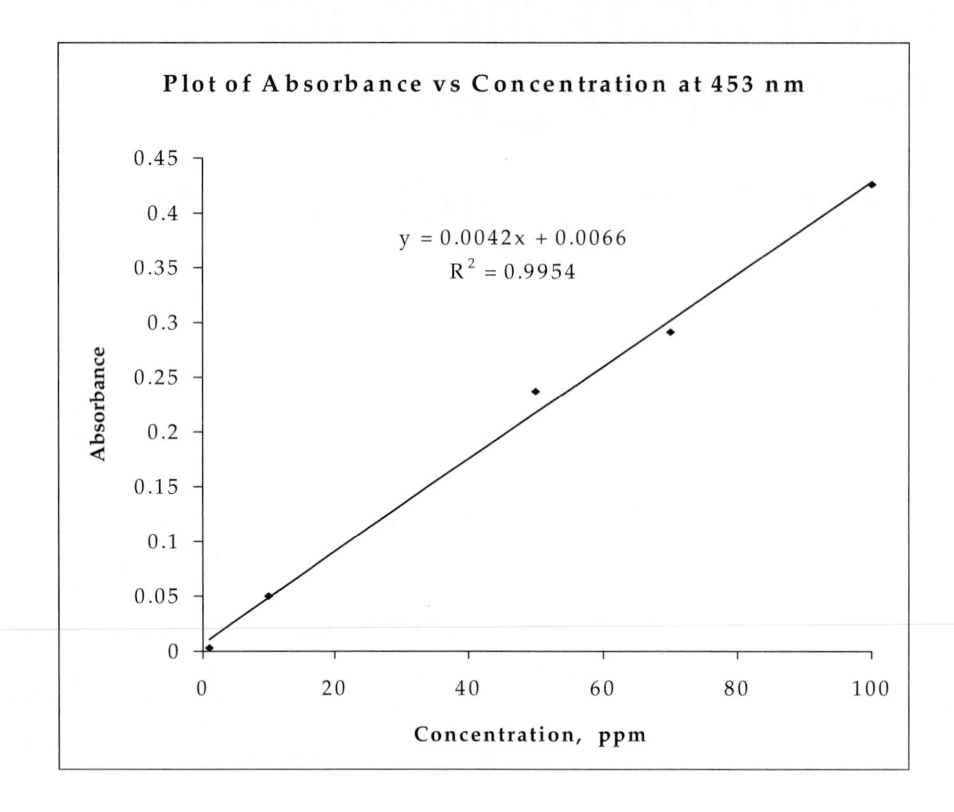

Instructions for Plotting Multiple Excel XY Thermograms for Freezing Point Depression

1. Open an Excel worksheet.
2. Go to File Open, open the MeasureNet tab delimited file containing temperature versus time data for the solvent. Click Finish.
3. Copy the first two columns (containing time and temperature data) in the tab delimited (*text*) file, and paste it into columns A (time) and B (temperature) in the Excel worksheet. Close the tab delimited file.
4. Go to File Open, open the MeasureNet tab delimited file containing temperature versus time data for the solution. Click Finish.
5. Copy the first two columns (containing time and temperature data) in the tab delimited (*text*) file, and paste it into columns C (time) and D (temperature) in the Excel worksheet. Close the tab delimited file.
6. Click on the "Chart Wizard" Icon.
7. Click on XY scatter plot. Click on the smooth line type.
8. Click Next. Click Series. Click Add.
9. In the *Name* box, enter the name for the curve (i.e., water or solvent).
10. The *x-values box* is for the X axis data. Click (with the cursor) inside the *x-values box*. Highlight all cells in column A (time data for the solvent).
11. The *y-values box* is for the Y axis data. Click (with the cursor) inside the *y-values box*. Highlight all cells in column B (temperature data for the solvent).
12. Click Add. Repeat Steps 9–11 for columns C and D (time and temperature data for the solution).
13. Click Next. Click Titles. Chart Title is the name of the plot, enter a name for your plot (i.e., Freezing Point Depression). The x-value box is for labeling the X axis (Time, s), and the y-value box is for labeling the Y axis (Temperature, °C) on your plot.

14. If you wish to remove the gridlines, Click on Gridlines and click on the axes that are checked to turn off the gridlines. This is a suggested cosmetic function, it is not necessary.

15. Click Next. The plot can be saved as a separate Excel sheet which can be printed. The plot can be saved as an object in the current worksheet. The plot will appear in the Excel worksheet beside columns A–D. If the file is saved in this manner, select and highlight a copy area box (i.e., highlight the plot area) under Page Set-Up to print the plot. *The plot should look similar to the one depicted below when you are finished.*

Alternatively, the plot may be copied and pasted into a Word document (i.e., a Formal Lab Report), and printed from within the Word file. To use this option, highlight the plot, click Copy, then go to the area in the Word file where the plot is to be inserted, click Paste.

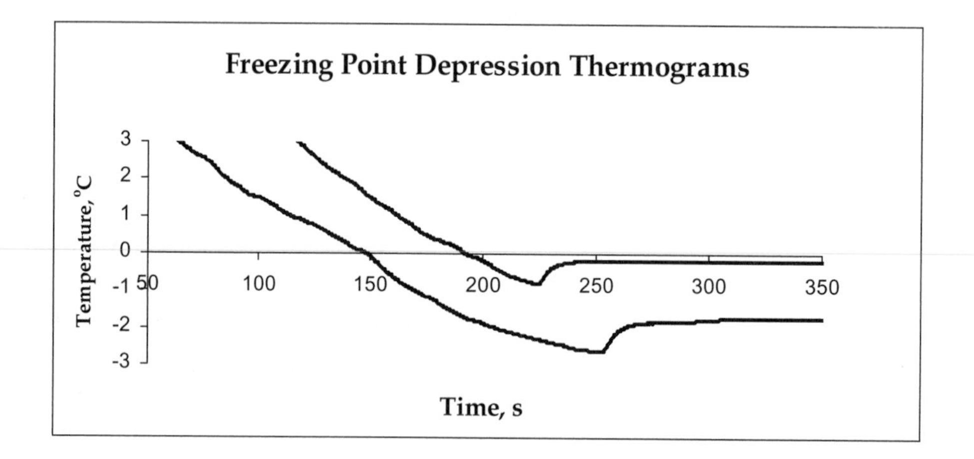

16. There are numerous other functions and options available to you to enhance the appearance of your plot. Trial and error is the best way for you to become proficient with Excel.

17. Scaling a Plot – You may scale the plot to exhibit only those portions of a scan that are significant to your experiment. For example, the time scanned occurred over the 0–100 second time interval, but you only wish to display the 30–60 seconds portion of the plot. Click on any number (0–100) on the time axis, the Scale Box appears. Press Scale, then enter a new *minimum* number (30 s) and a new *maximum* number (60 s), then click Okay. Excel will automatically produce a re-scaled plot.

18. Repeat Steps 1–17 to prepare additional thermograms.

Instructions for Plotting Excel Titration Curves

1. Open an Excel worksheet.
2. Go to File Open, open a MeasureNet tab delimited file containing pH versus time or volume data. Click Finish.
3. Copy the first two columns (containing time and pH, or volume and pH) in the tab delimited (*text*) file, and paste it into columns A (time or volume) and B (pH) in the Excel worksheet. Close the tab delimited file.
4. Click on the "Chart Wizard" Icon.
5. Click on XY scatter plot. Click on the smooth line type.
6. Click Next. Then highlight all cells in columns A and B that contain time or volume and pH data. An XY plot of the data will appear.
7. Click Next. Click Titles. Chart Title is the name of the plot, enter a name for your plot. The x-value box is for labeling the X axis (Time, s, or Volume, mL), and the y-value box is for labeling the Y axis (pH) on your plot.
8. If you wish to remove the gridlines, Click on Gridlines and click on the axes that are checked to turn off the gridlines. This is a suggested cosmetic function, it is not necessary.
9. Click Legend. Click on Show Legend to remove the legend, it is not required for a single curve.
10. Click Next. You can save the plot as a separate sheet which can then be printed. Alternatively, you can save it as an object in the current worksheet. The plot will appear in the Excel worksheet beside columns A and B. If you save the file in this manner, you will have to select and highlight a copy area box (i.e., highlight the plot area) under Page Set-Up to print the plot.
11. Alternatively, you can select the plot, copy, and paste it into a "Word" document, and print the plot from within the word document.
12. Scaling a Plot – You may scale the plot to exhibit only those portions of a scan that are significant to your experiment.

Instructions for Plotting Excel XY Emission Intensity Versus Wavelength Curves

1. Open an Excel worksheet.
2. Go to File Open, open a MeasureNet tab delimited file containing intensity versus wavelength data. Click Finish.
3. Copy the first two columns (containing wavelength and intensity) in the tab delimited (*text*) file, and paste it into columns A (wavelength) and B (intensity) in the Excel worksheet. Close the tab delimited file.
4. Click on the "Chart Wizard" Icon.
5. Click on XY scatter plot. Click on the smooth line type.
6. Click Next. Then highlight all cells in columns A and B that contain wavelength and intensity data. An XY plot of the data will appear.
7. Click Next. Click Titles. Chart Title is the name of the plot, enter a name for your plot. The x-value box is for labeling the X axis (Wavelength, nm), and the y-value box is for labeling the Y axis (Intensity) on your plot.
8. If you wish to remove the gridlines, Click on Gridlines and click on the axes that are checked to turn off the gridlines. This is a suggested cosmetic function, it is not necessary.
9. Click Legend. Click on Show Legend to remove the legend, it is not required for a single curve.
10. Click Next. You can save the plot as a separate sheet which can then be printed. Alternatively, you can save it as an object in the current worksheet. The plot will appear in the Excel worksheet beside columns A and B. If you save the file in this manner, you will have to select and highlight a copy area box (i.e., highlight the plot area) under Page Set-Up to print the plot.
11. Alternatively, you can select the plot, copy, and paste it into a "Word" document, and print the plot from within the word document.
12. Scaling a Plot – You may scale the plot to exhibit only those portions of a scan that are significant to your experiment.

Format for Writing the Procedure Proposal

A written Procedure Proposal and a Formal Lab Report are required for each self-directed lab experiment. One Procedure Proposal will be submitted per team of students to their laboratory instructor two weeks prior to performing the self-directed experiment. The graded and annotated procedure proposal will be returned to each team by the lab instructor one week before the self-directed lab experiment is to be performed.

The Procedure Proposal must be *typed* using either Microsoft Word, Word Perfect, Clarisworks, or some other word processing program. The manuscript must be *doubled spaced*.

The Formal Lab Report that is submitted after the self-directed experiment is concluded *must reflect* the corrections and revisions noted by your laboratory instructor in the Procedure Proposal.

The written format of the Procedure Proposal is provided below.

Title Page

On the first page include the name of the experiment, the names of all lab partners, and the date.

Introduction

In three to four paragraphs, describe the experiment to be performed.

Provide background information such as why the experiment is being performed, what the central question in the experiment is (thesis statement), how the experiment addresses the learning principles presented in previous experiments, and what plausible outcomes might be expected.

Experimental

Give a detailed description of the proposed procedure. Provide a step-by-step account of how the experiment will be performed.

List all relevant physical constants, properties, and mathematical and chemical equations.

List the specific reagents (including concentrations) and quantities of each reagent to be used in the experiment.

Thoroughly describe the apparatus to be used in the experiment and how it will be used. A sketch or drawing of the experimental apparatus may be useful.

You must include *all relevant calculations* for every step in the experimental procedure.

The experimental section should be written so that one of your classmates could perform the experiment. Do not make assumptions as to what "*everyone knows*." It is better to be too explicit than too brief.

Results

Prepare data sheets for recording data. Tables are an efficient method for recording and organizing data. You must record all measurements made while performing the experiment.

Bibliography

List all references used to prepare the Procedure Proposal (textbook, internet URL, or reference books).

Format for Writing the Formal Lab Report

One Formal Lab Report per team of students will be submitted for the self-directed experiment. The corrections and revisions noted by your laboratory instructor in the Procedure Proposal *must be reflected* in the Formal Lab Report. *Every team member is responsible for all the contents submitted in the Formal Lab Report*.

The Formal Lab Report must be *typed* using either Microsoft Word, Word Perfect, Clarisworks, or some other word processing program. The manuscript must be *doubled spaced*.

The format of the Formal Lab Report is outlined below.

Title Page

On the first page include the name of the experiment, the names of all lab partners, and the date.

Introduction

Include the revised introduction from the Procedure Proposal.

Experimental

Include the revised experimental section from the Procedure Proposal.

Results

Include all collected data in this section. The use of tables is an efficient way to report data. However, how data is reported is left to the discretion of the team.

Be sure to include *all* relevant calculations. Both Word and WordPerfect contain an equation editor that is easy to learn and use. The use of an equation editor for the reporting of calculations is highly recommended. However, hand written equations will be accepted on the condition that they are extremely neat and legible.

The answer to the experiment's central question is to be summarized in this portion of the report.

Discussion

Discuss the significance and relevance of the experimental results. Was there a clear and indisputable answer to the central question of the experiment?

Are the results accurate and precise?

Discuss possible sources of errors in this experiment. If necessary, propose procedural changes that might reduce these errors.

Conclusions

Summarize the experiment in one to two paragraphs. Did the experiment provide the data necessary to answer the central question of the experiment? Summarize the answer to the central question of the experiment (e.g., the identity of my unknown is or the concentration of the unknown is).

Bibliography

List all references used to prepare and perform the experiment (textbook, internet URL, or reference books).

Instructions for Recording an Absorbance Spectrum using the MeasureNet Spectrophotometer

1. *Be sure the spectrophotometer is turned ON before you set-up your workstation.*
2. Press the **On/Off** switch to turn on the power to the workstation.
3. Press **Main Menu**, then press **F5 Spectroscopy**, then press **F2 Absorbance**.
4. Press **SetUp** to establish scan parameters for the experiment.

   Press **F1** to set the scan limits. Use the ← → keys to move from **min** to **max** or from **X** to **Y**. An * marks the parameter selected to change at any given time.

   Absorbance data is recorded on the **Y** axis. Set the max intensity to 2.0, then press **Enter**. Leave the min intensity as 0.0, then press **Enter**.

   The wavelength, in nm, is recorded on the **X** axis. Unless otherwise instructed by your lab instructor, leave the *min* and *max* wavelengths set as the default values (200 nm *min* and 850 nm *max* for a UV-Visible source, or 350 nm *min* and 1100 nm *max* for a Visible source). (*Consult your lab instructor to ascertain the type of light source that will be utilized in this experiment*.) If a UV-Visible source is being used, the *min* and *max* wavelengths can be varied over the 200–850 nm region. If a Visible source is being used, the *min* and *max* wavelengths can be varied in the 350–1100 nm region.

   Press **Display** to accept all values. The workstation is ready to record an absorbance spectrum.
5. Add 3 mL of one of the solutions to a cuvette (*the cuvette should be 3/4 full* to ensure that the light beam passes through the sample). Dry the outside of the cuvette with a Kimwipe®.
6. Add 3 mL of the blank solution to a second cuvette.
7. **Go to the Spectrophotometer**. Be sure the Visible or UV-Visible source power switch is ON. Press **Station Number**, enter your Station

Number, then press **Enter**. The Spectrophotometer will display a "*Ready to Scan*" message.

8. Insert the light block (black cylinder or black plate) in the sample compartment, then press **ZERO**. The screen will display the "Ready to Scan" message once the "*Zeroing*" process is complete.
9. Remove the light block and lay it on top of the light source. Insert the blank solution into the sample compartment. Press **Reference**, the screen will display the "Ready to Scan" message once the "*Referencing*" process is complete. Remove the blank from the sample compartment.

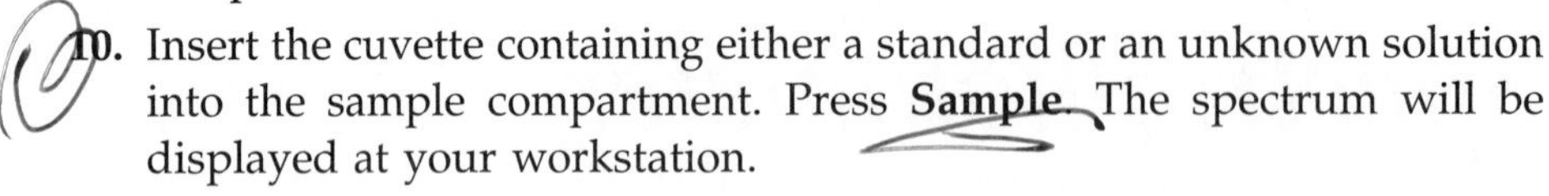

10. Insert the cuvette containing either a standard or an unknown solution into the sample compartment. Press **Sample**. The spectrum will be displayed at your workstation.
11. When the spectrum is complete, remove the cuvette containing the sample solution from the sample compartment.
12. **Go to your workstation.** If the plot is outside of the parameters you entered, (e.g., the absorbance was 2.5 and the maximum absorbance was set at 2.0), you may press **Setup**, then press **F2** (re-plot). Enter new minimum and maximum values, then press **Display** to accept these changes and automatically re-plot the absorbance spectrum.
13. Press **File Options**. Press **F3** to *save* the scan as a tab delimited file. You will be asked to enter a 3 digit code (any 3 digit number you choose). **Press Enter.**
14. Press **Display** to clear the previous scan.
15. Return to the spectrophotometer, repeat Steps 10–14 to record absorbance spectra for the remaining standard solutions or unknown sample. Save each spectrum.
16. Empty the contents of the cuvettes into the Waste Container. Clean the cuvette by thoroughly rinsing it with distilled water.
17. Save your files to a flash drive or email the files to yourself via the internet.
18. When absorbance spectra for all standard solutions and unknowns have been recorded, *return to the procedure in the corresponding laboratory experiment*.

Instructions for Recording an Emission Spectrum Using the MeasureNet Spectrophotometer

1. *Be sure the spectrophotometer is turned ON before you set-up your workstation.*
2. Press the **On/Off** switch to turn on the power to the workstation. Press **Main Menu**, then press **F5 Spectroscopy**, then press **F1 Emission**.
3. Press **SetUp** to establish scan parameters for the experiment.
4. Press **F1** to set the limits for the scan. Use the ← → keys to move from **min** to **max** or from **X** to **Y**. An * marks the parameter selected to change.
5. The **Y** axis represents the **Intensity** (brightness) of the emitted light. Set the maximum (*max*) intensity to 1500, then press **Enter**. Leave the minimum (*min*) intensity set to 0.
6. The **X** axis represents the wavelength of the emitted light in nanometers, nm. Leave the *min* and *max* wavelengths set at the default settings.
7. Press **Display** to enter the values into the MeasureNet system. The workstation is now ready to record an emission spectrum. (*If performing Experiment 10, go to Step 2 of the procedure in Experiment 10 before proceeding to Step 8 in Appendix E.*)
8. Go to the MeasureNet spectrophotometer. Press **Station Number** and type in your Station Number. Press **Enter**. The MeasureNet system will display *"Ready to Scan."*
9. Completely cover the end of the MeasureNet fiber optic cable with your finger and press **ZERO**. Hold your finger over the end of the fiber optic cable until the work station reads *"Ready to Scan."* (This procedure adjusts the spectrophotometer to a setting of zero intensity, i.e. there is no light.)
10. Press **Intensity**. Heat a nichrome wire in a Bunsen burner flame until orange light is emitted. Adjust the position of the burner relative to the

fiber optic cable until the intensity reading is somewhere between 2500–4000 (see Note). (This value only needs to appear on the MeasureNet screen for an instant, it need not persist). Press **Intensity**. A "*Ready to Scan*" message will be displayed.

NOTE: The Bunsen burner flame must **NEVER** be closer than 6 inches to the tip of the fiber optic cable. Excessive heating will DAMAGE the fiber optic cable.

---

11. Pour a small amount of distilled water onto a watch glass. Heat the nichrome wire until it glows bright orange in the flame (Figure 1).
12. Position the edge of the watch glass near the air intake of the Bunsen burner, tilting it slightly so that the water moves to the edge of the watch glass. Quickly immerse the hot wire into the water on the watch glass and *spatter* the water into the burner air intake (Figure 2).

Figure 1
Heating a nichrome wire in the Bunsen burner flame

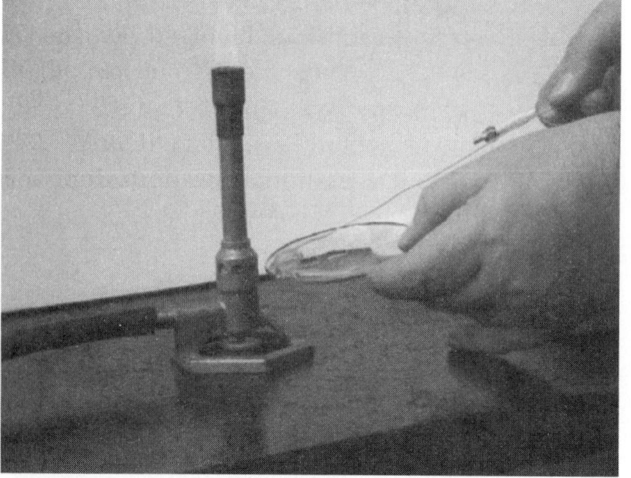

Figure 2
Spattering sample into the Bunsen burner flame

13. Repeat this process 3–4 times in succession. The gas flow carries the water up to the flame and rinses the burner of residual metal impurities from previous experiments.
14. Pour ~ 1 mL of a standard metal ion or an unknown metal ion solution onto a watch glass.
15. Heat the nichrome wire until it glows bright orange.
16. Positioning the watch glass with the sample near the burner intake, and immerse the hot wire into the sample to spatter it into the flame. Your lab partner must press **Sample** on the spectrophotometer to record the emission spectrum as soon as the hot nichrome wire contacts the sample (*it will sizzle when the hot wire contacts the solution*).
17. Should you record the color of the emission flame in the Lab Report for each known metal and the unknown metal solutions?
18. Pour the remaining metal ion solution in the Waste Container, and rinse the watch glass thoroughly with distilled water.
19. Return to your MeasureNet workstation. Examine the emission spectrum on your workstation's display. If the intensity exceeds the parameters you entered, (e.g., the actual emission intensity was 2500 and you initially set the maximum intensity to 1500) press **Setup**, then press **F2** (re-plot), enter new minimum and maximum values, then press **Display** to accept your changes. MeasureNet will re-plot the emission spectrum. (*This Step is optional.*)
20. Press **File Options**. Press **F3** to *save* the scan as a tab delimited file. You will be asked to enter a 3 digit code (any 3 digit number you choose). Then **Press Enter**. The name of your file will be saved as your station number plus the 3 digit number you entered. Should you record the file name in the Lab Report?
21. Save the file to a flash drive or email the files to yourself via the internet.
22. Press **Display** on the workstation to clear the previous scan, and to ready the Workstation to record another spectrum.
23. Return to the MeasureNet spectrophotometer. Repeat Steps 11–22 for each of the remaining standard solutions, or for an unknown water sample. Be sure to rinse the burner by spattering water through it 3–4 times after the emission spectrum of each known metal solution or unknown water sample spectrum is recorded. If the Bunsen burner or fiber optic cable are accidentally bumped out of alignment, redo Step 10.
24. *When all emission spectra have been recorded, turn off the MeasureNet workstation, and return to the procedure in the corresponding laboratory experiment.*

Instructions for Recording a Titration Curve Using the MeasureNet pH Probe and Drop Counter

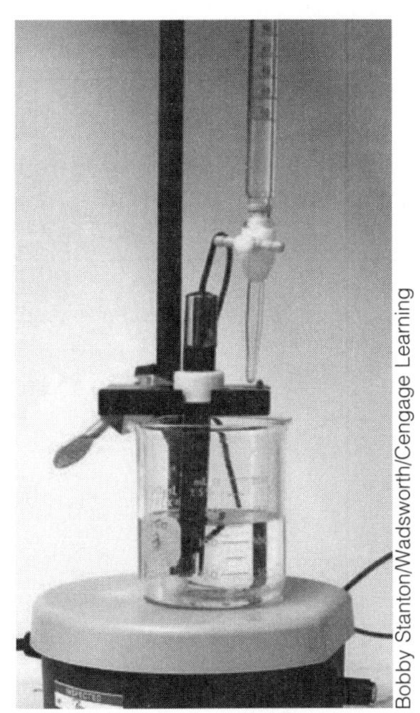

Bobby Stanton/Wadsworth/Cengage Learning

Figure 1
MeasureNet drop counter, pH electrode and buret assembly

1. Setup the MeasureNet drop counter, pH electrode, and buret assembly as depicted in Figure 1.
2. Rinse a 50-mL buret with distilled water, be sure to rinse the tip. Close the buret stopcock. Add 3–4 mL of NaOH solution to the buret. Rinse the buret with NaOH solution by tilting it on its side and twirling the buret. Drain the NaOH solution through the tip of the buret. Discard the NaOH into the laboratory sink. Be sure to flush the sink with copious quantities of water.
3. Close the buret stopcock. Fill the buret with 50.00 mL of NaOH solution. Check the tip for air bubbles. To remove bubbles, open the stopcock and quickly drain 2–3 mL of NaOH from the buret.
4. Set up your MeasureNet workstation to record a pH versus volume of NaOH added scan.
5. Press the **On/Off** button to turn on the MeasureNet workstation.
6. Press **Main Menu**, then press **F3 pH/mV**, then press **F2 pH v. Volume**.
7. Press **Calibrate**. The MeasureNet pH probe will be stored in a beaker containing pH 7.00 buffer solution. Measure the temperature of the pH 7.00 buffer solution (using a thermometer), enter the temperature at the work station, then press **Enter**.
8. Enter 7.00 when asked for the pH of the buffer, then press **Enter**. When the displayed pH value stabilizes (should be close to 7.00, but it does not have to be exactly 7.00), press **Enter**. Press F1 if a 1 point calibration (using pH 7.00 buffer only) of the pH meter is to be performed. Proceed to Step 10. If a 2-point calibration is to be used, proceed to Step 9. Your laboratory instructor will tell you whether you are to perform a 1- or 2-point calibration of the pH electrode.

9. Enter the pH of the second buffer solution (either pH 4.00 or pH 10.00 are suggested), press **Enter**. Gently stir the buffer solution with a stirring rod. When the displayed pH value stabilizes, press **Enter**.
10. Press **Display** to accept all values.
11. Remove the pH electrode from the buffer solution, rinse it with distilled water using a wash bottle over an empty beaker. *Gently* dry the tip of the electrode with a Kimwipe®.
12. Immerse the pH probe (supported by the drop counter) in the beaker containing the acid solution. If the tip of the pH probe (cut out notch of tip cover) is not submerged in the acid solution, add sufficient water to cover the tip of the probe (at least 1.5–2.0 cm of the probe tip should be submerged in the acid solution).
13. Insert a stir bar into the solution, and place the beaker on a magnetic stirrer. Turn on the magnetic stirrer to a low setting to gently stir the solution. The stir bar must not contact the pH probe when stirring. If a magnetic stirrer and stir bar are not available, continuously stir the solution with a stirring rod.
14. Position the buret filled with NaOH over the beaker of solution. The tip of the buret must be centered over the "*eye*" (notch) of the drop counter (see Figure 2). (**NOTE**: When the stopcock is open, drops of NaOH must pass through the center of the *eye* of the counter to be counted. The LED on the drop counter will flash red every time a drop passes through the eye and is counted).
15. Press **Start**. Read the volume of NaOH in the buret. Enter the initial volume at the workstation, then press **Enter**. If your buret is completely filled, and the bottom of the base's meniscus is on 0, enter 0.00 mL as the initial volume.

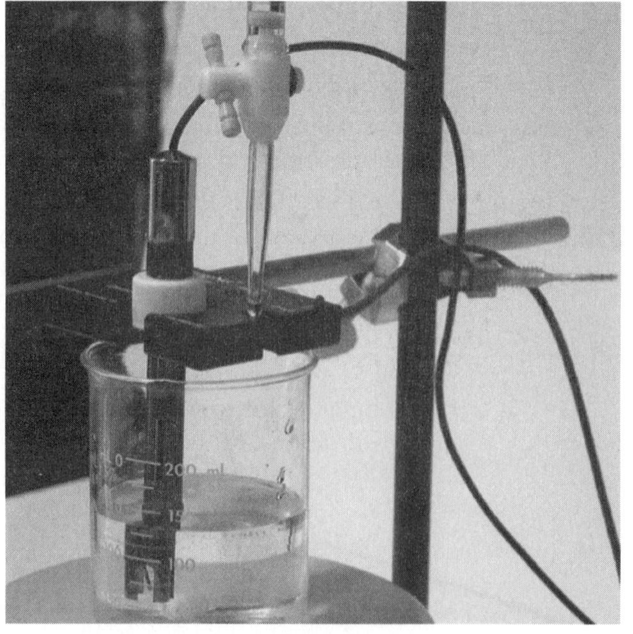

Figure 2
Buret tip centered over "eye" of drop counter

Bobby Stanton/Wadsworth/Cengage Learning

16. Press **Start.** Open the buret valve and begin adding NaOH drop-wise. Stir the solution continuously until the titration is completed. The base must pass through the *eye* of the drop counter. The red LED on the drop counter should blink every time a drop is counted. (*The flow rate must be slow enough that each drop is counted. The number of drops added will be displayed as counts on the station screen. The data will automatically be plotted as pH versus volume of NaOH added*).

    MeasureNet Plotting Information – MeasureNet plots the pH of the solution versus the mL of NaOH added if all aspects of the titration are done correctly. If a student makes a procedural error, MeasureNet defaults to drops, and the pH of the solution will be plotted versus drops of NaOH added. Procedural errors include: 1) drop rate that is too rapid; 2) using a buret with a chipped tip that drops oversized drops of NaOH; or 3) a student enters 50.00 mL as the initial volume, when the buret is completely filled, instead of 0.00 mL. (*The initial volume of NaOH must always be a smaller number than the final volume of NaOH in a titration, or MeasureNet will default to drops.*) Drops can be converted to mL using the following conversion factor: 20 drops of base = 1 mL of base.

17. When the titration is concluded, close the buret and Press **Stop**. (*The titration is finished when the sharp rise in pH begins to level off; this should be in the pH 11–13 range.*).
18. Read the volume of NaOH in the buret and enter the final volume of base in milliliters at the workstation. Press **Enter**.
19. Press **File Options.**Press **F3** to *save* the scan as a tab delimited file. You will be asked to enter a 3 digit code (any 3 digit number you choose) to name the file. **Press Enter.**
20. Ask your laboratory instructor whether you are to decant the reaction solution in the beaker into a Waste Container or into a laboratory sink. *Be sure to remove the stir bar with a magnetic rod from the solution before decanting the solution*. Rinse the pH probe and the stir bar with distilled water.
21. Return the pH probe to the beaker containing pH 7 buffer solution.
22. Press **Display** to clear the previous scan and begin the procedure for a new titration.
23. When you're finished with the experiment, empty the NaOH solution in the buret into the Waste Container.
24. Rinse the buret thoroughly with distilled water. Return the buret to its regular storage place.
25. **Turn Off** the power to your MeasureNet workstation when finished with the experiment.